THE FLOWER FINDER
VOLUME I

Volume One: Angel's Rest to North Lake

BARRY MALETZKY, M.D.

Flower illustrations by Shyla Villanueva

The Flower Finder
VOLUME ONE: ANGEL'S REST TO NORTH LAKE

Copyright © 2023 Barry Maletzky, M.D.

All rights reserved. No part of this book may be used or reproduced by any means, graphic, electronic, or mechanical, including photocopying, recording, taping or by any information storage retrieval system without the written permission of the author except in the case of brief quotations embodied in critical articles and reviews.

iUniverse books may be ordered through booksellers or by contacting:

iUniverse
1663 Liberty Drive
Bloomington, IN 47403
www.iuniverse.com
844-349-9409

Because of the dynamic nature of the Internet, any web addresses or links contained in this book may have changed since publication and may no longer be valid. The views expressed in this work are solely those of the author and do not necessarily reflect the views of the publisher, and the publisher hereby disclaims any responsibility for them.

Any people depicted in stock imagery provided by Getty Images are models, and such images are being used for illustrative purposes only. Certain stock imagery © Getty Images.

ISBN: 978-1-6632-5425-2 (sc)
ISBN: 978-1-6632-5424-5 (e)

Library of Congress Control Number: 2023911939

Print information available on the last page.

iUniverse rev. date: 07/08/2024

*Dedicated to Marjorie;
the best person I know.*

Contents

Acknowledgements　vii

Introduction　ix

Purpose and Scope　1

Angel's Rest/ Devil's Rest/Wahkeena Loop　8

February and March 9
April.. 10
May... 14
June.. 23
July and August 32
September and October............................ 39

Rock of Ages/Oneonta Loop　43

February and March 43
April.. 44
May... 47
June.. 50
July and August 52
September and October............................ 54

Nesmith Point Trail　57

February and March 57
April.. 60
May and June 65
July and August 74
September and October............................ 77

Munra Point　79

February and March 79
April.. 80
May... 83
June.. 86
July and August 88
September and October............................ 90

Eagle Creek　93

February/March 93
April.. 95
May... 98
June.. 100
July and August 102
September and October............................ 104

Ruckel Ridge and Ruckel Creek Loop　107

Up Ruckel Ridge 107

February/March 107
April.. 109
May... 112
June.. 114
July and August 119
September and October............................ 121

Down Ruckel Creek................................. 122
February and March 123
April.. 125
May... 130
June.. 135
July and August 140
September and October............................ 142

Nick Eaton Ridge/Gorton Creek Loop   146

Up Nick Eaton Ridge.................................... 147

    February and March 147
    April.. 149
    May..152
    June...159
    July and August............................ 166
    September and October............................ 171

Down the Ridge Cutoff and the Gorton Creek Trail 174

    February and March 174
    April.. 175
    May..176
    June... 179
    July and August............................ 180
    September and October............................181

North Lake (Wyeth Trail)   184

    February and March 184
    April.. 186
    May.. 187
    June... 193
    July and August............................ 199
    September and October............................ 202

Epilogue   205

Appendix   207

Selected and Annotated Bibliography   209

Acknowledgements

No work such as this is achieved by a single author. Credit must first be accorded to Shyla Villanueva, the main illustrator without which this book would be close to meaningless, and in addition, lack any color! Her illustrations literally bring the text to life. Of even greater importance, they help the reader identify any particular flower with far greater certainty than any long-winded written description ever could. Through countless and perhaps overly obsessive iterations, she persevered and tolerated my constant and undoubtedly incessant demands for perfection. The results bear witness to her artistic abilities. I also owe a debt of gratitude to Francis Linman, who laboriously perfected a number of the illustrations following Shyla's move to Hawaii, where she helps with a number of conservation efforts on the Islands. It is crucial to note that the many flower illustrations in texts and online are not necessarily truly representative of the wildflowers as they present themselves in the Columbia River Gorge. The Gorge's particular and unique ecosystems nourish our blooms with a variety of soil types and bacterial and fungal communities, thus rendering the exact pigments and leaf shapes somewhat different than in, for example, our Coast Ranges or the Oregon and North Cascades. Shyla's and Francis' illustrations are spot-on for how these blossoms actually appear on the many trails described herein.

I also owe more than a mere debt of gratitude to the professionals at iUniverse, not only for agreeing to publish such a lengthy text but in guiding me through the publication process. These experts helped to curtail my enthusiasm and condense overly wordy passages into more concise sentences. Among those who deserve more than a brief mention, I should single out Karen Stansberry for accepting the original manuscript, and helping to publicize it, Ellie Go for her skills in composing how the book would appear once printed, and Cindy Hervey for retyping the complete text. In addition, I must thank Leigh Allen at iUniverse for leading me through the sometimes complex task of actually publishing this text and Mae Subido for formatting this, at times, complex piece of work.

Were I to list my countless hiking and climbing companions who accompanied me on our many flower forays and who put up with my floral-centric company and frequent, unvarying expressions of "My word – did you just see that rare Yellow Bell?", this text would occupy too many extra pages to justify publication. To these friends, I must express a simple thank you for bearing my sometimes-boorish company but who also quite frequently spotted that Yellow Bell I probably would have missed. You know who you are – thanks to all!

Introduction

Note to Readers: *To render this manuscript into a more readable format, I have divided the text into two Volumes. This first Volume presents the wildflowers alongside Columbia Gorge trails beginning with the Angel's Rest/Devils Rest/Oneonta traverse and ends with the North Lake Trail, also known as the Wyeth Trail. In the second Volume, which I hope readers will also purchase, begins at the Mt. Defiance Trail and continues along the Oregon side before switching to the Washington side of the Gorge. Note that the Washington trails are reversed so that the trails are presented from east (Dog Mountain) to west (Hamilton Mountain). The general format, including the maps, illustrations and the Introductions, remains the same in each Volume.*

One advantage of such a division is that some hikers may prefer only certain trails encompassed in either Volumes One or Two. In my experience, a subset of hikers prefers a limited assemblage of trails and thus can choose either one Volume rather then the other. A second set, somewhat larger based upon my interactions with fellow hikers, takes advantage of many Gorge trails and may wish to purchase both volumes, though I must acknowledge, with some chagrin, the extra fee that entails. I hope, however, that whichever choice you make will ensure an appreciation for the unique botanical bounty encapsulated in the Columbia River Gorge's exceptional ecosystem, one that cannot be duplicated in any other location amongst our wild places across our planet.

Cresting a ridge just above the Nick Eaton Trail in the Columbia Gorge, I am breathless, as much from the scenery as from the exertion. I am immersed in a sea of violet-purple, satiny Grass Widows swaying in a kindly breeze – a choreography of color and motion elevating the spirit. What qualities in flowers energize us so? Color surely plays a role, as does form and even motion. Yet I believe there is more: an appreciation of the durability needed to survive in frequently adverse circumstances, and toughness packaged in a deceptively fragile and beautiful form. Discovering yellow Stonecrop apparently dangling, yet obviously thriving, on a north-facing cliff in the Gorge, or the dainty purple, white and blue Spreading Phlox thriving in abundance on meadowed heights exposed to the harsh westerly winds, we cannot help but admire the products of evolution, even if we tacitly admit that they were not designed expressly for our enjoyment.

Of course, not everyone is an ardent admirer of wildflowers. Indeed, I used to think of my interest in our native flora as trivial, almost an embarrassment. While colleagues were racing up some peak or scrambling to secure a favored campsite, I would often look up from my flower studies to find myself left behind and alone, but at least satisfied that I had identified some obscure variant of a tiny, probably unnoticeable and drab part of our native flora. Indeed, it sometimes seems that, to appreciate natural wildflowers, one must look, and see, with a small eye.

I started hiking the Gorge trails for exercise alone as I fancied myself a mountain climber. At first, I raced up these paths and, to be honest, took little heed of the native and varied habitats of the flora. But what became merely a conditioning playground soon turned into the realization that the Gorge was more than a wooded series of hills. Instead, I began to notice the micro-ecosystems and the vast diversity of flora within each. Soon, I was hooked; I just *had* to learn the names of these beautiful flowers, both in meadow and woodland, and, just as importantly, how each plant got there, why it thrived in its specialized location, and how it was related to other flowers; thus was born what has become, for me, an obsession to learn as much as I can about plant physiology, ecology and where each wildflower belongs in its particular genus and family.

You will notice some repetition in flower descriptions and botanical side-bars. This is unavoidable as many readers will not take on all these trails. For those hiking just a few, repeating illustrations and information is necessary. For others, enjoying the majority of these hikes, I hope they will not be too annoyed.

In truth, natural wildflowers can hardly compete with their human-engineered cultivated cousins of the urban garden, at least in terms of size and design. Why, then are wildflowers so enchanting, even comforting? There may be an evolutionary advantage to this appreciation – after all, our primate ancestors needed to be attracted to colors to anticipate what might prove appetizing. Perhaps, as well, we can admire them as beautifully formed by nature, rather than bred by our own species. And finally, also perhaps because we can appreciate the softer aura they lend to the rugged landscapes they inhabit, similar in spirit to our use of indoor plants to cushion the linear interiors of our homes and workplaces. There is truly splendor and fascination if one takes the time to simply look, rather than hurrying off to some far-flung destination.

Despite the comparison to garden plants, not all wildflowers are tiny, of course. The showy Tiger Lily of open woods, the gorgeous wild Iris of road-side banks and upper reaches of the western Gorge, and the deliciously fragrant Plume-flowered Solomon's Seals of low- to mid-elevation slopes are all fully capable of qualifying as garden plants. Moreover, many natural flowers, while individually inconspicuous, grow massed together on a single stem, often yielding an impressive display. The common Bear-grass, seen in clearings at higher Gorge elevations, is an example, its tiny fragrant white blooms arranged artistically on a three to four-foot spike. Other plants may spread by runners (branching stems extending roots as they march along the ground), producing fields of color. High meadows are often punctuated by such displays, with the blues of Lupine contrasting with the red and orange of Indian Paintbrush, while the faintly delicate violet of Phlox provides a lush backdrop for the bolder yellows of Cinquefoil and Buttercup. Even the lower woods boast such occasional displays, with Chocolate Lily decorating drier slopes and the daisy-like Broad-leaf Arnica providing bright splashes of yellow in darker forests.

But wildflowers do not serve a solely decorative purpose, even though their colors and forms attract the busy pollinating insects upon which their reproduction can depend. They also provide food and shelter for a host of birds and animals who, probably lacking an aesthetic appreciation of their beauty, have nonetheless come to rely upon them for sustenance. In addition, many plants, such as the showy Balsamroots yellowing Dog Mountain's upper meadows each May, possess sturdy root systems which anchor the soil, limiting erosion. Moreover, wildflowers often enrich the earth around them for other plants by providing a home in their root systems for fungi and bacteria which add valuable nutrients to the soil. Thus Indian Paintbrushes are often found thriving near Lupines, which provide nitrogen the Paintbrushes would have trouble extracting from the atmosphere on their own. Even all this ignores the powerful impact plants have had upon human civilization: It is the flowers and resultant seeds of grasses, such as wheat, rice, and corn, which have sustained us over millennia.

The Pacific Northwest provides a variety of habitats in which wildflowers flourish. Open meadows are often filled with a profusion of colorful species, thriving in the ample light. But deeper woods provide a good example of the compromises inherent in a plant's need to balance its energy investment with its capacity to reproduce and perpetuate its species. There is much less light available to woodland plants than to those inhabiting meadows; this may partially explain why there is a preponderance of flowers which are white and small in forests. These plants cannot afford to invest in energy-expensive flash and color as they need to create a large leaf surface to collect as much precious light as possible. But things are not always so easily explained by all-inclusive answers. Bright crimson Columbines thrive in the shade of tall Douglas Firs while tiny white Whitlow Grass only grows in sunny fields. Quirky evolution does not result in entirely logical or systematic arrangements, whether in the animal or plant worlds.

You will notice some repetition in flower descriptions and botanical side-bars. This is unavoidable as many readers will not take on all these trails. For those hiking just a few, repeating illustrations and information is necessary. For others, enjoying the majority of these hikes, I hope they will not be too annoyed.

Environment

We must recognize that much of what we think of as aboriginal is, in fact, second or third-growth timber, yet now fortunately preserved. Our world of the wild is mostly delimited these days by humans. The idyllic notion of an untouched wilderness, while enchanting, is an imaginary ideal. From the earliest *homo sapiens* to the crossing of an ice-bound Bering Strait, the primeval world is gone. Nonetheless, to be silent in the face of this oncoming calamity is to be an accomplice to it. When a tree falls in the forest, everyone hears it, and feels it.

We cannot allow bias to camouflage science. Our country, which gave birth to the idea of untrammeled wilderness, should not be among the first to destroy it. The denial of man-made global warming flies in the face of science; to deny it is simply intrinsically wrong. While popular opinion, abetted by the media, has focused on the loss of tropical forests, about a third of intact forest landscapes are still located in the boreal woodlands of North America, northern Europe and northern Asia. Satellite imagery has revealed a significant decline in these forests, mainly through manufacturing activities, chiefly logging, agriculture and mining. Not much of our world's spaces or species have escaped the impact of human agency.

We are heedlessly destroying many of our plants' ecosystems and crowding out their natural beauty or hunting valuable organisms into history. We have decimated hundreds of animal species. They say that domestic and feral cats kill more than one billion birds each year. To find a more predatory species, you only have to look in a mirror. Each photo, seminar or study sounds one more note in the cacophony of this distressing concert. The symphony we know as wilderness might well become this world's swan song.

Governments, policy makers, business leaders and we as citizens must view these changes not with resignation, but with revolt. There is no question that the grim fact of climate change has markedly altered the 4½ billion-year history of our planet; it is now the increasingly rapid *rate of change* that is alarming. It should be expected that a child fears the dark, but it is tragic when adults are afraid of the light. History may well judge the present generation's greatest sin as ignoring the advent of this crisis and doing nothing to prevent it.

Surely science and reason must issue an indictment: to not acknowledge a battle is to ensure losing the war

Indisputably, anthropogenic warming has been compared to a slinky going down a set of stairs – slowly, with many stops and starts. A more profound analogy stems from the seas: Its encroachment is not unlike a tide, imperceptibly rising yet unavoidable to notice as it reaches its crest. But unlike a tide, global warming will never return to its original position. Ignorance is no longer an excuse for inaction, nor are despair or complacency but writing a book by itself cannot rescue this wondrous wilderness, nor can strident voice alone; you have to be more than an anti-something. Speak up, get involved, talk with your senator and representative. Climate change has been so frequently noted that we may become inured to its perils. Act now to let your legislators realize that its impact on the wild worlds we adore requires out utmost attention.

While every botanical area claims a wealth of floral diversity, the Pacific Northwest is especially rich in its plant characteristics and patterns. The wettest coastal areas boast temperate-region rainforests unique in all the world. The Cascades' west slopes display both woodland and meadowed terrain bearing floral gems widely admired and studied in the botanical community, while the eastside regions produce a wholly separate and diverse array of plants better adapted to the drier conditions prevailing there. In a culmination of this rich floral profusion, the subalpine and alpine reaches of the Cascades harbor plants, some indigenous and rare, which not only survive, but appear to thrive under the most severe of climatic conditions.

A botanist finds years of enjoyable study under these skies; the casual hiker and mountaineer lacks both the time and interest to devote more than a few minutes to ponder the question of "what's the name of that flower?" As I ascend to Nick Eaton's ridge, then descend to Gorton Creek and climb out to the pinnacle of Indian Point, still lazily dreaming of fields of purple Grass Widows, I wonder if these nontechnical average folk (like me) would like to know the many answers to that question.

Note: The 2017 Eagle Creek conflagration, which decimated certain areas surprisingly as far east as the Wygant Loop may have altered barely a few, but not the vast majority, of the flower descriptions and locations for the areas covered in this text. Most of the common hiking trails, fortunately, will still remain intact and will soon recover. This is demonstrated by the late 20th century fire on Angel's Rest. Within a few years, the wildflowers not only recovered but this entire area showed an even more diverse display of wildflowers than before. Fires are a natural occurrence even within forests this far west; the Tillamook burn of mid-20th Century infamy, demonstrates nature's ability to recover, as does the devastation after the 1980 eruption of Mt. Saint Helens. As of this writing (summer, 2024), the 2017 Eagle Creek conflagration has still kept a few of the Gorge trails closed. However, most are now open, with only minor alterations. For example, the traverse from Angel's Rest to Devil's Rest has been modified by eliminating the brief round-about use of old logging roads, muddy even late in the season. It takes a more steep, but more sylvan and slightly more direct track to meet the original Devil's Rest Trail just below the top.

In addition, the approaches to the Mt. Defiance and Starvation Ridge Trails are slightly modified but quite easy to follow. The Wygant Trail itself is intact but the lower bridge over Perham Creek has been obliterated by a massive mudslide following the fire. In addition, the Chetwoot part of the Trail is blocked at the power line. The Nesmith Point Trail, the Ruckel Ridge/Ruckel Creek Loop and the Wyeth Trail to North Lake are as yet unopened. However, the Forest Service, along with volunteers from the Oregon Trails Association, the Pacific Trail Association and even the Washington Trails Association are all working to repair the bridges and re-open all these trails by the summer of 2024. Hopefully, they will be able to do so. When these trails become accessible, they will follow the same routes as in the past. Check the Gorge Trails websites or call the Forest Service at 541-308-1700 if in doubt. Also, read the introductions to each chapter herein for more specific information. Please continue to hike these trails as they may well become even more bountiful with wildflowers within the coming years.

It is also of note that two species of wildflower have flourished since the fire: the appropriately-named Fireweed and Varied-leaf Phacelia. While it has been known for centuries that Fireweed seeds can lie dormant underground for decades and then awaken into a spouting plant from the heat of a conflagration, I can find no reference to the same occurring for the Phacelia. I suspect that it also thrives after the heat generated by fires but it is equally possible that its predominance since the Eagle Creek blaze, especially on the Angel's/Devil's Loop and on Nick Eaton Ridge and Mt. Defiance, may be secondary to the openings left when many branches of the larger trees fell, allowing more light to reach the forest floor. A second factor could be the elimination of competing shrubs and flowers burned in the blaze.

Purpose and Scope

Often outdoor travelers ask about the name of a flower. But what is the value of learning simply the name of an object? Perhaps we want to know because it helps us order, classify, and store the information for later use. Learning the name seems to satisfy the unique human need to know. But there may be more: If you repeatedly ask, then remember, the name of a flower (saying it out loud while staring at it helps), you can begin to better appreciate how it got its name, what else it's related to, and a bit about its place in the local environment and its relationship to other plants which grow around it. You can begin to understand why daisies and asters usually don't grow in the forest (they need a lot of light); why Large-Flowered Blue-Eyed Mary isn't the same flower as Small-Flowered Blue-Eyed Mary, only larger (it's a completely different species); and why clones of female Meadow Rue grow in extensive patches without male plants nearby to fertilize them (they can replicate asexually). Learning the names is only the first step on a path toward greater knowledge and understanding, if the learner desires.

Based upon this premise, the present work attempts to help hikers and mountain walkers to identify the many wildflowers which they may encounter in their adventures on trails and routes in the Columbia Gorge of western Oregon and southwestern Washington (see Volume II). The locations described are within a reasonable drive from the Portland metropolitan area, although they may certainly be easily reached from further afield. All of the trips encompassed in this manual can be accomplished in a day by average hikers. While the majority of locations are adjacent to a trail and, hence, the flowers can be easily viewed as part of a hike, some wildflower locations will be described on frequently-used short scenic by-paths as well, because these have become so popular in recent years.

While some outdoor enthusiasts have deplored the burgeoning numbers of mountain explorers in the 21st Century as inevitably leading to ecological degradation, others welcome this popularity as providing opportunities to improve an understanding of the need for environmental preservation and to enhance an appreciation of the natural environment, values which grow in importance as our media-oriented world becomes increasingly focused on an urban milieu.

Although a number of wildflower identification books are available (see bibliography), none pinpoints the exact location of flowers in relation to distance on a trail or route, altitude, proximity to a landmark, and geographic position. This guide attempts to identify not only what a particular flower looks like, but where you can find it, and what it is doing there in the first place – that is, its relationship to its surroundings. In doing so, however, I recognize that most outdoor travelers are not botanists who plan trips in order to discover wildflowers. Rather, for most folks going out to enjoy the wilderness, coming upon a field of colorful flowers in a meadow or on a rock outcrop, or a single bloom in an otherwise dark forest, is simply an additional delight, a bonus of their outing.

Nonetheless, for many hikers, learning a bit more about the name of the plant, to which family it belongs, and how it is related to more familiar plants, might add to the enjoyment of the outdoors. It is for those individuals that this book is intended. This manual is thus not expected to be easily read from start to finish, but rather as a field guide to be consulted when planning a particular trip.

Those interested in learning more about wildflower identification should, by all means, borrow or buy, and read, the texts referred to in the bibliography.

I expect that most outdoor adventurers will either print out the section they are undertaking on any particular trip (for example, the April section on Angel's Rest or the May section for Dog Mountain) and carry just that section on their hike. It would take the most intrepid of hikers to carry this full volume in their pack for just a single trip! Others might read the section which pertains to that day's adventure either before or after their trip, trying in the first instance to recall what they have read or, following the hike, remember what they have seen. Most Gorge hikers repeat trips from time to time and thus, any of these approaches should help the amateur flower enthusiast to recognize many species and to learn about the whys and wherefores of their location and, once learned, be able to generalize flower names and information to additional Gorge hikes.

This book is organized by geographic areas in two volumes and then by trail (generally west to east on the Oregon side and vice-versa on the Washington side). Descriptions of flowers to be found are given as the traveler would ascend a trail, meadow or peak, and hence directions are referred to as "right" or "left", as these are most easily and rapidly ascertained; compass directions are sometimes included, however, if any ambiguity might arise. Locations of wildflowers are most often indicated as elevations (altitude in feet) as many mountain travelers now carry wristwatch or GPS altimeters and these are most easily consulted quickly to check reasonably accurate locations. Distances (for example, "1/4 mile from the trail-head" or "50 yards from the stream crossing") are also sometimes offered, especially when altitudes do not vary greatly along a route.

Whenever possible, landmarks are also noted. Thus, fluffy masses of purple Ball-head Waterleaf are noted on the Dog Mountain Trail at 2½ miles from the trailhead, just as one emerges from deep woods at 2,200' and enters an open brushy area, and also just before and after the large right-hand switchback at 2,250'. Most people don't carry distance measuring wheels with them on outdoor excursions, thus the emphasis on pinpointing locations by altitude and landmark. Elevations and mileages are based upon USGS topographic maps and on a wrist altimeter (Suunto); hence variances of 5% to 10 % must be allowed.

In addition, wildflower locations are displayed for each trail on an associated topographic map for each chapter and number-keyed to the map. For example, where the Angel's Rest Trail approaches 1,600' at a trail junction, a number **5** appears in boldface corresponding to the notation "**Point 5**" in the text where flowers at that elevation are described. Of course, not all flowers are described in detail for every trail and area (although most flowers on each route are mentioned) as this would lead to needless duplication. Rather, significant flowers are selected within each trip for more detailed discussion. By reading the description for any trail, one can gain an appreciation of the prominent flora to be encountered; by reviewing many of these descriptions, the reader will begin to more easily recognize the majority of wildflowers in our botanically-rich area.

Elevation is often of more significance when ascending these trails than mileage. One thousand feet of elevation gain may feel like several miles when, in fact, it has only been one – witness for example the Munra Point Trail as an example. In using this book, the hiker should refer to the maps at the beginning of each chapter and the **Points** noted on them, which correspond to the **Points** in the text in boldface, in order to locate more accurately where specimens of a species are likely to be found.

Most GPS devices now provide easy-to-read elevations with built-in altimeters, and smart-phone apps for altitude are also available. Combining altitude, **Points** on the maps associated with **Points** in the text, and landmarks mentioned, you will be able to judge fairly accurately your position on the trail and the wildflowers present there.

I have chosen not to employ exact GPS locations for the flowers as most hikers would be unlikely to whip out their GPS device and take a reading for each flower description. Moreover, GPS readings can be difficult to obtain with exact accuracy in heavily forested areas so common in the Gorge.

Driving, hiking and route directions are not provided as excellent guides (particularly by the Lowe's, Schneider and Sullivan– see the Bibliography) exist and the present work concentrates on where and when to find the flowers, not the trail-head. Hopefully, you will not get too lost if you follow those other comprehensive references.

Wildflower locations are based upon field notes and tape recordings from 1973 through 2024. Because flowers in the Northwest appear in succession over a season beginning in February and ending in early November, the month, and occasionally even the week the flower can be observed are specified in the text. But beware: Season by season, the same flowers may not always bloom during the same week, or even month, due to annual changes in climate. Moreover, some wildflowers, particularly annuals (which die and must re-seed each year), may not set seed and re-bloom the following year in the same location. Even perennials (which should re-bloom in succeeding years) seem at times to flower at will, without our capacity to fully understand, and therefore predict, all the factors affecting their distribution. Remember too that flowers that bloom during one month may well still be in bloom the next. The lilies and candy flowers of April may well persist throughout May, but in this book, they may not always be repeatedly mentioned month after month. That said, there is sufficient recurrence and persistence in the plant world to render the vast majority of location and month citations reliable year after year.

Obviously, not all flowers to be found on any particular trail or meadow can be enumerated here. Thousands of plant species exist in the Northwest and some, such as the obscure willow herbs found in urbanized parts of trails, the microscopic mitreworts, or the minimally-flowered bedstraws lost amidst the jungle green of valley floors, are intentionally omitted. In addition, certain less common or less conspicuous plants, such as the almost unnoticeable twayblades, are not mentioned. Instead, this guide focuses on the more obvious, colorful and, hopefully, interesting species. I hope no one is offended.

In addition, I have focused on the times of year when these trails and routes will be both most popular and most colorful. For example, most hikers frequent many of the trails in the Columbia River Gorge in April through late September, either because the weather is more conducive to outdoor travel then, the flowers are at their peak during those months, or these folks are trying to get into condition for longer backpacks or climbs for the upcoming summer months. In contrast, the trails and routes at higher elevations, such as up Mt. Defiance or on the heights of Nesmith Point, receive more attention during late June through August, when receding snow levels allow the gradual reappearance of the high meadow blooms unique to the Cascades.

Please note that the separation in each chapter into subchapters denoting each month, is artificial, though useful in signaling which species might be most prominent during that particular month. In fact, floral species do not adhere to our human-manufactured calendar. Thus it would be wise, if you are planning a trip up, for example, Mt. Defiance or Dog Mountain in June, to read both the May and July/August Chapters as well. Many plants last through several months in the Gorge. This is especially true for the months of May and June as well as for June through the early months of summer and equally of note should your trip occur in the middle of any particular month.

In the trail descriptions, two types of references to the flowers exist: One simply mentions the presence of a particular species and its location, with perhaps a minimal description of the plant. If this flower is not described more fully, you may look it up in the flower index, which will provide the page where a more detailed description will be provided along with its Latin name in parentheses; an illustration of that flower will also appear on that given page. Thus, if you encounter a flower without a full description, or one with merely a brief mention, you will be able to find a more complete description and illustration by checking the index. Occasionally, a prominent species may be described quite fully in several chapters. Please note that the color illustrations are not drawn to scale; a tiny plant may appear as large as a huge shrubby one in the drawing but dimensions, such as height, are provided in the text.

I have often included sidebars, which I hope will be of general botanic interest. These most often attempt to expound upon issues mentioned in the adjacent text. In fact, a reader might be interested in scanning all the sidebars in the book even if she or he were not planning on tackling any particular trail. The themes of a few sidebars are repeated because many folks may wish to skip chapters on trails they never plan to visit. These sidebars, I hope, will prove of interest to all nature-loving Gorge visitors, regardless of which trail is undertaken. Moreover, one could simply be entertained by flipping through the pages to see all the colorful illustrations, although my hope is that, in reading the text, the hiker will gain a greater appreciation of a flower's significance within the Gorge's complex ecological systems.

In contrast to most wildflower authors, I have chosen to include the description of some weeds where they constitute a conspicuous or important component of the environment. Weeds carry an ignoble reputation, undoubtedly deserved in part due to their sometimes harmful effects on crops as well as their often ungainly appearance. Moreover, weeds are most commonly adventitious, introduced from far-flung sources, usually by human plan or error. Examples include seeds attached to the clothing of travelers, those accidentally introduced into grain shipments from overseas, or those imported for agricultural purposes. They are therefore considered artificial.

Still, some weeds are quite attractive: Most people admiring the millions of white Oxeye Daisies lining Northwest highways from May to July, with their golden centers and glistening white rays, scarcely realize that they are not native plants. Moreover, some weeds can be beneficial (St. John's Wort), while others are quite harmless (Henbit, Self-Heal). Dispersal of plants by animals, including humans, has occurred for millions of years over many continents, both ancient and present-day. While some plant introductions can be disastrous (Kudzu in the Southeastern states, for example, or the Himalayan Blackberry in the Northwest), we should probably not discriminate on the basis of origins or beauty alone – after all we are all, in a sense, immigrants.

I realize that smart phone apps can now allow hikers to merely point their phones' cameras at a plant and the app will attempt to recognize it. For those not too interested in that flower's evolution, why it is where it is, its relationship to other plants in its micro-ecosystem, or other facts about the species, these may prove adequate. I have employed these apps on many occasions and found several problems with them, especially related to the Columbia River Gorge. First, they rarely comment on anything other than the name of the plant itself rather than its relationship to its environment. Secondly, they are, more often than any botanist would like, inaccurate. A third problem is that they ignore many species commonly found in the Gorge. Yet a fourth issue is that a photograph is intrinsically inferior to an illustration due to conditions of light and shadow, proximity, sense of scale, leaf and petal venation, etc. A fifth problem is that most apps ignore the majority of weedy species comprising a host of interesting plants not uncommon to the Gorge. Unfortunately, these apps, while seeming to bring flower identification into the 21st Century, may do more harm than good.

The standard system of biological nomenclature is employed throughout this work, with the common name of a flower given first, followed by the scientific Latin name in italics. The scientific names employed are those in most common use today. However, it is in the nature of science to change. For example, recent botanic work has suggested that the genus *Smilacina*, comprising the very common False Solomon Seals, should be merged with *Maianthemum* and the latter genus name be used. However, because the former genus name is still the most frequent one hikers will see in flower books, I have chosen to adhere to that name in this, and in most such cases. This practice may anger forward-thinking botanists, but will hopefully help aspiring naturalists not get too discouraged when looking up plants in different texts. An appendix at the end hopefully corrects this discrimination with both prior scientific names (thus more frequently encountered) and updated scientific names alongside.

Although scientific names are well-established and unique, common names differ not only as to name, but in spelling as well. I have, in the main, used the common name spellings as presented in Russ Jolley's *Wildflowers of the Columbia Gorge* as this book is in more frequent use than more technical works, although the spellings in *Pojar and MacKinnon* could just as easily be substituted. Occasionally, I have provided alternative common names, at times in parentheses, as these common names and spellings are in frequent use in our part of the country. The differences are minor; for example, the former reference lists *Tellima grandiflora* as Fringe Cup, the latter as Fringe cup. These differences would be of interest to only the most obsessive taxonomists.

In the text, I have occasionally omitted a common name modifier if the species being described exists mainly in the Gorge and has been mentioned repeatedly. For example, at times I refer to "Cliff Larkspur" as "Larkspur" and to "Small flowered Prairie Star" as plain "Prairie Star" because the other larkspurs and prairie stars in the Gorge are far less common in the areas described for these trails. Moreover, these species' distinctions are only realized upon microscopic examination. I have also chosen to capitalize the common names of these plants, in opposition to most other texts, because it is the wildflowers *themselves* we are featuring. This amateur text emphasizes these names and I believe highlighting them with capitals helps them stand out and be better remembered.

Why employ impossible-to-pronounce, scientific names (even in parentheses) in an unspoken language? Because common names can vary widely, and in an idiosyncratic and unpredictable fashion based upon diverse local usage: The Johnny Jump-ups of Maine are Stream (or Wood) Violets to us in the Northwest; the pretty yellow butterweeds of the Midwest are known to West Coasters as senecios. This taxonomy forms the alphabet for the biological sciences because these scientific names remain as constants in the changing world of folk names. They consist of a genus name first (a genus is a group of related species within a family), then the species name second (a species is a group of organisms which can successfully breed among themselves). Thus there are many lupines (the genus), but only one Broad-leaved Lupine (*Lupinus latifolius*); of the five or so Indian paintbrushes which inhabit the Northwest, just two, Common (or Tall) (*Castilleja miniata*) and Harsh (*Castilleja hispida*) are repeatedly encountered by most back-country travelers. While the majority of hikers will be satisfied to learn the common name only, the scientific names are provided to ensure clarity and precision. If and when you can learn the scientific name, you can more easily grasp the relationships (in an evolutionary as well as a physical sense) among our many wildflowers.

We thus name and classify objects in the natural world so as to provide some order to the multitude of organisms populating our planet and thus to transform raw facts into usable knowledge. Nonetheless, science and understanding aside, one of life's modest pleasures can be to stoop and examine a flower, and to know its name and its place in our universe. A riot of discoveries awaits. So go hiking and adventuring, but take a friend or two. This world is too beautiful to fit into just one soul.

Please note that this is *not* a trail guide. Directions to trailheads, descriptions of stream crossings, trail junctions, rougher parts of the trails and the like, while given brief and occasional mention here and there, are best gleaned from careful readings of the trail guides to this area by the Lowe's, Schneider, and Sullivan as noted in the bibliography, as well as by diligent study of the many topographic maps widely available for this area.

It may prove unnecessary to remind most readers of this book to respect the Gorge, a National Scenic Area well worthy of that designation. Off-trail travel and bushwhacking may be fun for some adventurers but these cherished woodlands should be honored as they are, rather than trampled for the fleeting pleasure of pioneering new routes. Stick to the trails – that's what they're there for and by all means avoid cutting switchbacks, especially on the Angel's Rest and Dog Mountain Trails. You may believe you're gaining ground on your companions by taking the steeper cuts rather than remaining faithful to the main trail but, believe it or not, actual measurements have shown that you are not only not saving time, but are adding to an unsightly gash and contributing to what, in the rainy months will become a watery channel eroding both the cut itself and the trail below.

Many of the hikes described herein are strenuous and the introductions to each hike should be perused to learn about the particulars of specific difficulties for each trip, regardless of location or season. Heed the "**CAUTION**" section preceding each trip and be prepared. It will prove most helpful to acclimate each season with the easier trips first, such as the Angel's/Devil's/Wahkeena Loop, Eagle Creek, or Hamilton Mountain before tackling the 4,000 - 5,000' workouts of Nesmith Point, Nick Eaton Ridge or especially Mt. Defiance and Starvation Ridge. Do not be surprised by Delayed-Onset Muscle Soreness (DOMS) 1-2 days following your first few hikes; a gradual approach early in the spring will greatly enhance the pleasure of the more strenuous outings later on as your body adapts to the demands of hiking. Alternative activities, such as biking or skiing, or even gym workouts, do not always transfer to the specific parts of muscles required for hiking. Go slower than you think you can at first to avoid unnecessary agony thereafter.

Another note of **caution**: Despite historical references to the use of some native plants by early explorers, prospectors or Native Americans, picking or molesting any plant in the Columbia Gorge National Scenic Area is prohibited by law, although tasting a few edible berries would not be considered a crime. Believe what you read about the ancients but please respect the sanctity of this glorious outdoor museum.

A final note of **CAUTION**: I AM NOT A BOTANIST. This work undoubtedly contains errors, many of which I hope will be corrected by readers more astute than I. Please communicate your opinions and observations to me through the Publisher.

The Flower Finder: Purpose and Scope

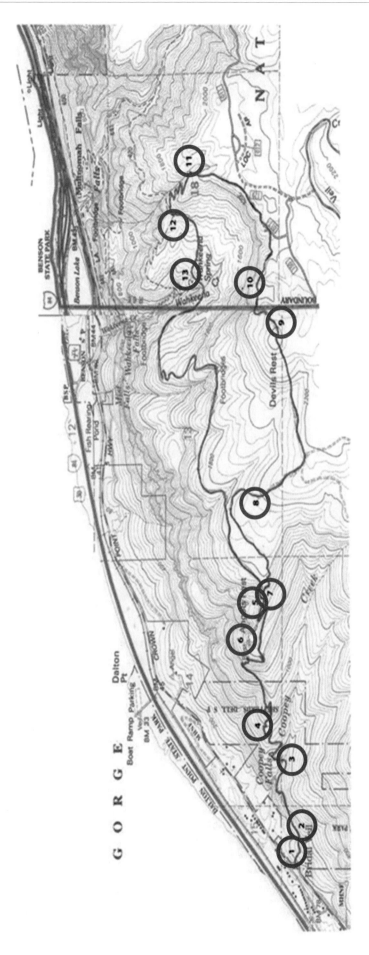

Angel's Rest/Devil's Rest/Wahkeena Loop

Perhaps angels really have descended from on high to rest their haloed heads upon this outcropping of basalt, but you have every right to feel a bit closer to heaven after scaling the Angel's Rest Trail, not only because you'll be 1,600' above sea level, but because the view over the Columbia River and out past Crown Point can be awe-inspiring. On the way up, however, be prepared to encounter neighbors, cousins, dogs, and a host of colorful characters on what will seem at times to be as much an escalator ride at the mall as a wilderness hike. Angel's Rest and Dog Mountain remain the most popular trails in the Portland hiking area, testified by the need, a number of years ago, for the Forest Service to build a second parking area just up the road. Both are usually filled on weekends from April through September so arrive early. By noon, both lots may be full and a slew of cars lie perilously parked off the old highway before and after the parking lots, half of them infringing on the paved road itself.

Fortunately, hikers come and go, but the land endures. Still, solitude isn't the only reason to hike in the Gorge; if being alone is paramount, try this hike in the off season – the views are still spectacular on a crisp early November morning or a misty March day (and there are still flowers to see). Better yet, go in the evening hours to see the vermillion sunset glow and the lights of Portland. Or extend the hike, as described here, to Devil's Rest, then loop back via Trails 420C, 420, and 415 to escape the crowds and earn the bonus of some extra elevation and exercise. Fortunately, the wildflowers have not seemed to mind the population explosion; perhaps they even enjoy all the admiration as year after year they continue to impress experienced hikers and newcomers alike.

Indeed, this Angel's/Devil's/Wahkeena circuit provides the most wildflower exposure for the least effort of any trip in this book. Share the flowers with others (as long as you don't trample them) – it's probably a good sign that so many people of all ages are at least experiencing the outdoors – and climb Ruckel Ridge in the Gorge if you really can't tolerate running into cousin Ernie yet again.

CAUTION: While even small children and dogs make the ascent of Angel's Rest comfortably, there are two rock steps near the top which may require a helping hand for young children. In addition, both at the top of Angel's Rest and on the first stretches of the turnoff to Devil's Rest, youngsters should be watched and not allowed too close to the eastern edges of the trails, where steep drop-offs await the unwary or the overly adventurous.

Note: Since the Eagle Creek fire of 2017, a small section of the trail between Angel's and Devil's Rests has been altered. Instead of routing the path through a series of muddy old logging roads, the new path, easily followed, steers slightly away from the original at about 2,000' of elevation, then regains it but not before you lose about 100' in ½ mile. It then rejoins the original trail, then encounters a junction. Watch for a small sign (on an alder tree in the forest) straight ahead. Then stay straight ahead, the middle of three branches, following the arrow pointing to Devil's Rest. This climbs, steeply at times, to the junction with the way path to your left to reach the top of Devil's Rest. While this newer section may seem a bit more precipitous than the original, it is easily traversed and avoids the muddiest of sections for which the old trail was infamous.

February and March

One encounters wildflowers as soon as the trail, newly groomed from a modern parking lot (our **Point 1**), begins, entering deep woods in the first quarter mile, then emerging onto several open rocky swaths at about a half mile from the trailhead. Even as early as mid-February some wildflowers (well, really weeds) appear: the ubiquitous Hairy Bittercress (*Cardamine hirsuta*) in openings in the woods (between **Points 1** and **2** on the map), and especially at the rocky swaths (**Point 2**) around 290' to 350' of elevation, where its tiny white flowers nod inelegantly from their 5" to 10" stalks. This Bittercress has a native cousin, Little Western Bittercress (*C. oligospermia*). These plants apparently have few redeeming values as they are neither pretty nor odiferous. Little Western Bittercress is more common on higher ground; it looks almost the same as its hairier sibling.

Hairy Bittercress

Various forms of bittercress prosper in the Northwest's wet climates and the Hairy species is possibly the most ignoble of them all, with its straight up-shoots, straight-up seed pods (technically called *siliques*) and almost unnoticeable four-petalled flowers. However, the leaf pattern of this unfortunate flower is actually quite beautiful – with tiny scalloped leaflets protruding at various angles followed by a similarly-shaped but larger leaf at the end, the whole shaped like a lyre – the ancient musical instrument. Both Bittercresses, small as they are, are still typical of the Mustard Family – four petals almost forming a cross and pods as seeds. They both appear to thrive in urban as well as forested environments, obnoxiously intruding into our yards, gardens, and lawns from February through April or May.

These plants apparently have few redeeming values as they are neither pretty nor odiferous. However, in a human-centered way, the bittercresses do offer one benefit – munch on a leaf or two and enjoy a pleasing taste of bitter salad greens, especially pleasant on a hot dry day. In fact, the bittercresses are not too distantly related to the Water-Cress (*Rorippa nasturtium-aquaticum*) you add to salads, and they contain similar compounds. While you may eradicate these with impunity in your gardens and yards, please leave them intact in the wild.

Common Chickweed

Just above the second rock-strewn opening find the equally weedy Common Chickweed (*Stellaria media*), with its trailing, branching stems and similarly tiny white flowers, found in one patch on the right (south side of the trail) in a breach at approximately 390' elevation.

By all means, take the side trails left before you reach a plateau; these are paths to an overlook of Coopey Falls and, this time of year, the Falls are spectacular. You will have two opportunities to do so as there are two marked side trails, one at about 100 yards after crossing the second rocky stretch at around 380' of elevation, the second a few minutes further on.

Trillium

This early in the season, one can't be too picky about the size and glamour of the flowers you can find. In fact, the three species noted above are also common garden weeds.

Beyond **Point 2**, few wildflowers are in bloom this early except, in milder years, Trillium, as described below. Still, you should consider trekking forward from **Point 5** to Devil's Rest and even beyond to the Wahkeena Trail. Candy Flowers will accompany you, along with a few Stream Violets as you climb to Devil's Rest, a rock outcrop on both sides of the top, then continue down the Wahkeena Trail on your left after the Devil. Some snow patches may persist through March but these are usually easily negotiated. You will come to two trail junctions along the way; always take the left-most turn to remain on the correct route back to the Angel's Rest Trail and the descent to the parking lot. You may have to brave some fog, wind and showers this early in the year but for the active Gorge adventurer, there is no such thing as bad weather.

April

From the beginning of the trail, right by the stairs leading up from the road across from the parking lot, note the Forget-me-nots charming the drab asphalt with their yellow-eyed blue blossoms. Up the actual trail, by late March to early April it is reassuring to encounter the wonderful white blossoms of Trillium (*Trillium ovatum*), somewhat like encountering an old friend. Well known and admired throughout the West, Trillium, also known as Wake-Robin because of its early appearance, grows not only in forested areas of what we consider wilderness, but in many city parks, urban woodlands, and not a few home gardens. Its cheerful triumvirate of three spotless petals astride three pointed green sepals reminds us that the rains may soon be memories and that spring has arrived (even though it may still be raining). Look for Trillium in the first 500 yards of the trail, around **Point 1**, wherever some light enters the forest; by late March it begins to bloom and by early April the blossoms are at their peak.

Notice especially how new Trillium plants emerge in the spring with their petals rolled up, much like a large sheaf of paper. This habit, common with many northern latitude plants, enables the delicate young petals, and the vulnerable reproductive organs inside them (such as stamens and pistils), to remain warm and protected during the chilly days of early spring. As the plant matures, the blossoms gently unfold to expose petals and sexual parts to the warming atmosphere and to visiting pollinating insects, but unless you're watching one of those fast-forward films on the Nature Channel, you'll likely miss the movie.

Trillium makes another fascinating transformation later in the season, when its petals begin to show some streaks or spots of burgundy during late April or early May. In this low forest setting at the beginning of the Angel's Rest Trail, this change can even begin in early April. By mid-May, Trillium blossoms look like some forest elf has dipped them into a deep Cabernet Sauvignon, their purple blooms just as striking, if a bit withered, as the fresh white petals were in April. For those who may mourn their passing, simply climb higher – throughout the Cascades in Oregon and Washington, Trillium will continue to sparkle well into June in open woods and clearings at mid-to-subalpine elevations.

Hooker's Fairy Bells

Although *T. ovatum* is easy to recognize because of its relatively large flowers, many people are still surprised to learn that it is a certified member of the Lily Family, a not-too-distant relative of the showy lilies, tulips and daffodils growing in your yard. A moment's close inspection, however, reveals the similarities: three petals, three sepals (the usually green leaf-like supporting structures under the actual petals), and parallel veins on the leaves – well, they're not actually parallel in a physics-textbook way as they do reach a common point at the tip of the leaf, but they certainly run parallel for much of their length and are quite different than the net-veined leaves typical of most other families.

> While most outdoor travelers assume that early-season plants are coaxed to bravely emerge above ground by the increasing warmth of the spring-time sun, many of our early bloomers are actually responding to the longer daylight hours or, in a fascinating inversion of human intuition, the fewer hours of darkness at night. Daylight levels can vary but the hours of nighttime darkness

are far more reliable. Plants like the bittercresses can thus be induced to sprout at odd times of the year by simply exposing them to varying levels and durations of light. Conversely, some plants can be made to wither and die by altering various conditions of warmth and light. Thus humans have exerted some control over the republic of plants with beneficial effects for us in terms of greater food productivity and profitability. Whether we will also reap any ill effect, only the future will tell.

In fact, lilies are among the few families that feature leaves of this type, often thin and grass-like. So too are irises, orchids, and the grasses themselves. This large order of plants, technically termed the *monocotyledons*, are thought to be among the most ancient of the angiosperms, the group we commonly know as flowering plants. Among these primal families are some of the most conspicuous and earliest-blooming flowers we encounter in Northwest mountains and forests.

Another lily dwelling in the early woods on this trail, and especially prominent between **Points 1** and **2**, is the pleasantly-named Hooker's Fairy Bells (*Disporum hookeri*). Despite its name, this lily is not as showy as Trillium; nonetheless, a close look reveals its dainty and beautiful architecture. The plant itself looks more like a bush than a simple flower, with multiple leafy branches up to 3' tall. Dangling from the tips of these branches, and almost hidden by the taper-pointed leaves, are creamy white couplets of bell-like flowers with protruding golden stamens, completing the image of bells with clappers. Listen hard as you might, they will still not ring, although for all the world, they look like they are just about to. You may need to reach under the leaves and turn the flowers upward to reveal the blooms fully but the effort will be well-rewarded as you can then appreciate the exquisite shape of these flowers, which frequently decorate the deep woods between the beginning of the trail and on up to 450' of elevation. Look for them again in deeply-wooded areas after crossing the bridge over Coopey Creek, and between **Points 3** and **4**, 500' to 1,300'.

Hooker's Fairy Bells (and the similar Smith's Fairy Bells (*D. smithii*), which are more abundant in the Oregon Coast Range, devote more surface space to leaves than flowers; thus they are able to thrive in the light-poor gloom of the deep forest. They avoid being drenched by shedding the profuse rain they receive from their taper-tipped leaves ("drip-tips"). Evolution has prepared them well for their existence, and persistence, in the wet Northwest. I have spotted just one example of this species, along the road on the downward loop of Hamilton Mountain.

Smith's Fairy Bells

A number of other interesting lilies thrive in glorious profusion by mid-April along the first mile of the Angel's Rest Trail. Many of these, and their peculiar characteristics, will be more fully described in later trail and outing chapters. Of special note are the strangely speckled purple, brown, and yellowish-green nodding bells of the Chocolate Lily (*Fritillaria lanceolata*). This 1½' to 2' tall plant is particularly prominent on the right side of the trail in the rocky swaths at 250' and 320', **Point 2**, and will be more fully described in the Starvation Ridge Chapter. Stoop and sniff these foul-smelling blooms at your own risk and see why they attract garbage-loving flies as pollinators.

Among the lowland plants flourishing within the emerald maze of these west-side forests, note the low lavender blossoms of Oaks Toothwort, often drooping but always welcome as a harbinger of spring. These are prominent in open woods before and after the bridge over raging Coopey Creek, **Point 3**, but also brighten slopes from 850' to the areas just before the trail junction at **Point 5**.

No species, however, is more plentiful yet more obscure in the first half-mile of trail than the Pacific Waterleaf (*Hydrophyllum tenuipes*). Masses of these moisture-loving 1-3' tall plants populate shady groves alongside the trail from its very origins up to 300', appearing in bloom from early April through early

Chocolate Lilly

May. This waterleaf, with its dull buff- to lavender-colored fuzzy flowers held in compact clusters, thrives especially in the earliest woods between **Points 1** and **2**, particularly on the right side of the trail, then again around the bridge at **Point 3**. They display the typical habit of most waterleafs to extend their stamens (the male parts) well beyond their petals, giving the impression of tiny antennae announcing their availability to neighborhood insects.

Look too for the common (and happily sweeter-smelling) Plume-flowered Solomon's Seal (*Smilacina racemosa*) also called Western Solomon Plume and False Solomon's Seal. Although its individual cream-to-white flowers are tiny, like many other plants with insignificant blooms (at least to us), it displays them in panicles dwelling in rich profusion at the ends of its 1½' to 3' tall main stem, thus earning notice without investing excessive energy in any one blossom. Note this plant's sheathing elliptical leaves with parallel veins, a sure sign it is a card-carrying member of the Lily Family.

Pacific Waterleaf

Although there are plenty of these Solomon's Seals in the woody areas of this trail right from its start all the way to the first trail junction at **Point 5**, it seems to avoid open areas and to prefer slopes, arching out gracefully to conveniently offer its sweet fragrance to hikers passing by. It is easily noticed just before and after the bridge at **Point 3** (480'), especially on the left bank as one ascends above the stream.

The "False" in one of this flower's other common names is an example of the manner in which many plants attain descriptive monikers that do little to help us amateurs understand their place in the floral universe. "True" Solomon Seals were first named in Europe; their roots bore surface scratches that resembled old Latin document seals. Our Northwest plants are of a different genus entirely, but they resemble these old-world specimens and, since they were only later discovered by scientists, they are therefore called "False" Solomon Seals, although they look real enough to most of us to merit a more distinguished name – no plant of which I am aware is capable of this type of deception.

Star-flowered Solomon's Seal (*Smilacina stellata*) shares its genus and many of its characteristics with its larger cousin *S. racemosa*. Its 5 to 10 white starry blossoms atop a 1-2' tall, but arching, stem begin to appear by mid-April and persist through May and into early June, particularly at higher elevations. However, on the Angel's Rest Trail, this Solomon's Seal is neither as abundant nor wide-spread as its larger cousin. You can find its cheerful stars bursting forth to brighten the deepest woods within the first ¼ mile only. Much more plentiful elsewhere in the Gorge (see Nick Eaton Ridge, for example), it still offers here a tantalizing hint at the amazingly rich floral bounty waiting above.

Proceeding up the trail from the parking lot, one encounters two types of Candy Flower. The first, and most common, Candy Flower (often called Siberian Miner's-lettuce), is *Montia sibirica*. Growing just 6" to 12" tall, its blooms are nonetheless notable for their sprays of 5 to 10 flowers with 5 slightly notched white petals streaked delicately with fine pink lines and its paired heart-shaped upper stem leaves. Although this plant prefers damp habitats, it can be found growing in masses on the Angel's Rest Trail in woods and openings where any amount of moisture can collect, but especially between 200' and 750', that is between **Points 2** and **4.**

Notice these appealing blooms just before and after the bridge, **Point 3**, particularly on the left side of the trail and again between 750' and 1,000' of elevation, just after the first set of open switchbacks at **Point 4** and then delightfully scattered in openings in a second set of switchbacks and up to the trail junction at **Point 5**, 1,600'. These openings have partially resulted from a forest fire; indeed, the charred trunks of some

Candy Flower

remaining trees bear testimony to this long-ago blaze.

Yet a second type of Candy Flower, called somewhat unimaginatively Little-leaf Candy Flower (*Montia parvifolia*), prefers rocky substrates and at times seems to actually be growing directly out of a basaltic boulder. Actually, it thrives on the moss and dirt which collect on these rocks. Its flowers are a bit less showy than those of regular Candy Flower, and its alternate upper stem leaves are definitely narrower, yet it can nonetheless be relatively easily observed by the casual visitor on the right, or upper, side of the trail amidst the rocky swaths at about 300' elevation, around **Point 2**.

Even if you don't immediately notice its flowers, the leaves of Pacific Waterleaf provide plenty of botanical interest, even for the non-specialist. These 5-7" long leaves, often covering large swaths of rich and moist bottomlands, are doubly toothed: Their main blade is divided into five to seven deeply-incised segments which themselves bear jagged teeth. As with many of the deep forest plants which must survive without access to much sunlight, these leaves, which

Little-Leaf Candy Flower

function like vegetative engines converting light into energy, spread their surface mass horizontally to intercept as many of the sun's precious rays as possible which reach the forest floor. Pacific Waterleaf may have received its name from these leaves as often you can notice light blotches on their surface; some floral historians believe these reminded early botanists of watermarks on the endpapers of ancient books.

Just past the second rocky opening at about 350-400', two signs point left for overlooks to Coopey Falls, quite spectacular at this time of year. These are brief side-trails well worth the minimal effort. The first occurs just past the rocky opening, while the second sign to the overlook follows closely thereafter

Beyond **Point 5** most blossoms on this loop must await May to flower, although by late April, many of the species described below may well be in bloom, especially at lower elevations. Watch particularly for Cliff (Menzies') Larkspur, 1-2' tall with blue-purple blooms, just emerging in the openings after 800' of elevation. Reach the top of Angel's Rest, **Point 6**, at 1,640'. (Note that some texts list the top at 1,620' – I suppose it's where you're standing that makes the difference.)

However, not all floral bounty is lost after returning from Angel's Rest and taking the left-hand trail to Devil's Rest, at **Point 5**. Candy Flower travels with you much of the way to Devil's Rest and will also be found right at the top of the Devil's rocky head. The yellow flower sometimes accompanying it will be Stream Violet.

After another mile, a three-way sign points left to an alternative way back to Angel's Rest (brushy) and a left points to a connector to the Wahkeena Trail. Stay straight ahead for Devil's Rest. Another ¾ mile brings you to the side trail to the top of Devil's Rest. You may return the same way you ascended (muddy at times) or, as I recommend, take a left down at the junction with the Devil's Rest side trial to descend to the Wahkeena Trail and note that the white Trilliums you've encountered below 1,000' are now stained a wine-red while those on this plateau, around 2,000' along the Wahkeena Trail are still gleaming white and are still populating open forest.

A new species in the Mustard Family (though it doesn't much resemble our notions of what a Mustard should look like) appears in late April. Turn left and start to descend from the plateau. American Wintercress, 1½ - 2' tall, appears, usually in solitary stands at around 1,800' and again after you pass the two signed cutoffs (always take the left-most turn at both) to return to the Angel's Rest Trail). This plant is not often seen trailside in the Gorge but it does appear more frequently on patches just after Triple Falls on the Eagle Creek Trail. It holds a cluster of small yellow flowers at top, typical of the Mustard Family, with four petals in a cross-like pattern.

An even more colorful plant, Scouler's Corydalis (*Corydalis scouleri*) cannot be missed nor forgotten once found and admired. It can be found about 20 minutes after passing Wahkeena Spring, on your left in bunches, then in about another 10 minutes' walk along this mostly level part of the trail. A 3-5' tall relative of the many Bleeding Hearts now opening alongside the Wahkeena Trail, it sends forth a ladder-like array of flaming pink spurred flowers, each 1-2" long atop its tower-like stem. Its leaves resemble those of Bleeding Heart but are a bit softer and rounder around the edges. It prefers moist substrates and, indeed, can be found growing either on the banks of, or in the middle of, many low-level western Gorge streams. A garden variety comes from California, bedecked with yellow flowers which mimic those of the Scouler's species and is often referred to as "Golden Corydalis".

Most of the remaining April flora consists of Candy Flower, happily inhabiting both sides of the Wahkeena Trail all the way back to **Point 5,** where you turn left to return to the Angel's Rest parking lot.

May

The flower-wise outdoor traveler knows that the month of May marks the height of wildflower season in the lower elevations of the Gorge. Many of the plants described above are still in bloom between **Points 1** and **3** in early May. Just past the second rocky opening at about 350-400', two signs point left for overlooks to Coopey Falls, still a raging creek through June, spectacular and deafening at times. These are brief side-trails well worth the minimal effort. The first occurs just past the rocky opening, **Point 2**, while the second sign to the overlook follows closely thereafter

Indeed, there are too many flowers, even on a single trail, to fully describe here, but, of the many plants growing in happy profusion in the lower woods and openings on the Angel's Rest Trail, one hallmark species demands recognition: the gorgeous larkspur, known as the Poison, Tall, or Trollius-leaf, Larkspur (*Delphinium trolliifolium*). The only way to miss seeing these 2'-to-5' tall plants with deep blue-purple flowers would be to close your eyes as you cross the bridge at **Point 3**, a somewhat dangerous undertaking. Often beginning in early April and continuing through late May, these glorious plants display blossoms both before and after the bridge to rival that of most garden-variety delphiniums. Indeed, they have been employed as seed-stocks in hybridizing a number of garden delphiniums. Note the spur at the back of each flower; this appendage holds nectar for pollinating insects, who may be less impressed than we are with the flower's beauty, but appreciate it for more practical reasons. You will gaze in wonder at the largest groupings of this marvelous delphinium in all our mountains.

Tall Larkspur

Six or seven species of larkspur thrive in our mountains, and all share the same structure: The spur, and what appear to be the petals, are actually sepals, modified leaves which underlie, and form the supporting structures for the petals. In this case, the petals are smaller, often a paler blue, and sit in the center of the showier sepal display. This arrangement is not unusual for flowers in the Buttercup Family (the nearly unpronounceable *Ranuculaceae*). Yet it is hard to believe that larkspurs are buttercups, at least of a sort. In fact, many families which contain familiar garden plants, such as the roses and heathers, include species that appear so morphologically distinct that most aspiring naturalists cannot understand their relationships. While first-time botany students may believe that this system was devised simply to confound them, botanical classification is, in reality, never based on the appearance of a flower or the architectural structure of a leaf, as these features are exceedingly ephemeral, at least in an evolutionary sense. Shapes or colors of petals can change over a relatively brief period of time based upon environmental shifts and pollinator choice.

Instead, more consistent features of plants must be considered in determining what is related to what. Taxonomy within the floral kingdom is most often determined by the obscure structures of a plant's sexual parts, such as its number of stamens (male organs), the shape and position of its pistil and ovary (female parts), and, also crucially, the type of its seeds. While of less beauty and interest to the typical wilderness visitor, these characteristics display the permanence and consistency across species, genera, and plant families to warrant the scientific accuracy needed to classify biological organisms. Too bad – there is no easy observational way to group all these beautiful flowers into meaningful families; the small and strange catchflies really are close relatives of the powerfully beautiful carnations and the ubiquitous Fireweed is botanically a first cousin to the evening primroses blooming in many of our gardens. Dissimilar floral structure does not preclude a close relationship.

Above the rocky openings, proceeding beyond the bridge at **Point 3**, the gorgeous display of the Tall Larkspur continues in purple profusion up to about 750'. Just before the trail emerges from deep woods to begin its first series of more open switchbacks, plentiful Western Meadowrue (*Thalictrum occidentale*) lifts its sprays of inconspicuous flowers 1½' to 3' above the ground in moist fairly open settings – inconspicuous only if you fail to look closely. Upon more intimate scrutiny, the hiker may be surprised and amused by the male flowers of this dioecious (female and male parts on separate plants) species, also a Buttercup Family member. Drooping fascicles of yellowish anthers hang listlessly like the fringe on antique lampshades, supported by long purple rods called filaments. These create filmy drooping chandeliers, an effect quite unlike that of any other plant you are likely to encounter in the Northwest. The spray-like purplish female flowers, smaller and far less showy, may feel jealous of their male counterparts, although they have no reason to for

Bleeding Heart

they can perform a trick no male plant or animal could ever accomplish: When no male plant is nearby to fertilize the female, it can reproduce asexually, producing a host of clones. Thus one encounters masses of female plants, especially on the left in the first ½ mile after the bridge, without a male in sight. Similarly, however, and more difficult to explain, clusters of male plants also seem to hang together; analogies to the human species, nonetheless, are hazardous.

There is nothing inconspicuous about the Bleeding Hearts, also growing in masses along with the Meadowrues, as much because of their familiarity as their beauty. In fact, their scientific name, *Dicentra formosa*, means "beautiful two-spurred flower", an apt description. This common woodland plant grows as a semi-shrub 1' to 2' tall, capped by dainty drooping sprays of pink heart-shaped blossoms. These can frequently bloom as early as mid-March but, fortunately for the fair-weather hiker, they persist even at lower elevations of the Gorge through early June in most locations.

Local nurseries have created garden varieties of Bleeding Heart, which can be found growing in many city gardens. However, please don't try to transplant this plant, or any other, from the woodlands to your home yard. Not only is this practice illegal in most areas described in this book, it is almost always pointless. Fragile native plants have survived in a balanced ecosystem which most often cannot be duplicated in a garden environment; transplants are usually doomed to wither and die, a just reward for those pilfering the wilderness.

By early May, many shrubs are in bloom, providing eye-level color without the need to stoop and examine. The brashly flamboyant 4' to 6' tall Red-flowering Currant (*Ribes sanguineum*) displays its pendant clusters as early as mid-April in some locations. Look for one particularly healthy specimen on the left at 450' just before the bridge at **Point 3**. Don't eat the currants, which form later in the summer, though – they're plain awful.

Much more common on this trail is a Rose Family shrub, Serviceberry (*Amelanchier alnifolia*). This 10'-to-20' bush carries leafy clusters of sparkling 1" to 2" five-petalled white flowers in early spring and serves as a cheerful reminder that better weather is on the way, even if its petals often look a bit ragged. Although the berries were dried and eaten by local Native American tribes, I wouldn't advise it unless you appreciate bland mealiness in your lunch. While Serviceberry occurs on the Angel's Rest trail above 500' in a variety of settings, it requires a good deal of sunshine, and hence can be found abundantly rounding the switchbacks at **Point 4** and in masses just beyond the trail junction at **Point 5** and again at the very top, **Point 6**, where it forms a forest of white, frayed-looking blooms from late April through late May.

Red-flowering Currant

By the second week in May, so many flowers are in bloom that we must delay their detailed descriptions to later chapters. Particularly note the star-shaped yellow blossoms of Broad-leaf Stonecrop (*Sedum spathulifolium*) on the right at **Point 2**, seeming to grow right out of the rocks in these bouldery areas. These succulent plants, cousins of the common Chicks and Hens of our rock gardens, store water in their fleshy leaves to help with survival during the summer droughts in their well- drained and rocky environments. (Not widely appreciated in our marine-influenced climate dripping with moisture is that the greatest challenge for survival facing our flora is the lack of water during the summer months.) Also note the cheerful, but drearily-named Field Chickweed, with its five white petals so deeply notched that there appear to be ten petals on each flower. Look for this close relative of the wooly mouse-eared garden plant, Candytuft, in the rocky openings and switchbacks at 800' to 900' near **Point 4**.

Fringe Cup

The lowland woods, between **Points 1** and **2**, also host a number of less showy, but nonetheless attractive white-flowered species in May, including the charmingly-named Columbia Windflower, with its neatly trifold whorl of leaves midway up the stem and its five-petalled inch-wide flowers held a foot high above the forest floor greenery. Here also find the appropriately named Inside-Out Flower (see below), with its oddly inverted shooting-star blossoms lighting up the forest at regular intervals. Look too for the dull white-to-pink tiny and oddly-shaped cups of frilly petals marking Fringe Cup (*Tellima grandiflora*), up to 3' tall, in moist woods and openings, particularly before and after **Point 3**, but also just after **Point 5**. Note its similarity to another Saxifrage family denizen of the damp, Youth-on-age (*Tolmiea menziesii*), whose tiny tubular flowers tend towards a brownish-purple color hard to describe. This moisture-loving plant is particularly abundant around streams so look for its spikes of frilly flowers especially around the bridge at **Point 3** and note, also, its resemblance to the common houseplant called the Piggyback Plant. There's a good reason for the similarity – they're the same species –although the housebound variety has been manicured for indoor windowed environments.

Broad-leaf Stonecrop

Western Corydalis

Why do so many of these deep forest dwellers have whitish or dull-colored flowers? Perhaps they need to invest more of their energy in producing broad surface areas for their leaves, but I also suspect that in a dark forest, stark white petals may contrast with the jungle greenery as well as, or better than, more deeply-colored blooms. This may also be true of foliage. Rather inconspicuous plants of the under-story, such as Rattlesnake Plantain and the Pyrola's, both described in later chapters, also have white markings, although on their ground- hugging leaves rather than on their flowers, perhaps in order to advertise their presence to pollinating insects creeping on the forest floor.

Youth-on-age

Fortunately, not all forest dwellers are dull or white. The bright crimson Columbine (see June, below), growing here on the left as you cross the second talus slope at **Point 2**, and the flaming pink spikes of spurred petals on Western Corydalis (*Corydalis scouleri*) just before reaching the bridge at **Point 3** form glorious exceptions to this rule. The Corydalis, related to Bleeding Heart, can grow to 5' tall and displays pink-spurred flowers in a ladder-like arrangement progressing up its stem. Also note the low-lying but cheerful white Woods Strawberry blossoms growing in profusion at the openings at around 300' at **Point 2** and again around the switchbacks between **Points 4** and **5**. While these bloom in May, check out the delicious miniature strawberries growing close to the ground which follow the flowers from mid-July through September.

Proceeding up the trail, and amidst a garden show of splashier Stonecrops and Serviceberries, the hiker will encounter, at 750' on the left in the open switchbacks at **Point 4**, the strange half-bush of Varied- leaf Phacelia (*Phacelia heterophylla*). With its silky-coated leaves with arrow-like protuberances at their bases, and crowded dull white arching plumes of flowers, it is easily ignored. Perhaps that's why it bears the ignominy of lacking a common name.

Varied-leaf Phacelia

Nonetheless, there is something unusual about its presence on the Angel's Rest Trail as this is west of where it likes to live. Preferring drier conditions, it may survive this west- side habitat only in these dry, well-drained situations, another reminder that soil types and conditions can sometimes determine plant locations. Since the 2017 fire, it has been much more prominent alongside trails on the Oregon side of the Gorge, perhaps spurred on by the heat of the fire's effect on increasing its germination rates or by its preference for the openings made available by the blaze.

A bit more colorful, although certainly tinier, are the intricate blossoms of Small- Flowered Blue-eyed Mary (*Collinsia parviflora*). Above 1,000', these tiny blue flowers, just one to several inches off the ground, sparkle like sugary pieces of cerulean sky sprinkled from above onto vernally-moist slopes and meadows wherever even modestly filtered light can reach the forest floor.

Small-flowered Blue-eyed Mary

There is not only strength in numbers, but beauty as well, for, although each individual two-lipped blossom of this intricately shaped flower may be relatively inconspicuous, there is no ignoring a field of watery blue Mary's waving in the sun. Just such a show rewards the hiker at 1,350', between **Points 4** and **5** in the bouldery areas before the trail junction. To be even more impressed with this genus, explore (with great effort!) the upper reaches of the Munra Point Trail for a soul-melting sight of fields of this flower's

Martindale's Desert Parsley

grander cousin, Large-Flowered Blue-eyed Mary (*C. grandiflora*). Whoever Mary was, she must have been of uncommon beauty!

Here, both before and beyond the level rocky sections of trail after **Point 4**, Tall Larkspur re-appears. This beauty, once thought to be found only at very low elevations in the Gorge, has progressed in recent years to decorate portions of the Angel's Rest Trail at these higher elevations, perhaps an indication of global warming.

Large-flowered Blue-eyed Mary

Because the Angel's Rest Trail isn't particularly steep, the hiker should have the energy to enjoy some of the plants that bloom near and at the very top, **Point 6,** 1,640'. It's hard to miss the Nootka Rose bushes beginning after you pass the switchback at 1,500' and wend your way across the rocks that serve as a trail. These, our largest wild roses, can exceed 4' in height and are prominent on your right in the openings after the rocky pathway gives way to smoother earth and after the last switchback at 1,600'. While Serviceberry (see above) is prominent by early May, a smaller flower, Martindale's Desert Parsley (*Lomatium martindalei*) is of equal interest. Yes, its small yellow flowers in compact heads barely reach above the rocky substrates in which it thrives, and its lacy leaves, while a nicely blue-colored variant on green, are not particularly notable. However, what's so interesting about this plant, found growing right off the rocks at the top from May through June, is its family affiliation.

Look closely to note how the leaves resemble those of the common garden carrot, and how the flowers are carried in umbrella-like clusters at the ends of their inch-long stems, just like the common and weedy Queen Anne's Lace of August roadsides and the less common, but equally weedy, huge Poison Hemlock of damp waste places in our urban environments. This resemblance stems from the fact that all these plants belong to the Carrot Family (*Umbelliferae*), but please don't pull them up to eat the roots – many are highly poisonous. And don't worry about why a *desert* parsley is growing in some of the wettest places known to meteorologists – its originally-named cousin was a denizen of scrubland east of the mountains. So often, the first species in a genus is named after a particular characteristic that is then embarrassingly absent from closely related species. Thus, blooms in the Pink Family are often white or red, and sunflowers don't always favor the sun.

One further top-side plant deserving mention at Angel's Rest is the woolly-headed Field Pussytoes (carrying the somewhat cruel scientific name, *Antennaria neglecta*). Although for some folks, Pussytoes is rather too cute a name, it is fairly descriptive of the 7" to 12"- tall, dull white, fuzzy but separate heads of this plant, which look sort of like what you would see if you held up the ends of your pet cat's furry foot. It generally blooms from mid-May through early July in openings throughout the Gorge. At the top of Angel's Rest it grows in swarms, a not atypical habit, as it is another dioecious plant, and hence can reproduce asexually. Many of the patches which grow here, at **Point 6**, are female clones; the great advantage of this trait for all the six or so species of Pussytoes in the Northwest is that these plants can continue to thrive even in the absence of pollinating insects. Indeed, many *Antennaria* species grow in exposed areas where the climate may prove too nasty for many pollinators.

Field Pussytoes

For those not satisfied to simply soak up the Gorge-wide views at the top of Angel's Rest, taking a left at the trail junction, **Point 5**, hopefully after first visiting the top, is an attractive option, for, by mid-May, an additional floral bounty awaits the hiker who proceeds on to Devil's Rest. Proceeding up Trail Number 415 a short distance, one encounters the Nootka Rose again, on your left this time, along with masses of the tall bushy Thimbleberry (*Rubus parviflorus*), unmistakable with its fuzzy 5-8" maple-shaped leaves and crinkly white 1½" flowers dangling in long-stemmed terminal clusters with petals as thin as tissue paper. *Parviflorus* means small-flowered in Latin, so somebody either wasn't that observant or was comparing this *rubus* to some other giant in its clan, but, compared to most wildflowers, Thimbleberry's flowers are fairly large and obvious. You must wait, however, until later in the summer to fully appreciate this thorn-less relative of garden roses and strawberries, for it is only by early to mid-August that you will be able to contribute to the ongoing and heated controversy about the berries' taste.

Thimbleberry

Many authors have slandered these noble red fruits as insipid; please ignore these prejudiced opinions as stemming from gastronomic ignorance – or more likely from sampling the fruits from suboptimal specimens. If you hit it right, a bright red Thimbleberry is an explosive mouthful of raspberry-like fruit on the tongue, a well-earned reward for all your hiking efforts; two or five or ten are even better.

Now, Lupines are beginning to appear here as you tread (carefully) along the wobbling rocks along the trail, minding the severe exposure (and expansive views) to your left. Just before a left turn about 5 minutes up the trail, note on your right the delightfully-named Blue-eyed Grass, an Iris Family relative with six petals of the most appealing blue atop 1' tall stems.

Beyond the junction where Trail 415 ("Wahkeena Trail") turns left, **Point 7**, and where you will keep to the right (the sign which pointed to "Devil's Rest" has been AWOL since the 2017 fire)) you cannot avoid taking notice of the clover-like leaves of the Wood Sorrel (*Oxalis oregana*) carpeting the forest floor. By early May this plant sends forth its 1" five-petalled flowers held just above the leaf cover by slender stems in seemingly random patterns. Never abundant, the flowers are nonetheless quite agreeably pleasant, with their white petals penciled with a series of fine pink lines. Some botanists believe these serve as landing strips, guiding insects into the reproductive parts of the blossom, just as in many violets, where a series of colorful lines and dots signal, like landing lights for our 747's, the best places to alight.

Don't be content however to merely admire the gorgeous green carpet of leaves and the dainty white flowers of this *Oxalis*. Munch on a leaf or two and be pleasantly surprised by the puckery refreshment this plant has supplied to many a thirsty traveler. You can even pop a few leaves in a salad, but don't overdo this treat – oxalic acid in large doses wrenches the gastric lining enough to supply a sour ending to most trips.

Wood Sorrel

This plant requires a lot of west-side moisture, so much so that it cannot be found even 10 miles east of Angel's Rest. If you have the time, study the leaves of this fascinating plant, common in the Coast Range and appearing all the way from Washington to the Redwood forests of northern California. These leaves are alive, bending and folding to protect themselves from too much sunlight or rain. Have patience – it can take 5 to 30 minutes for the leaves to fold, and even then, they do so only under the right conditions. Scientists still debate all the reasons why leaves of several species are so active, but we amateurs can simply enjoy the irony that one of the main features distinguishing plants from animals, mobility, is not absolute.

As you proceed up toward Devil's Rest, a magnificent forest of firs and alders awaits, the latter occasionally arching over the trail, almost forming a French *allee* through which you are easily guided. After about ½ more miles of mostly mild up's and down's, the Forest Service has slightly altered the path since the 2017 fire, (our **Point 8**), which now briefly sidesteps the old logging roads which used to lead to a steep, but muddy trail to Devil's Rest, and instead now heads directly toward the top. Watch for new signs in the forest which direct you to either a left turn on a different, but brushy, trail back to Angel's Rest or a right turn to short-cut your way to the Wahkeena Trail, thus bypassing Devil's Rest or, my preference, take the middle branch directly to the Devil. Steep at times, it takes you through pleasant second-growth Douglas Firs whose bases are bedecked with Wood Sorrel and yellow Stream Violets. In just half a muddy mile, this branch deposits you just below the side path on your left leading to the top of Devil's Rest, our **Point 9**. This branch, while steeper overall than what you've already encountered, has the benefit of avoiding what had been the muddiest of approaches to any point in the Gorge, so be thankful even if you're occasionally laboring up the precipitous newer path.

Both distant views and flowers are wanting here, at the top of the Devil, though Little-leaf Candy Flower is prominent on the rocks at the top in June. You can take solace, however, in the extra exercise gained and in the array of May-time flowers which await if you descend Trail 420C to make the loop back to Angel's Rest, a worthwhile endeavor, for here you will sample the best of the west-side floral bounty which spring can offer. To accomplish this loop, proceed down Trail 420C (it's a left just below the top) heading northeast and east along the rim above Wahkeena Creek.

Although deep woods are encountered, enough light reaches the forest floor to prompt the growth of several shade- loving plants, especially the ubiquitous Candy Flower. Here and there, however, between **Points 9** and **10**, find the wonderfully weird Inside-out Flower (*Vancouveria hexandra*), whose flowers seem to have been designed by a punch-drunk committee. This 1-3' tall plant carries compound leaves springing up from just above the ground (called basal leaves) whose three-lobed blue-green leaflets are pleasantly heart- to egg-shaped and which have led some sleep-deprived botanists to call the plant "Duckfoot". However, it's the floral anatomy which elevates *Vancouveria* to let's-remember-this-plant status. Look closely and you will see that the ½" flowers really *are* turned inside out. The six dull white petals are reflexed and flare outward like little parachutes, resembling a garden cyclamen or shooting star. This remarkable design may better expose the pollen-bearing stamens and receptive stigmas to visiting insects. Thus these strange riches are not without their robbers.

Inside-out Flower

As you progress along the rim, descending to about 2,170', you come upon several short side trails to the left (the one providing the best views is our **Point 10**, the third side-trail encountered) which reward the hiker with expansive views from the cliffs rimming Wahkeena Basin to the river far below and the far-flung peaks of the northern Gorge in Washington. At the first of these, about ½ mile from the top of Devil's Rest, you will find by late May the first of what will be several Indian paintbrushes in the Gorge, at least of this particular species. Easily recognized by genus, paintbrushes are devilishly difficult to pin down by species, even for trained field botanists.

> All this does seem arcane but there is a practical aspect to this knowledge: Don't try to transport paintbrushes from the wild to your city garden. They will fail miserably, as will many other wilderness plants, because we most often cannot duplicate the exact mix of soil and nutrient conditions necessary for these hardy yet unnervingly demanding flowers to thrive outside of their natural environments. Yes, there are examples of wilderness flowers gracing urban landscapes – Trilliums abound in Portland and Seattle parks and Twin Flower likely grows as a lush groundcover over a few garden walls in your neighborhood. For the most part however, you will not be successful in transplanting flowers without advice from a native plant specialist, and not without seriously fracturing the law against removing native plants from the wild.

Fortunately for us amateurs, there are just two or three common species to learn in our area, and in the Gorge, only two are usually present in the areas most folks would visit. The first of these to bloom in the spring, and the ones mainly here, are the Harsh Indian Paintbrush (*Castilleja hispida*). This species generally grows on low-elevation bluffs and is actually more common than the second species, variously termed Common or Tall Indian Paintbrush (with the lilting Latin name, *Castilleja miniata*). This latter is found at the first lookout, although it is more plentiful at many open grassy sites throughout our area from May through August. This Paintbrush is taller (usually to 2' tall) than the Harsh variety (generally ½ to 1' tall), smoother, and its leaves and scarlet bracts are sharply pointed; Harsh Paintbrush, with its stiff bristles on the stem and three to five-lobed leaves and bracts is clearly (at least to botanists) a different plant. While many people are familiar with paintbrushes, not many realize that their glistening flashes of red, scarlet, magenta, orange, yellow, and even white do not grace the flowers at all; these splotches of "paint" instead color the leafy bracts which hide the true flowers. These are rather inconspicuous greenish tubes inside the bracts which shelter the plant's reproductive organs.

Many flowering plants advertise their attraction to insects not with flowers but with leaves (see Rattlesnake Plantain or White-vein Pyrola in future chapters for examples) or pigmentation of their bracts. A good summer example is Bunchberry, a close relative of Dogwood trees.

But yet another surprise awaits those interested in plumbing the recondite depths of paintbrush physiology. These apparently placid inhabitants of all our wilderness meadows are, in reality, barbaric parasites, attacking other peace-loving plants which were minding their own business, and raping their root systems, thus robbing them of precious nutrients. Lest we get carried away with anthropomorphic parallels, however, we should realize that floral inhabitants of an ecosystem evolve in step with one another, and nature's checks and balances generally hold sway. Thus fields of purple-blue Lupines (see June, below) are commonly punctuated with the crimson of the very paintbrushes which feed off their roots – the lupines provide nutrients, especially nitrogen, to their hosts but what, if anything, paintbrushes provide in return is still debatable. Such potentially beneficial associations are sufficiently common in the botanical world that they might provide a worthy model for the political arena.

Between **Points 10** and **11** along the rim you will continue to meet only the occasional Inside-out Flower and a few examples of Candy Flower in the wetter areas. However, once you begin the descent down the switchbacks, between **Points 11** (do not go straight here, but turn left) and **12** (the junction with Trail 420) and from 1,950' to 1,600', expect to be entertained by a typical show of west-side floral profusion, and unfortunately, at times the insect equivalent of Fort Knox. Some of the flowers found here will be described below, under June, such as the spectacular five-trumpeted Scarlet Columbine, the spiky white blossoms of False Bugbane thriving at streams, and the graceful single white 1- 2" blooms of Columbia Windflower sprinkled too sparingly in deeper woods. Of particular note here, especially in wetter spots on the right at 1,700' to 1,600', are the luscious blue-purple blooms of the Poison Larkspur described above. Just after the point where Trail 419 enters on the right, at 1,560', Western Meadowrue and Bleeding Heart, described above, put in their almost obligatory appearance. All along, watch for, and remember, the tissue-thin white 2" blooms of Thimbleberry, for, by August, you will be the hero of your hiking group as you point out where to collect and consume the tasty berries.

On the descent from **Point 11**, you come close to a spring, then-level out and pass two signed trail junctions. Be sure to always take the left-most branches to remain on the loop which will lead you back, past Wahkeena Spring and eventually to **Point**

Harsh Paintbrush

5, the junction with the Angel's Rest you had originally ascended.

Continue along Trail 420 to Wahkeena Spring, between **Points 12** and **13** (the junction with Trail 415) pass samples of Candy Flower, Youth-on-age and Fringe Cup, as well as others to be described in later chapters, such as the pretty yellow Stream (Wood) Violet and the tiny but shiny Little (Small-flowered) Buttercup. One shrub with notable flowers, however, demands to be identified here – the immense Cow Parsnip (*Heracleum lanatum*). Specimens growing up to 10' tall are not uncommon, although the plants encountered along this section of the trail, and just before and after Wahkeena Spring and in the wetter areas along Trail 415, are commonly 5' to 9' tall. These giants are topped in mid-May with umbrella-like clusters of tiny white flowers. There is prominence in numbers, as these many small flowers constitute clusters up to 1' in diameter, making these united "flowers" the largest floral display in our mountains.

Cow Parsnip

Heracleum means Hercules, signifying the mass and strength of these carrot relatives, while the *lanatum* part means woolly, a reference to the hairy stems and leaves. While its floral display is impressive, with 15 or 20 "umbellets" emanating from a single point on top of the huge stem, the leaves of Cow Parsnip bear equal attention. One to 2' broad and long, they resemble huge spiky maple leaves, and originate at a base on the stem which is inflated and winged. In fact, the entire plant looks like it was transplanted from a tropical jungle, although it appears to thrive in our temperate climates from sea level to subalpine locations, and can be found without too much trouble from Alaska to the middle of California. While we may not appreciate it, much of our Cascade flora can be found in our neighboring state to the south, but, according to Northwest botanists, our specimens are much superior.

The casual hiker/botanist will almost certainly notice the Carrot Family (*Umbelliferae*) resemblance in Cow Parsnip plants, and may be tempted to sample a specimen to gain some moisture or to supplement a trail salad. This may work with Cow Parsnip plants, and bovines seem to enjoy them, but they are rather tasteless unless boiled in butter or fat, as the Native Americans did. Prudence is advised, however, when eating any native plants, especially in the Carrot Family. While many such plants are edible, several Cow Parsnip look-alikes are almost as tall yet are deadly poisons, although they usually grow in lower elevation urban and wetland locations.

Chief among these are the ferny-leaved and purple-spotted Poison Hemlock, imported from Europe and famous for poisoning Socrates, and the stout Douglas' Water Hemlock, with its thrice-divided lance-shaped leaves and thickened stems. People have been known to mistake the ferny leaves of these plants for the related Carrot Family plant, edible parsley, or their flat and abundant seeds for fennel or anise, similar plants and also members of the Carrot Family. Should they do so and consume any parts of these species, they may encounter, at times, fatal results. It's best to not sample plants in the wild unless you find yourself in a survival situation and are certain of your greens; flowers don't molest us, so it's probably well advised to leave the flora alone.

As you traverse deeper woods here the hiker will find frequent specimens of Sweet Cicely (*Osmorhiza chilensis*). While this plant can grow up to 2½ -3' tall, it usually assumes an inconspicuous role in the understory, with its rather ordinary toothed leaflets twice divided into three's and its white spray of tiny short-lived flowers. In fact, the most notable feature of this plant is probably its needle-like blackish seeds, which begin to form as the flowers fade in late May and early June, and which can reach 1" in length. It is good to know

Sweet Cicely

this species however, for it is ubiquitous, if never abundant, under conifers in all our deep forests. Its common name, Cicely, derives from a Greek word meaning a carrot family plant with medicinal value, and the "sweet" portion stems from the flavor of its root, which tastes remotely of licorice. A similar plant grows in Chilean woodlands and this was once thought to be identical to our own Sweet Cicely but botanists, ever hopeful of sowing trouble, have now determined it to be a slightly different species.

About 20 minutes of trail-walking past Wahkeena Spring, note the pink ladder of blooms ascending the 4' tall Scouler's Corydalis in a damp area on your right. These close relatives of Bleeding Hearts are not commonly seen from Gorge trails so take a moment to appreciate their lovely architecture and their resemblance to Bleeding Hearts.

You now traverse along fairly level open woods and brushy slopes on Trail 415, encountering many flowers to be described in later chapters or below, including the fascinating Scarlet Columbine in moist areas in the two miles before Angel's Rest, around 1,300' between **Points 12** and **13**, Columbia Windflower in the shade of deeper woods between **Points 13** and **7**, and Broad-leaf Lupine (see below) just beginning to show its spikes of wooly blue blooms in more open areas, especially in the mile after **Point 13**.

Along the way back to the Angel's Rest Trail, and above a brief rocky opening at about 1,900', multitudes of Candy Flower decorate the trail-sides. You eventually reach the junction with the Wahkeena Trail at **Point 7** and shortly thereafter, the junction with the main trail, our **Point 5**, back to the highway (left) or to the top of Angel's Rest (right). Also note, scattered here and there in happy abundance just before and after **Point 5**, the low-growing cheery white petals of Woods Strawberry promising tiny but luscious fruit by July through September.

Two species of wild strawberries grow in our mountains: the more common Woods Strawberry described above and the Broad-petal (or Virginian or Wild) Strawberry described in the Dog Mountain Chapter. Both are closely related to the *Frais du Bois* of European woodlands for which continental chefs and their customers pay dearly. Enjoy!

Turn left at **Point 5** to return to the Angel's Rest Trailhead and view once again the tremendously diverse republic of wildflowers populating this west-side botanic wonderland.

June

The spectacular outdoor garden show on the Angel's Rest Trail continues this month unabated. Many of the same plants will last through mid-July, so both spring and summer hikers should really read both subchapters as significant overlap occurs. Entering deep woods at the start of the trail, **Point 1**, one immediately encounters many of the same plants seen in May and described above, still in glorious bloom. These include the Inside-out Flower, Candy Flower, Pacific Waterleaf, Youth-on Age and Thimbleberry. One newcomer, in more ways than one, is the European import Wall Lettuce (*Lactuca Muralis*). This 2½-3½' tall somewhat gangly invader sports ½" yellow flowers unique among the 22,500-member Sunflower family (the *Compositae* or *Asteraceae*) in producing *only* 5-petalled flowers. All other sunflowers, such as the asters and daisies, have a variable number of petals. (Actually, these "petals" are entire flowers themselves, called ray flowers; the numerous disc petals in the centers of these daisy-like plants are also entirely separate flowers.) It's not only the petals that make Wall Lettuce unique; the shape of its broadly lobed leaves is also distinctive and difficult to describe, even for literate botanists, with flanges on the sides and a 2" to 3" terminal lobe shaped like a maple leaf. Indeed, these Rorschach-blot leaves and yellow blossoms are abundantly interspersed among the understory of many coniferous Gorge forests, especially in areas which have been logged and replanted with Douglas Firs.

Although its common name is lettuce, and it *is* a close relative of garden lettuce, Wall Lettuce has a taste rather like what you would expect a plant to taste like – sort of green and leafy but not really appetizing. It can grow from Columbia River bottomlands up to 4,000' in elevation and appears to have displaced many native wildflowers. However, we can still reflect upon its spread from what may have been a few itinerant seeds hitchhiking on the pants of early immigrants or stealing away in the hold of an 18th-Century sailing vessel, to its present abundance and persistence in a foreign environment. Our country has always welcomed immigrants but many outdoor enthusiasts in the conservation movement are now regarding this *Lactuca* as about as welcome as Himalayan Blackberry and English Ivy and are calling for its eradication. This will prove a tedious task given its prevalence, often in areas not easily accessible.

A more attractive native plant, easily recognized by many outdoor travelers, is the Red Columbine (*Aquilegia formosa*). *Formosa* means beautiful in Latin, an apt name for this plant, with its five yellow cup-like petals surrounded by five scarlet sepals that are reflexed (pointed backwards) to form spurs while at the same time spreading outward in graceful arcs. These reminded early Greeks of a quintet of doves (*Aquilegia*) in flight; more modern observers may think of flashing red and yellow jet fighters in a Star Wars movie. Gently turn over the 1' flowers to more fully appreciate the remarkably beautiful symmetry of their floral parts, which almost seem to be taking flight, and marvel again at how distant evolutionary events could have produced such a pleasingly intricate architecture.

Red Columbine

Look for these 1' to 3' tall plants at the rocky openings around **Point 2** at 250' to 300' of elevation. Although Columbine enjoys moist settings, it also can be found in relatively dry areas such as this, perhaps because of the abundant rainfall on west-side slopes. While the showy flowers of this species draw the most attention, don't ignore the leaves of this striking plant. Resembling those of Meadowrue (see above), they are a pretty blue-green, twice divided into three's (tri-pinnate), and, being rather thin, flutter pleasingly in the slightest breeze. Many Northwest plants growing in the partial shade of conifer forests possess thin leaves, perhaps to rapidly absorb as much precious light as possible.

Several by-paths have now been opened by the Forest Service just after **Point 2**, the second rocky opening, at about 380'. These are clearly labelled as viewpoints to Coopey Falls, and bear the brief side trip as, for some reason (perhaps spring-fed sources above?), the Falls seem almost as turgid and full as earlier in the spring.

A number of the flowers beginning to bloom in May, and described above, continue to color the slopes and woods between **Points 2** and **4** through June, including the Inside Out Flower; Poison Larkspur by the crossing of Coopey Creek, **Point 3**; Cow Parsnip in wetter areas, especially between 350' and 900'; Bleeding Heart, also in wet woods at those elevations; and Sweet Cicely between **Points 3** and **4** in deeper and drier woods. By mid-June through mid-July, the woodlands between 600-1,000' contain the greatest concentration of Columbines lining any trail in the Gorge. These gorgeous, complex, yet easily-identifiable flowers usually are modestly displayed in random locations, usually isolated or restricted to just a few specimens. But at these elevations one cannot take but a few steps before encountering groves of these gorgeous blossoms, almost as if planed by a professional landscaper. Perhaps Nature hertself, at least here, is the perfect gardener.

Accompanying the Columbines is another easily recognized Aster Family plant, Annual Daisy. With white-to-pinkish petals and yellow centers, the small daisy flowers appear to be mis-matched by the overall size of the plant, whose stems can reach over 6' tall.

One additional conspicuous denizen of lower-elevation grasslands and open woods can usually only be seen beginning in early June, and then only at select Northwest locations, such as here, on the drive to Mt. Hood flanking Highway 26 for 5 miles west of Sandy, and around the southern flanks of Mt. St. Helens: the Oregon Iris (*Iris tenax*), also known as Oregon Flag. This gorgeous garden-sized iris begins to display its purple, yellow and white 2" blossoms at about an altitude of 1,000', by the immense boulder on the left of the trail (at around **Point 4**) and beautiful specimens continue to appear in grassy areas and open woods up to the trail junction at **Point 5** and even beyond to **Point 7** at around 1,750'. *Tenax* is Latin for tough – its leaves were used by early tribes to weave hefty baskets for carrying fish.

Yes, this Iris is related to the common irises seen in all our springtime gardens, but it does not transplant well so it should not be molested. Indeed, this is probably as close to a garden plant as you will encounter in the wild; some hikers have trouble believing this Iris is not adventive – that is, a garden escapee. Several apparent wildflowers really *have* eloped from the laundered nature of our city gardens to become established in our wilds (see Rose Campion in the Mt. Defiance Chapter, for example). However, Oregon Iris is a certified native, and thus bears testimony to how widely a genus can travel on the wings of a successful floral design, as irises appear on all northern continents. Such far-flung dispersion is not unusual for the Monocots, a group of flowering plants which include the Lilies, Orchids, and Grasses, perhaps because they were the first such plants to appear and because their genetic make-up allows abundant flexibility.

Oregon Iris

Bronze Bells appear *en masse* on your right at the first set of rocky openings at **Point 4**, about 800'. This is the largest congregation of these plants I have seen in the Gorge. They sprout 1' stems, usually bending toward the trail, and are capped by a series of purplish-brown bells as flowers, most often drooping downwards. Although they look for all the world like some Saxifrage Family sibling of Youth-on-age, they are in fact a member of the Lily Family, as signaled by their thin, parallel-vein leaves and six petals. They continue in these early openings to 900' of elevation and can also be found in several openings above 1,400'.

Negotiating the switchbacks as one continues up the trail this month, between **Points 4** and **5**, the floral biota is nearly continuous. The hiker will pass many flowers in the relatively open switchbacks and on the rocky substrates of these north-facing slopes (between 1,300' and 1,600'), each to be more fully described in later chapters, including the ubiquitous carrot-like Yarrow, the unmistakable maroon-dotted orange Tiger Lily, and two species of purple penstemon (Cascade and Fine-tooth).

In woods and openings, Parsley-leaf Lovage (*Ligusticum apiifolium*) has begun to appear in large numbers between 1,000' and 1,500'. While not as colorful as some of the other flowers near Angel's Rest, this Carrot Family plant is still attractive, with its fern-like lacy leaves gracing a 1-3' tall stem and its compound "umbel" (an aggregation of diminutive flowers packed together in one flower head, all arising from the same point). Indeed, this lovage (and its higher elevation sibling, Gray's Lovage – *L. grayii*) form the prototypical Carrot Family profile, with fern-like leaves and compound umbels of tiny white flowers. This design, copied by many plants outside the Carrot Family as well, ensures that even tiny flowers don't get ignored by potential pollinators. As with many animal species, there is strength in numbers.

Parsley-leaf Lovage

Among this floral abundance, one yellow daisy-like flower stands out not so much for its size or showy plumage, but for the simple purity of its color and form – the Woolly Sunflower (*Eriophyllum lanatum*). Everyone's favorite wild daisy (well, at least mine), this 1-2' tall shrubby plant frequents the rocky slopes between **Points 4** and **6** and can especially be enjoyed just before the left turn at 1,400'. Its cheerful soft yellow rays and deeper yellow-orange central disks, set off above its tangle of grayish wooly leaves, seem almost too perfect, as if the plants were conceived by an impressionist artist and painted against the dull gray background of its rocky habitats. Woolly Sunflower can be found sparkling in many Northwest locales, but nowhere does it exert a happier effect than on the Angel's Rest Trail, perhaps because, so close to the city, one encounters such an engaging patch of the wilderness.

The true botanist describes the pure and simple Woolly Sunflower as a "loosely tomentose perennial with leaves glabrate or glabrous above; the lower spatulate or oblanceolate to obovate and pinnatifid; heads loosely corymbose; phyllaries loosely floccose or glabrate; the pappuspalae 4-10, usually obtuse and arose and not exceeding 2 mm in length" (from Abrams and Ferris, a technical text). Scientists need to employ what appear to be obscure and confusing terms in order to be precise; indeed there are thirteen woolly sunflowers in North America and nine or ten varieties of our species alone, differing in just a few miniscule technical details, such as the length of a seed head or the hairiness of the leaves.

Wooly Sunflower

Fortunately, for us amateur hill walkers, value and enjoyment lie in simply appreciating the color and form of a flower, and perhaps in learning its name.

As one proceeds through the rocky areas from 1,400' to 1,600' and nears the top of Angel's Rest in June, additional discoveries await. Rough Wallflower advertises its head of multiple orange-yellow blossoms around the rocks between **Points 4** and **5**.

The small but fascinating Bronze Bells throws sprays of its fringed bronze tubules just after a right turn at the opening that begins at **Point 4**, then again at the second set of openings beginning around 1,300', while 3-4' tall Nootka Rose appears just before the left turn at the trail junction, **Point 5**, and again at the very top. One rose relative to note here in June, and appearing in many dry open habitats in the Northwest from low to middle elevations, is the Birchleaf Spirea (*Spirea betulifolia*). In rocky openings from 1,400' to the top, this plant crests 1'-tall stems with a single flower head of dull white blooms, a pale equivalent of the many rosy-colored sprays on the shrubby spireas familiar to Northwest gardeners. In addition, this Spirea appears to thrive within a wide elevational range; as is so often the case, wildflowers seem hardier than their more glamorous city cousins.

Strictly botanical names often lack the originality and folksy charm of common ones: witness Wake-robin for Trillium and Johnny Jump-Up for Stream Violet. Often, botanists select a name based upon some characteristic already familiar to them from another species. Thus the leaves of this Spirea do indeed resemble birch leaves, and the leaves of many wintergreens are often pear-shaped (and hence called *Pyrola's*, which means pear in Latin). Confusion comes about, however, when new, related species don't quite fit the first descriptions. Thus, not all wintergreens have pear-shaped leaves and not all bittercresses are sour. Academic botanists, however, have never been known for their mirthful originality, but then again, amateurs like us probably aren't that obsessive to learn all the details.

Birchleaf Spirea

Up on top of Angel's Rest, **Point 6** at 1,640', you can see in all four directions and the view is superb, but don't forget to look around and down as well. The trees on the left (west) are Black Hawthorne (*Crataegus douglasii*), named, as are so many of our plants, after the early intrepid floral explorer David Douglas. Its fuzzy white flowers, borne in clusters at the ends of the stems, and its oddly-shaped leaves with blunt large lobes but saw-toothed margins, are unmistakable. Although its fruits, ½" maroon little

Black Hawthorne

apples, won't appear for another 1 to 2 months, its stinky flowers can be noticed and smelled through the early summer in many Gorge environments. While it may not look like it, this 10-30' tall shrub or tree is a member of the Rose Family.

When enjoying the view at top, also look for Nootka Rose (looking more like a typical rose bush), Yarrow, and the occasional Orange Honeysuckle, the last occurring just before the top. Even at this relatively low altitude, in this exposed location some flowers grow that are more common at higher elevations, including yellow Martindale's Desert Parsley and the Round-leaf Bluebell, (*Campanula rotundifolia*), also called the Bluebells of Scotland. These Bluebells are commonly found on ridges at higher elevations but are somewhat unusual in the Gorge. Cheerfully ringing in the wind, these low- growing but strikingly-colored bells are indeed found in Scotland, and in many Eurasian countries as well. Don't confuse your botanically challenged hiking friends with a literal translation of their Latin species name, which means "round-leaf". Look as hard as you may, by the time this plant blossoms, its round basal leaves have withered, and all you will find are puny narrow lance-shaped stem leaves, furthering the continuing tension between the early scientists who named these plants and those of us who merely follow and try to understand.

Round-leaf Bluebell

For those hardy enough to continue up to Devil's Rest, further flower fields await. From 1,600' to 1,900', **Points 5** to **7**, many May plants are still in bloom, including Thimbleberry and the Oregon Iris, while others, such as Wood Sorrel, have gone to seed. Most outdoor enthusiasts disdain plants which are past their flowering glory days, and probably for ample reason – withered and brown just aren't as glamorous to our eyes, perhaps conditioned by evolution to notice, and be attracted to, youth, vigor, and color. But it is still of interest to glance at the architecture of a seed to appreciate the intricate manner in which nature sculpts its shapes to accomplish a purpose. The football-shaped seeds of this sorrel are ideally adapted to withstand the intense moisture of its west-side habitats; similarly, the freakishly long style (female organ) of the common wild geranium, called Filaree, looks like it belongs to a larger flower because it is so out of proportion. However, it is designed (not really – it just evolved) to penetrate the ground once it has detached itself from the plant so that germination can be assured.

Seeds can also help us identify plants because most biological specimens are classified by their sexual parts. This is particularly true of the Pea Family as exemplified by the Broad-leaf Lupine (*Lupinus latifolius*) so commonly found along this stretch of the trail. All lupines share the same basic seed structure, which looks suspiciously like a pea pod because that's what it is. Open one of these pods later in the season (the seeds follow the flowers by several weeks) and you will discover five to six "peas" cozily encased inside. While you may taste one of these, it is unlikely you will eat more – they are the definition of insipid. Good thing, as many botanists suspect that overconsumption could make you ill.

Broad-leaf Lupine

There are around eleven species of lupine in our mountains and most outdoor visitors will readily recognize the sweet pea-shaped blue flowers of this genus. Believe it or not, there really are five petals, although to confuse you, two are fused. An upright "standard" sits at the back of the arrangement, flanked by a pair of lateral "wings"; two inner and lower united petals form a "keel". While this seems like a random arrangement to us, pollinators must

Creeping Buttercup

love it – there are thousands of species in the Pea Family, which forms the third largest plant family in the world, after orchids and asters. Success may not follow our modern notions of spare functionality. Broad-leaf Lupine is our most common lupine and the one impossible to ignore in mountain meadows and open woods. It grows up to 2½' tall and can be found at almost all elevations.

Lupine means wolf in Latin, but no one knows for certain what the association is with this genus. While lupines may seem aggressive and even predatory, the opposite is true: They enrich the soil for other plants by capturing nitrogen, so plentiful in the atmosphere but so sparse in mountain meadows, and thereby "fixing" it into the ground. They then convert this element to a form that can be utilized by other plants. They accomplish this task by harboring bacteria in nodules on their roots, thus underscoring the often-helpful associations between organisms which make our planet work. Were it not for the blue lupines in our mountain meadows, the wonderful contrast of red Indian paintbrushes might be lost, as those Snapdragon Family plants cannot manufacture their own nitrogen. Despite what we thought in high school, chemistry can be beautiful.

The trail beyond **Point 5** continues with mild up's and down's for about ½ mile until it meets the new trail, designed since the 2017 fire, though it deviates just a bit from the original path, bypassing several muddy sinks in an old logging road. As you progress first gently up, then slightly down the trail, you have the opportunity to spy at **Points 7** and **8** at around 2,000' elevation, bright dabs of yellow color here and there thanks to the ubiquitous European weed, Creeping Buttercup (*Ranunculus repens*). Please don't dismiss this cheerful flower just because it's classified as a weed. While it does grow rambunctiously in gutters along city streets and in lawns wherever sufficient moisture can drench its roots, Creeping Buttercup also provides some magical splash to otherwise drab landscapes, such as the washed-out logging roads cutting through Northwest forests.

Buttercups and cinquefoils are often confused, but our Northwest cinquefoils (Rose Family) are usually higher-elevation plants with a similar appearance but quite botanically different. Most of the time, it's good to simply remember that, as the name implies, the petals of buttercups *look* buttery – they shimmer and shine with a waxy sheen and they are never notched.

Creeping Buttercup

Repens means creeping and this buttercup (there are around nine other buttercup species in our wild-lands) does extend runners, called stolons, which put forth roots at each node about every few inches. Thus patches of these shiny yellow flowers are often seen over extensive tracts of ground, a not unpleasant sight. Creeping Buttercup can withstand an abundance of moisture, a good thing in the Northwest, and is thus found in damp ditches and fields. Not so with many other flowers, even those common in our maritime environment, as many plants are susceptible to root-rot fungi and require adequate drainage; good examples are the lovely Twin Flower and the Evergreen Violet.

Don't simply admire the bright flowers of this buttercup; its leaves are also remarkable. Each dark green leaf actually consists of three leaflets. In fact, what we amateurs call a leaf is often just part of an overall leaf: The entire conglomeration of leaflets in Creeping Buttercup is technically the "leaf 'as this is what emanates from the main stem while all three leaflets hang from the secondary stem. The three leaflets are themselves lobed and toothed, making for a lacy appearance, but what makes these leaves truly unusual are pale green stripes which grace each leaflet, producing a sea of multi-colored green upon which float punctuated splotches of waxy yellow blossoms, not a bad show for what most people consider a troublesome weed. If we have to have weeds, they could be much worse.

The trail now leads to a three-way junction, fortunately signed. Take the middle branch; the right provides another way back to Angel's Rest but has not, as yet (2024) been cleared of annoying brush. The left path leads directly down to the Wahkeena Trail but bypasses Devil's Rest. The middle branch loses some elevation, but regains it shortly as the trail, steep and muddy at times but quite direct, features many Stream Violets, Wood Sorrels and Candy Flowers. A new summertime plant can now be appreciated, at about 2,000' elevation: False Lily of the Valley. This charmer can be found in deeper forest areas, its relatively broad bright leaves topped by cylindrical spikes of delicate white blossoms. The trail, after 1/2 mile, leads you to a side trail to your left heading shortly to the top of Devil's Rest (no views, unfortunately) at 2,450', **Point 9**. Along this brief stretch, note the Foxglove on your left.

At the boulders marking the summit you might admire the intense concentration of Little-leaf Candy Flowers making love to the boulders scattered about the bump called the top. This sibling to the more common Candy Flower, or Siberian Miner's Lettuce, also displays the typical pink candy-cane streaking on its white petals and often seems to grow right out of the rocks punctuating Gorge trails such as Nick Eaton and Ruckel Ridge at middle elevations (around 1,250' to 1,800'). In actuality, this member of the Purslane Family prefers the mossy substrates which coat these rain-pounded boulders, and they are able to derive minerals and other nutrients directly from the moss. In return, they also anchor the moss more firmly to the rock, thus returning the favor.

Regardless of whether you are a devil or merely think you are, you may take a well-deserved rest at this summit, but do try to take a left once down from the Devil's Rest side trail and complete the loop described above in May as the scope and intensity of the flower show only increases. Proceeding down Trail 420C and along the rim the hiker continues to encounter masses of Inside-out Flowers, Candy Flowers, and Star-flowered Solomon's Seals. Do take the short left turn at **Point 10**, 2,170', for a nicer lunch spot than the top, with views north down to the river and out to the Washington side of the Gorge.

Along this rim, large Western Hemlocks thrive alongside Douglas Firs. Scattered on the ground, you can recognize the Douglas Fir cones, with their raggedy mouse's tail emerging from beneath each scale; they are giants compared with the many marble-sized brown cones, which seem out of proportion to their giant Western Hemlocks above. Explorers of more alpine domains find that the more diminutive Mountain

Tall Bugbane

Hemlocks, generally 60' shorter than their Western kin, grow paradoxically much larger cones, perhaps necessary in their mainly harsher environments. Before human logging operations turned first-growth forest into second- and now third-growth woodlands, this stretch of forest consisted mainly of Western Hemlocks; loggers preferred the faster-growing, and thus more profitable Douglas Firs.

> As we travel through the Gorge's magnificent forests, most of us realize that much of what we encounter are landscapes touched by the hand of humankind. Logging, roads and trails are examples of man's passing, but this should not diminish our appreciation of the giants of our woodlands, the Douglas Firs; Western Hemlocks; and the wonderfully shredded bark of the Western Red Cedars, seeking out moist substrates and lining our creeks and rivers. But blowdowns and logging have left many huge stumps in view from all

our trails. Once hulking behemoths, these remnants may appear as dead as the multicolored leaves littering our trails each autumn. Yet, surprisingly, nothing could be further from the truth. Neither the fallen leaves or the apparently lifeless stumps have expired.

Yes, these stumps will never re-grow into full-fledged living trees as before. Nonetheless, they form part of a complex underground system of nourishment, unseen and underappreciated. In fact, recent research reveals that, despite each tree appearing as a distinct entity above ground, beneath the soil, a vast array of interconnecting root systems, fungi, and bacteria comprise a system evolved to provide sustenance to many of the plant and tree species visible above-ground. A stump's roots may interconnect with many others of their brethren still alive; they may actually fuse with the root systems of similar species. Such links constitutes a nexus that allows an exchange of water, carbohydrates, minerals (especially nitrogen and phosphorus), and even micro-organisms which can provide essential elements upon which a tree depends. Studies employing sensors have demonstrated that a stump's roots continue to provide such nutrients to their above-ground kin's root systems, especially at night, when such essentials may be at their lowest ebb.

Does the stump actually gain any benefit from such a nexus? Perhaps; from an evolutionary point of view, it is helping its close kin perpetuate its species, even though, by itself it cannot reproduce. These complex underground alliances are becoming increasingly recognized as essential to the health of all forests and, indeed, all plant (and fungal) growth. What we can see is merely the top-most layer of an interconnected ecosystem which could, without exaggeration, be considered a single gigantic organism, alive without apparent movement or junction, but one which allows our entire earth, and our own species to not only survive, but to thrive as well.

Down the switchbacks, our **Point 11**, many of the same blooms continue to display their cascades of bright color as in May, including the Poison Larkspur and Red Columbine. One new June addition to note, however, is a clump of the strange and spiky False Bugbane (*Trautvetteria caroliniensis*), growing on your right at the stream crossing (rude wooden bridge) approximately one mile from Devil's Rest. This 1-3' tall plant, rarely abundant, stands out nonetheless for its 5-10" long leaves with five to ten deeply-cleft lobes all emanating from the same point. (These are called palmately cleft, as opposed to many leaflets which branch from the petiole in opposite rows, like a feather and are thus termed pinnately cleft.) Of even greater note to casual hikers who come upon this unexpected tropical-looking plant are the strange spiky white flowers. False Bugbane has no petals; instead, startling starbursts of stamens explode from a central point like small silent firecrackers getting ready for the Fourth of July. (Check the Starvation Chapter for an illustration)

It is quite a treat to discover this uncommon flower here and even more so to learn about its place in our green universe. False Bugbane is in the Buttercup Family even though it hardly bears any resemblance to the more classical beauties of typical buttercups or columbines. Family kinship does not rely as much on appearance in the floral, as opposed to the animal, world, but instead relies upon the reproductive parts of plants. Indeed, the stamens of False Bugbane are similar in structure to all buttercups. Many other plants in this large family, not too distant from the roses, lack true petals and instead, what appear to be petals are, in fact, sepals (as in the Columbia Windflower) or congregations of stamens, as in this bugbane. Fortunately, we can leave that distinction to the botanists and simply enjoy the immense diversity within our floral communities.

If you are interested, there is a true bugbane, called Tall Bugbane (*Cimicifuga elata*), growing at a saddle just below the talus field on the Heart-break Ridge Trail on Table Mountain. This route is not recommended here, however. It is somewhat similar to the "false" species described here. The "bugbane" part of these names evidently refers to their supposed insect repellency. Anyone however who has ever tried to employ these plants to ward off the hordes of mosquitoes or biting black flies of the Cascades in July and August suspects that that is how the "false" was actually derived. They may be anemic in that regard, but most hikers and climbers feel the same about the commercial preparations available as well.

Trail 420C ends around 1,550' at the junction where Trails 419 and 420 meet, between **Points 11** and **12**. Just before this junction, you will have descended around three switchbacks and come close to a stream at two of these. Just after the first signed juncture, the striking Queen's Cup has by now begun to announce its presence with its 1' tall six-petalled white flowers while further on, Columbia Windflower, Meadowrue, and occasional Bleeding Heart reward the hiker along Trail 420 before arriving at Wahkeena Spring. This spring, remarkable for the volume of water pouring forth from underground, provides sufficient moisture for several wetland plants to thrive, including Little Buttercup, Stream Violet, Candy Flower and Youth-on-age.

As you proceed along the level after Wahkeena Spring, then climb up the switchbacks aiming to return to Angel's Rest, look for the dainty white to pink low-growing flowers of the appropriately named Star Flower (*Trientalis latifolia*). Barely a foot tall, this plant still demands notice not only because the inch- wide flowers are so cute, but because they seem to float above the greenery due to their thin, flexuous, and almost invisible, stems. A good plant to know, Star Flower is a frequent inhabitant of Northwest forests from June through mid-summer. Although most botanical texts describe its petals, which can number from five to nine, as pink, I see them as mostly white, but with a blush of pleasing pink often at their tips.

Star Flower

Heading back to Angel's Rest on the loop you encounter a second junction, this with Trail 415 at **Point 13** at 1,560'. Between here and **Point 5** many of the same flowers described above under May are still in bold bloom, including the brilliant purple of the Oregon Iris in grassy areas especially on east-facing slopes, balanced nicely by the tall blue spikes of Lupine. These colorful denizens of open areas are particularly prominent in the mile before the trail junction at **Point 5**. Candy Flower and Stream Violet grow jointly in lush arrangements at the wetter locations, while Woods Strawberry frequents those open slopes where the grass hasn't grown so tall as to obscure its low-lying leaves.

Columbia Gorge Groundsel   *Foxglove*

Several outliers as you ascend gently, unusual in this area and to be later described, include a single bush of Spreading Dogbane on the right at 1,550' one mile before the trail junction; ¾ mile before this junction, a single specimen of Foxglove (*Digitalis purpurea*), is also on the right. A most beautiful import from Europe, and now widely spread across our continent, Foxglove's 3-5' spike of stocking-like pink and white spotted blossoms is to be found everywhere along west-side roads from June through August, but only an occasional plant climbs up into real wilderness areas. Apparently, *Digitalis*, the source for the heart medicine and a frequent inhabitant of city gardens, requires some human assistance to disperse but this one must have been a climber.

Two yellow Aster (Sunflower) Family plants can be found along this stretch of Trail 415 in the mile before the junction with the Angel's Rest Trail. One is Broad-leaf or Mountain Arnica, a 2' tall plant with daisy-like flowers which supply surprising bursts of color in all our gloomier mid-elevation forests. The second, the Columbia Gorge Groundsel (*Senecio bolanderi*) is not only more scarce, but in this particular variety, it comes close to being an endemic species found almost exclusively in the Gorge, and more particularly on north-facing slopes in deep woods on the south side of the river. Although lower-growing than many plants in the Aster Family, this Senecio is not difficult to identify (as are many others in this huge genus, with over 1,000 maddeningly-similar species), because it is the only Senecio growing in the forest at these low to middle elevations. Look for this 1-2' tall plant in the 1½ miles before Angel's Rest between **Points 13** and **5** at around 1,600' to 1,400'. Its deeply-lobed stem leaves are surmounted by a cluster of 1" daisy flowers with rays and central disks that are both a deep yellow.

> We should be proud of our endemic plants, true Northwest natives. While many of these flowers are not technically endangered, their limited range pointedly reminds us of how fragile our ecosystem can be; destroying even a small fragment of the Columbia Gorge National Recreation Area might ruin the opportunity we now have to enjoy such narrowly-distributed citizens of our biota as the pink Columbia Gorge Daisy, decorating the wet and overhanging cliffs at Oneonta and Latourell Falls; the Wooly Hawksbeard, residing on rocky outcrops near the Indian potholes on the Ruckel Creek Trail; and my personal favorite, Columbia Kittentails, whose purply spikes of 6-12" flowers are largely to be found gracing north-facing slopes on the Oregon side of the gorge, particularly on the Mt. Defiance and Starvation Ridge Trails but also across the River, in the early spring upper meadows of Dog Mountain.

Along this stretch of trail, several flowers mentioned above can be seen, including Sweet Cicely and Fringe Cup. Another Aster Family plant common here, especially in openings on the left in the mile before the trail junction at **Point 5** is Scouler's Hawkweed. Although related to daisies and asters, hawkweeds look more like dandelions, with their strap-like yellow rays and milky juice if the stem is broken (take my word for it). In fact, there are so many members of the Sunflower clan (technically named the *Compositae* or the *Asteracae*) that botanists have had to divide the family into subfamilies called "tribes" so as to impose some nomenclatural order. Under these tribes come the many genera, followed by zillions of often very similar species differing only in the tiniest of technical details. There are often loud disputes in the botanical world about these differences. All in all, this is a noisy and disruptive family, although often with beautiful individuals. Good thing then that they don't all live in the same neighborhood!

July and August

While most flower-wise hikers believe that the major flower show in the Columbia Gorge occurs in spring, don't neglect these trails in the summer as many plants just come into bloom at this time. If you can tolerate the heat, summer is an opportune time to sample some members of the Lily and Sunflower Families just beginning to emerge, summoned by the warmth of this season. Moreover, stands of several flowers usually seen as single blooms now can be spied in masses uncommon on other trails in spring or summer throughout the Gorge. These include gatherings of Columbines, Annual Daisies, Cooley's Hedge-nettles, Foxgloves, Birch-leaf Spireas, and Scouler's Bluebells. Some of these can be seen by late June as well.

Even the parking lot for the Angel's Rest Trail displays some interesting plants, so before starting out, look for the small purple spikes of Self-heal in the ditches alongside the road, a plant with supposed medicinal properties, which looks and acts like a weed but may actually be a native American. Another parking-lot denizen, and an all too common weed around urban and rural landscapes, is Common St. John's Wort, a 2-3' tall yellow-flowered invader reputed to help those with mild depression (the controversy rages) but causing phototoxicity in cattle attracted to its succulent blossoms. (As an aside, *wort* is the old English name for plant and thus appears with charming regularity in the botanical nomenclature.)

Still out through the early summer months are the numerous Columbines, bequeathing their beautiful symmetry after the bridged crossing of

Coopey Creek at **Point 3** and seemingly teasing hikers with its charm in a variety of settings, both sunny and partially shaded all the way to the top of Angel's Rest and beyond. A congregation of Columbines and Annual Daisies continues their colorful presence as well, especially in the woods after the bridge and up to the openings around 1,300'. As you attain the plateau at 500' and just after the marked side-trails to viewpoints to Coopey Falls, be alert for Cooley's Hedge-nettle (*Stachys cooleyae*) on your right after about 50 yards on this relatively level break in the ascent. This Mint Family species may not smell too minty but its square stems and deep red-to-purple tubular flowers atop 3' tall stems have been used by Native Americans as a pick-me-up tonic and a treatment for healing skin infections. It can also be found, should you undertake the full loop, about 5 minutes after crossing the log bridge once you pass Wahkeena Springs itself, also here on the left.

Those elephant-sized ears you see on both sides of the trail belong to Western Dock, a weed from Eurasia now invading low-lying wet areas throughout city and trail in the west. A more appropriate, and frequently-used common name is Broad-leaf Dock. They are topped by a spike of reddish-brown flowers, each quite tiny on its own but, in aggregation, an apparently tasty sight for pollinating flies.

Past the bridge over the now-dwindling Coopey Creek, our **Point 3**, be prepared for the largest congregation of Scarlet Columbines you will come across not only in the Gorge but in all our mountains. Generally, Columbines stand in splendid isolation but here, between 600-900', just after you make a right switchback after the bridge, the Columbines are massed together as nowhere else which I have seen in all our Northwestern mountains. Usually on the uphill side of the trail, they put on a spectacular show from late June throughout the summer. They are accompanied by large gatherings of Annual Daisies, some exceeding 6' in height.

Up the trail, both before, then following the lookout opening at 900', Wooly Sunflowers share their yellow with the common urban weed, Common St. John's Wort, here not as obnoxious as it is in the city. These yellows are nicely counterpointed by the low-growing purple of Self-heal.

Fine-tooth Penstemon

One delightful summer flower to learn is the Nodding Onion (*Allium cernuum*), which creates swaths of pink in all the open areas on the Angel's Rest Trail from the rocky talus slopes at around 300', **Point 2**, all the way to the top at 1,660', **Point 6**, from early July through early August. It is frequently paired with Woolly Daisy and white, fuzzy Birch-leaf Spireas in all these openings. Look for Nodding Onion especially at the rocky switch-backs between **Points 4** and **5**; don't forget to look closely in order to appreciate the subtle yet rich nuance of the five to ten tiny pink bell-shaped flowers seeming to nod in embarrassment from a single point atop the 1½' drooping stem.

Nodding Onion

All our onions are indeed closely related to the common kitchen plant, as can be deduced by a sniff test. The onion smell is strongest at the root, but don't pull up the plant just to spice up your trailside salad – you can buy better-tasting onions at the grocery store. Still, the odor is unmistakably oniony, especially so in the midst of acres of this plant around 900' to 1,400'. The scent even precedes the flowers' appearance and persists after blooming, a lingering reminder of the pink fields coloring the rocks of early summer. For Nodding Onion aficionados (are there really any?) check out the upper slopes of Table Mountain as well at this time of year for another such show. Also see the Munra Point Chapter for a description of another onion, the Taper-tip Onion

(*A. accuminatum*), which seems to prefer the drier meadows further east.

From the beginning of the trail, and scattered in wooded areas all the way to the trail junction at **Point 5**, Candy Flower appears in floating white masses, with its candy-cane pink-striped flowers cheering up the forests. Other flowers, described above in June, continue to delight the summer visitor, including Red Columbine, scattered in the margins of openings particularly between **Points 3** and **5**, and the Inside-out Flower, found in deeper woods especially between 300' and 1,000'. At the very beginning of the trail, watch for Youth-on-age and, also in the first ¼ mile, the ubiquitous Robert Geranium, an introduced but highly decorative wild geranium providing a riot of pink color in open woods throughout the Gorge from June through August.

Several new flowers just begin to make their welcome appearance in late June or early July, and will happily remain throughout the summer. The Fine-tooth Penstemon (*Penstemon subserratus*) is one of these. Growing 1- 2' tall on sandy banks and gravelly ridges from **Point 2** to the top (300' to 1,640'), it clearly shows off the hallmark characteristics of the penstemon genus, a division of the Snapdragon Family, with its bluish-purple tubular flowers, rich with subtlety, and its opposite pairs of pointed leaves. Look closely inside the tube to see the oddly beautiful arrangement of the four fertile stamens and one wooly stamen resting on the throat of the floral tube (technically the corolla). One feature which sets this penstemon apart from the 80 other species in this genus are the very fine teeth of its leaves; most penstemons have entire (untoothed) or more coarsely-toothed leaves. One of these, the Cascade Penstemon (*P. serrulatus*), also scattered on this part of the trail, will be described further in the Hamilton Mountain Chapter. Often, miniscule technical details differentiate one species from another, particularly in large genera, making life a bit more difficult for us plant-identifying amateurs.

Also new in late June to early July is a common Pea Family plant, the Nevada Pea (*Lathyrus nevadensis*), bearing pink to mauve sweet-pea like flowers at the ends of 1-3' long clambering vines ending in grasping tendrils. The leaves are compound, that is they are composed of opposite smaller leaflets, usually 4 to 10. They are even- numbered and hence lack a leaflet at the tip of the stem. Most peas are weak-stemmed and scramble over other vegetation, using their neighbors for support. This, and the similar Leafy Pea (*L. polyphyllus*), often do so, although they can stand up to support themselves if no helpful plant next door takes up their cause. These peas are close relatives of the commercial Sweet Pea (*L. odoratus*), grown for its fragrant flowers, and the very common, but beautiful weed, the Everlasting or Broad-leaf Pea (*L. latifolius*), which brightens lowland waste places and roadsides (frequently on I-84, I-5, and Washington's Highway 14) with blazes of pink through July to November. It also lines the pockets of unscrupulous collectors, seen cutting this plant and selling it off as Sweet Pea at Saturday Market in Portland and Pike Street Market in Seattle every summer. No need to bother about this practice – the flowers are just as luscious even though they smell a bit rancid.

Leafy Pea

Nevada Pea

As you reach a plateau after the two rock-bound openings at **Point 2**, and just past the new signs pointing to an overlook at Coopey Falls, note on your right a Mint Family plant not too commonly found near trails in the Gorge, Cooley's Hedge-nettle. Its bright pink flowers seem almost too small for such a large bush as they cap stems extending 2-3½' in the air. Each tiny blossom consists of two fused petals above an extended tube, the bottom comprised of three fused petals and marking a landing platform for pollinating insects seeking sugary nectar hidden deep within. Rarely found in bunches, and always near water, this Hedge-nettle will reappear later in your trip – see below.

A white-to-pink daisy is also scattered in the open areas between 500' and 1,250', the Annual Daisy (*Erigeron annuus*). This Aster Family species has 2'- to 5'-long trailing stems and grows at relatively low elevations. This family is notorious for complexity and even seasoned botanists must sometimes resort to a key to properly identify its members. However, here, from just before the bridge over Coopey Creek at **Point 3**, and after this crossing all the way to about to 1,250', these daisies are easily identified. Often seen elsewhere with a pink tinge to its ray petals, these are more white here than pink. Their arching stems crowd over the trail in places and their abundance here during the summer months, particularly between 500-900', represents the densest concentration of these Aster Family species you will ever encounter in any Gorge location.

Erigeron annuus

At last count, 80 species of penstemon had been identified in the Northwest alone. This genus has probably been successful due in part to its floral architecture: The flower tube contains nectar hidden at the back of its throat. To reach this prize, a pollinating insect, often a bee but sometimes a moth or even a hummingbird, must crawl or poke its way past the four pollen-bearing stamens. Once these pollinators accumulate pollen from one flower, they can unwittingly deposit it on the stigma (female part) of the next flower they visit, again lured by the reward of a sugary nectar treat. Brevity forbids a lengthy description of how these flowers avoid self-pollination, but it does have something to do with the fact that the style, which supports the stigma, does not bend downward into the insects' way until after the flower's pollen has been deposited on an insect, thus ensuring cross-pollination. The fifth stamen (hence *Penstemon*) often wears a hairy beard, giving rise to another common name for these plants as Beardtongues, and is thought to function as a guard to protect the ovary from marauding insects. Members of this wily but beautiful and abundant genus have evolved a number of botanical tricks to become widespread in our mountains, a fortunate thing for the mountain visitor, and especially for the many members of the American Penstemon Society - yes, even flowers have spawned organizations.

The floral landscape progressing up the Angel's Rest Trail in summertime continues to impress with a diverse array of splendid plants. Following the second rocky pitch, at about 400', first one sign, then a second 5 minutes later, directs you to viewpoints over Coopey Falls. The side-trails, both to your left, are brief and worth the slight detour as Coopey Creek, even in summer, remains a powerful stream and its falls are among the finest in the Gorge.

Two types of Stonecrop, the Broad-Leaf and Oregon species, cling bravely to the rocks on the right at **Point 2**, while Large-leaf Avens sends up its pretty yellow 1" flowers atop 2' tall stems scattered in moister woods from **Point 3** to about 1,100'. Parsley-leaf Lovage adorns wooded areas from 300' to 1,100' and Yarrow can be found here and there in open areas ascending the switchbacks at **Point 4**.

Farther up the trail, the towering pink spires of Fireweed blaze in openings at around 1,450', while the diminutive flowers of Small-flowered Blue-eyed Mary at 1,300' provide a pleasingly robin's-egg contrast to the spikes of two red paintbrushes, the Tall or Common Indian Paintbrush and the slighter, shorter and hairier Harsh Paintbrush, both described above in May.

A few plants deserve to be more completely examined on the upper reaches of Angel's Rest as they occur elsewhere less commonly. At 1,300', just after the second set of open switchbacks begins, note on the right by the large rocks, two specimens of the shrub Mock Orange (*Philadelpus lewisii*), which grows here to 7' tall. Several additional examples of this proud plant can be discovered to the right of the trail junction at **Point 5**. Its presence here is a bit remarkable as it usually prefers drier east- side slopes and thus provides a rare summertime treat this far west. Its leaves are egg shaped with three prominent "parallel" veins, but it's the summertime flowers which command attention, with 1-2" glistening white petals radiating from a yellow center studded with bright stamens. This shrub is sufficiently showy to be included in some Northwest gardens, although it's almost identical twin, the common garden Mock Orange, is more commonly planted. The flowers actually have a pleasant fragrance, although the resemblance to any citrus plant is a bit of a stretch. I doubt the plant is trying to mock anything.

Native Americans used the strong yet supple wood of this shrub, a distant relative of the common garden hydrangea, for bows, arrows, and sewing needles. They never were forced to contemplate the meaning of the strange Latin name. The ancient Greeks assigned the name *Philadelphus*, which means brotherly love, to a Mediterranean tree with the appearance of a Mock Orange relative. Many years later the generic name was applied to related new-world plants, none of which were even remotely close to the original Greek tree – so much for the logic of scientific nomenclature.

> Botanical keys are employed to help identify families, genera, and species of plants. They often begin with distinguishing characteristics of a family, such as the location of the flower's ovary (above or below a supporting receptacle) or the number and shape of its seedheads. The key then progresses, in algorithmic style, to narrow down the choices, such as whether the flower is symmetrical, whether the leaves are arranged alternately or opposite each other on the stem, or whether the plant has sticky glands or is relatively smooth. As you can see, these keys become progressively more detailed and technical, sometimes to the point of requiring microscopic examination to determine an exact species. Because not too many hikers and climbers I know carry a microscope or even a hand lens on their excursions into the wilderness, and because keys are cumbersome and heavy, they are not included in this Guide. For those interested, the bibliography contains several resources.

A number of other colorful flowers are scattered above 1,000' near the trail. The large and common shrub Ocean Spray, with its bouquets of dull white blossoms, can be found at 1,250' on the right, while Birch-leaf Spirea (see June) is frequent in the switchbacks particularly around 1,350'. Scouler's Hawkweed, with its dandelion-like yellow flowers, is scattered in open areas as you proceed across the rocky stretches here, around **Point 4**, often accompanied by white-topped Yarrow and the Wooly Sunflower, described above in June. Note that, in most of these openings, Queen Anne's Lace predominates as the tall white flower clusters are beginning to curve inward, forming the "bird's nest" shape as it proceeds toward its seedling stage. Above about 1,350', however, though this Lace is still present, it is overtaken in prominence by Pearly Everlasting, also with white clusters atop 3' tall stems. These, however, hold multiple small flowers with yellow disk flowers at each small blossom's center. Living up to its name, these blooms will persist throughout even the early winter months.

Near the very top of Angel's Rest at **Point 6**, 1,640', a number of flowers described for June will remain throughout the summer. These include the Columbia Gorge Groundsel (groundels are also called butterweeds or senecios), the low-growing yellow Martindale's Desert Parsley, and the delightful Round-leaf Bluebell. These two latter species can be found gracing the rocks at the exact top. A diminutive relative of these Bluebells, Scouler's Bluebell (*C. scouleri*), described in the Mt. Defiance and Hamilton Mountain Chapters, grows under the forest canopy somewhat lower down on this trail, between 800' and 1,250', displaying blue-white bell flowers with spreading lobes, much like miniature starbursts of color amidst the green understory.

Continuing along to Devil's Rest during the summer months by turning right (east) at **Point 5**, many Thimbleberry flowers will be out from **Points 5** through **7**, and, by August, don't neglect the tasty dull red berries, the equal of any wild fruit in our mountains. Look too for the continuing display of Round-leaf Bluebell scattered on both sides of the trail, though predominantly on its wind-ward (right) side. These bells seem to favor ridges and other exposed locations, perhaps in hopes of their clappers ringing peals which only insects may hear; if so, such sounds are beyond the ken of human ears. Here also dwell blue Lupine, and the simple but elegant blossoms of the Little Wild (or Baldhip) Rose, one of our daintier and more common wild roses. Along this first part of the Devil's Rest side trail, the Cascade Penstemon is accompanied by a clutch of Tall, or Common, Paintbrush, with linear rather than forked leaves (as on the Harsh Paintbrush, more commonly found along Gorge Trails).

Scouler's Bluebell

The almost-level trail from **Point 5** continues through alder-bedecked woods, accompanied by an army of Candy Flowers with an occasional Stream Violet, then begins, after about 3/4 mile, a dip of about 200', then a steeper ascent through deep forest on a newly-refreshed trail, (since the 2017 conflagration), our **Point 8**, where a few Bleeding Hearts are still in bloom at this time of year, while False Lily-of-the-Valley is thrusting up its 6" spike of tiny white flowers, particularly between 1,900' and 2,100' in deeper woods. Also, at about 1,950', just at a left turn in the trail, an opening reveals an astounding sight: an astonishing display of hundreds of Foxgloves. These garden escapees were represented just a few years earlier by a single specimen near the top of Devil's Rest but here, they have taken advantage of a meadow laid open by the 2017 conflagration and paint this meadow various shades of pink, purple, red and white.

You must lose about 200' of elevation, then regain it to come upon a new sign, small and a bit confusing on an alder tree straight ahead. This provides an option to descend directly to the Wahkeena Trail to your left and to loop back to Angel's Rest to your right, but I recommend proceeding straight ahead to achieve Devil's Rest. Steep at times, the newly broadened trail, muddy in spots, even in late spring and summer, brings you to a side path just below Devil's summit. Take a short left to reach this rocky, but view-less perch, our **Point 9,** at 2,450'. No great outlook exists here but the boulders invite some scrambling and a convenient water and lunch-time break. Peer over the north side of the topmost rocks to see a wonderful display of Round-leaf Bluebells, a sight to make any Scotchman proud. Also, just below the rocks to your north dwell Small-leaf Candy Flowers, thriving off the moss-covered rocks just below the summit.

Descending from Devil's Rest, after about five minutes' walk, note on your left, a mass of white hanging flowers atop 2' tall stems which hold arrow-shaped toothed leaves. This is Western Rattlesnake-root (*Prenanthes alata*), rare in the Gorge. In fact, at least along trails, this plant can be seen only here and further down, on the switchbacks descending from the 2,000' plateau, and again about 10 minutes' walk up from Wahkeena Springs. This plant is not listed as endangered but if so, it must be hiding well away from the major Gorge trails as I have only seen it on this loop.

While the gorgeous Tall Larkspurs have mostly gone to seed, you will still find many Columbines, especially down the switchbacks before **Point 10**, and, just emerging from their long winter's slumber, the delightful Queen's Cup, with its six strap-like white petals illuminating forest floors throughout the Northwest. This Lily Family member can be most easily observed along the rim at 2,000' and through the woods up to **Point 11**.

Later in August, Foam flower can be seen, primarily on your left, all along the 2,000' plateau. It is usually the three-leaved variety *trifoliata*, with each leaf holding three leaflets joined before the middle of the leaf. But a few here seem so deeply incised, with the leaflets so sharply delineated, that they appear to be three separate leaflets, cut almost all the way to the stem. Thus they could be considered to be the variety *lanceolata,* or "Cut-leaf Foam Flower". However, the vast majority of Foam Flowers you will see in the Gorge are the *trifoliata* type.

As you descend from the plateau, you will encounter a series of switchbacks, the third of which brings you to within sight of a stream. A few feet further on the trail, spot a number of examples of Cooley's Hedge-nettle on your right but then, a repeat performance of the unusual and infrequent member of the Aster Family, a congregation of white tubular blossoms. It is again Western Rattlesnake-root. Atop 2' tall multiple stems droop white heads of ray-only flowers, perhaps nodding in deference to the summer's heat. This site is the sole location in which I have witnessed more than one such bush of this recondite species.

As you descend toward Wahkeena Spring, you cannot help but notice that you are accompanied almost continuously by blooming soldiers guarding the shoulders of most of the trail – Candy Flowers. They cheer you on, despite their diminutive size, just by their sheer numbers, their pink candy-striped petals and the knowledge that they've persisted through April's showers all the way to summer's swelter. Don't forget, as well, that they may still be present throughout the fall, a testament to their perseverance regardless of what Nature may throw at them in terms of moisture and heat, or lack thereof. Occasionally accompanying the Candy Flowers, pink tubular blossoms of Turtlehead poke their 3' long arching sprays out over the trail, low but quite visible; note that these are so close to penstemons in structure that some botanists list them in the Penstemon genus.

About 10-20 minutes past the Spring, you will come across an amazing site: a congress of Cooley's Hedge-nettles in such abundance that it would appear both the House and Senate have been called back from their summer recess. Their bright pink two-lipped flowers, though tiny, compensate through their sheer numbers, all the more surprising as, though moisture is omnipresent through all the Gorge, there is no large stream or creek nearby. Look closely at their minty architecture, described earlier in this subsection, as you will not encounter this species in any other trail-side location of which I am aware in any of the Gorge trails described herein.

A rocky opening a half-mile from **Point 13** may still exhibit some White-flowered Hawkweed and Turtlehead throughout the summer and also affords a nice viewpoint for a rest and a view across the Columbia to Table Mountain. Just 10 minutes from this talus-strewn meadow, and a few minutes after a left turn in the trail, sufficient moisture persists throughout the summer months for Cooley's Hedge-nettle to again display its bright pink tubular flowers at shoulder height on your right. Just before a rocky opening permitting a nice view of the river, and a great snacking spot, a field of Scouler's Hawkweed thrives, while at the opening itself, Lupine and Fine-tooth Penstemon predominate. Scouler's Bluebell awaits a left turn above the rocky opening you just passed, while Birch-leaf Spirea, Meadowrue and Bleeding Heart complete the journey back to **Point 5** and the reunion with the Angel's Rest Trail. Along the way, Fireweed is blazing its hot pink blossoms atop 6' tall stems in many of the openings you will encounter along this stretch of trail.

Thus, the advent of summer does little to diminish the floral galaxy of plants to be enjoyed at Angel's Rest and beyond. While spring may boast a greater number of species, summer brings forth a different, but equally interesting array of blooms on this easily accessible west-side hike. Enjoy this area of the Gorge in July and August but don't forget your water bottle in the heat. These plants are tougher than you when it comes to surviving the summertime drought!

September and October

Hiking the Angel's Rest Trail in the fall does not mean sacrificing wildflowers for the burnished colors of autumn. While the larger spectacle of *en masse* floral drama may be at its end, a more subtle, and even quirky, display of plant life persists for the discerning Columbia Gorge visitor at this more tranquil time of year. An added benefit are the colors of the Maple and Alder trees and their scattered multi-colored leaves resting on the forest floor, especially below 2,000'. Shades of golden yellow, orange and brown create a kaleidoscope of hues beneath your boots. Most folk think these leaves have "turned" from green to their autumn shades. In fact, all that's occurred is that the green chlorophyll fades away, leaving the autumnal colors bare underneath.

Starting from the parking lot, note on the south side of the road the common weed, Lady's Thumb (*Polygonum persicaria*). It thrives in ditches alongside many byways throughout the West, sending up a 4-6" curled spike of reddish flowers above 1' tall stems. Unfortunately for the lover of vivid color, however, deep woods in autumn are rather barren. Nonetheless, those who relish the obscure could inspect the unusual seeds of forest plants now abundant with potential life for the next year. The Inside-out Flower, whose strangely inverted white flowers were so prominently displayed throughout the spring and summer, now shows its brownish seeds as tiny curved packets within a ½" black follicle which ultimately will burst open and spread them as far as possible. This plant is especially abundant in woods during the first ½ mile and between **Points 4** and **5**. Look too for the lustrous black seeds of Candy Flower, scattered in masses in deeper forests from the bottom to the trail junction at **Point 5**.

Even more noticeable, and more valuable for the hungry hiker, are the "seeds" of Thimbleberry, found particularly at the start of the trail, and between **Points 5** and **7**. Seeds of the Rose Family shrub, of which this luscious plant is a member, are encased in a fleshy envelope we call a fruit. Many fruits within this family, such as the common orchard apples, pears, and cherries, are delicious. The edible parts are actually the ovary and surrounding tissue from which the seeds are meant to derive nutrients. Many of these, such as Thimbleberries, remain for the picking and eating into September in our mountains, making fall hiking that much more of a pleasure. However, not too many other berries in the Gorge are edible, although almost none are poisonous. Best not to experiment unless you are certain of what you are consuming.

> Most folks tend to ignore plants as they go to seed; indeed "going to seed" is a derogatory term implying a downhill progression toward uselessness. In the botanical world, however, nothing could be further from the truth. Seeds, however withered and plain they appear to our superficial gaze, form the well-springs for future generations of each particular species. To the plant, undoubtedly lacking our nervous system's attraction to color and form, seeds are simply an integral and necessary cog in the cycle of reproduction. Fallen to earth, seeds know they must bury themselves in dirt, but then generate a shoot, ever growing upwards in order to reach the light of day. Seeds, however, cannot foster a new plant's growth without first spreading beyond their parents' home turf so as to spread more widely and not compete with the mother plant. Seed dispersal is a subject about which entire books have been written. Common mechanisms which Northwest plants employ include gusty winds, the splash of frequent raindrops, and the eating habits of a variety of birds and animals, who consume fruits and seeds, then deposit them relatively unchanged in their far-flung wanderings as droppings. Apparently, plants aren't squeamish when it comes to their survival.

When examined closely, many seeds may seem strange, but none so much so as the long cork-screw-shaped style of Filaree, of the Geranium Family. All such species have long pointed styles; see these here at the beginning of the trail on the still-flowering pink Robert Geranium. The capsules of the related Filaree, more common to the east, which form as the pretty pink flower petals drop to the forest floor, are surmounted by a 1½" long spirally-twisted tapering rod which, for all the world looks like a carpenter's wood screw. While its form may seem whimsical, this seed head performs an act of germination as unbelievable as a Grade B science fiction movie. As this helical rod drops to the ground in the fall, intermittent rains swell the style while during dry spells, it shrinks. This alternate expansion and contraction actually enables the screw-shaped seed to twist itself into the ground, much as you would screw a light bulb into a socket. It thereby ensures that a new plant will take hold. Evolution makes no small plans.

After balancing your way up the two rock-strewn openings at 250-300', our **Point 2**, note two new signs, the first followed closely by a second, both promising views of Coopey Falls. While the Creek and Falls are in autumn's decline, they still merit a visit on these short side-paths.

Not all floral color is absent during the fall season at Angel's Rest. The above-mentioned Robert Geranium still sports an occasional pink blossom in the woods between **Points 1** and **3**, while Wall Lettuce (see June, above) waves its ½" yellow five-petalled flowers above its weirdly-incised leaves, at times clear through November. Wall Lettuce blooms as well in higher woods between **Points 3** and **4**.

Autumn is also the time for daisies and asters to thrive, especially the latter. Annual Daisy (see July and August) often retains its pinkish flowers throughout mid-to-late October, beginning just after the bridge, **Point 3**, and especially in swarms between 800' and 1,250'. Two other Aster Family plants continue to bloom throughout the fall: Pearly Everlasting, with its papery yellow and white flowers almost appearing as perpetual buds; and the omnipresent Yarrow, with its head of small white flowers pressed tightly together and lacy leaves which look like ferns. They can be found in all the rock-strewn open areas between **Points 4** and **5**. Neither of these plants resemble our common notion of what a daisy or sunflower is supposed to look like. However, the multifaceted *Asteracea*, also known as the *Compositae*, or Sunflower Family, throws an array of sizes and shapes at the botanist, at times without much apparent order. Scientists have attempted a better understanding of these plants through classifying and sub-classifying them, but nature can still, at times, defy categorization.

One other flower that may still be in bloom in September is the Large-leaf Avens, growing sparingly in open forest just before and after the bridge at **Point 3** and between **Points 6** and **7**. It persists usually where some remnants of moisture remain throughout the fall. Its 1-3' tall leafy stems are surmounted by 1" bright yellow flowers and, even though this plant, a member of the ever-present Rose Family, can be found in weedy roadside areas in our cities, it nonetheless serves as another pleasant reminder that traversing the mountain wilds in autumn can still produce pleasant floral surprises – not to mention frequent spells of cool, clear, and much-appreciated weather.

As you gently ascend the switch-backed trail above 1,100', a variety of openings allow vistas of the River below. Note the striking clutch of Canada Goldenrod on your left between 1,350' and 1,400'. Strangely isolated, this batch comprises the sole example of the common flower along this entire loop. All Goldenrods are members of the Aster Family, and examples such as the Annual Daisies noted as you ascend this trail, and the Goldenweeds of higher meadows in the Gorge (see the Nick Eaton, Ruckel Creek and Dog Mountain Chapters) give testament to the longevity and hardiness of these Sunflower relatives. Many members of the Aster Family have evolved to synchronize their late-summer to early-fall blooming times with their preferred insect pollinators, cold-loving insects more robust than many of us hikers. Many are disseminated by the wind (think of the fluffy dandelion seeds) and, once pollinated, can take their time awaiting autumn's frequent gusts to ride the air to far-flung, and sometimes more propitious, locations. It would not be unusual to spy Annual Daisy in bloom up until late October.

After the left turn just short of Angel's Rest, our **Point 5**, and continuing throughout September, and even into mid-October, Candy Flower accompanies you through the woods almost to the top of Devil's

Rest and, indeed, surrounds the top-most rocks as well. Note that the loop trail dips a bit, then begins a gradual ascent to a junction; the trail to your left leads down to the Wahkeena Trail, the one to the right takes you on an alternative path back to Angel's Rest, but I recommend proceeding straight ahead The trail dips a bit more, then climbs steadily to the viewless top of Devil's Rest. This guarantees greater exercise and a longer trip around the Loop, if you're up for it

A few Steam Violets can be seen along this way accompanied by the ever-present Candy Flowers, in masses through September and into late October as well. Descend the brief side-trail from the top of Devil's, turn left and after just five minutes' walk on the Wahkeena Trail, a gathering of Western Rattlesnake Root can be spied on your right. You descend to a plateau, at about 2,000', which features three side trails left to lookouts over the River. Take the third; it has the most open view and rocks shaped well for resting and eating. As you descend from the plateau, after the first left switchback, you will hear a stream. The following three switchbacks all harbor masses of Cooley's Hedge-nettle, bright pink Mint Family flowers atop 3' tall stems.

Note that all along this way, and even after descending to the trail junctions and past Wahkeena Spring, Candy Flower persists as a constant companion, even into late October. Foam Flower still rings its tiny white blooms masquerading as bells throughout the mostly-level path along the plateau and, in occasional openings, Fireweed still may be in bloom through early September; later in the month almost all these specimens will showcase their fluffy white seed-heads; these often litter the trail, so numerous that, in some sections, it's as if you're walking on a bed of wooly white wisps rather than a trail itself.

As you descend to Wahkeena Spring, numerous Candy Flowers are scattered all about. After 10 minutes from the Spring, you will again encounter Cooley's Hedge-nettle throughout September. By October, just the remnants of its seeds, as well as those of Fireweed, survive but never fear, they will bear a new floral display the following summer. However, the most prominent flower here is a continuation of the masses of Candy Flowers you've passed on your way up to Devil's Rest.

Continuing along the up-and-down trail past the Spring (mostly up it always seems to me), Varied-leaf Phacelia predominates, a legacy of the Eagle Creek fire of 2017. Its presence along many Gorge trails is much more abundant than before the conflagration perhaps owing to the opening of light and the lack of competing shrubbery. It is also possible that, as with Fireweed, its seeds are activated by the heat of a fire but I can find no reference to this in the many botany texts I've come across, though I certainly haven't digested them all.

Fireweed still stands tall as you make your way back to the Angel's Rest Trail, **Point 5**, then descend to the parking lot, almost always entirely filled even in the fall (along with the upper lot as well, even on weekdays let alone weekends), testimony to the popularity of Angel's Rest hikers and the trail's proximity to Portland.

Overall, the Angel's Rest/Devil's Rest/Wahkeena Loop described here will introduce the interested outdoor traveler to a vast botanical domain which forms a typical west-side display throughout the growing season. Moreover, all this is within an easy drive from the Portland metropolitan area. As well, this trip is often snow-free earlier than other higher-elevation trips described in later chapters both in early spring and late autumn

It also requires less effort than many alternatives in the Gorge – not a bad combination of advantages to explain the burgeoning popularity of this premier Columbia Gorge outing.

Barry Maletzky, M.D.

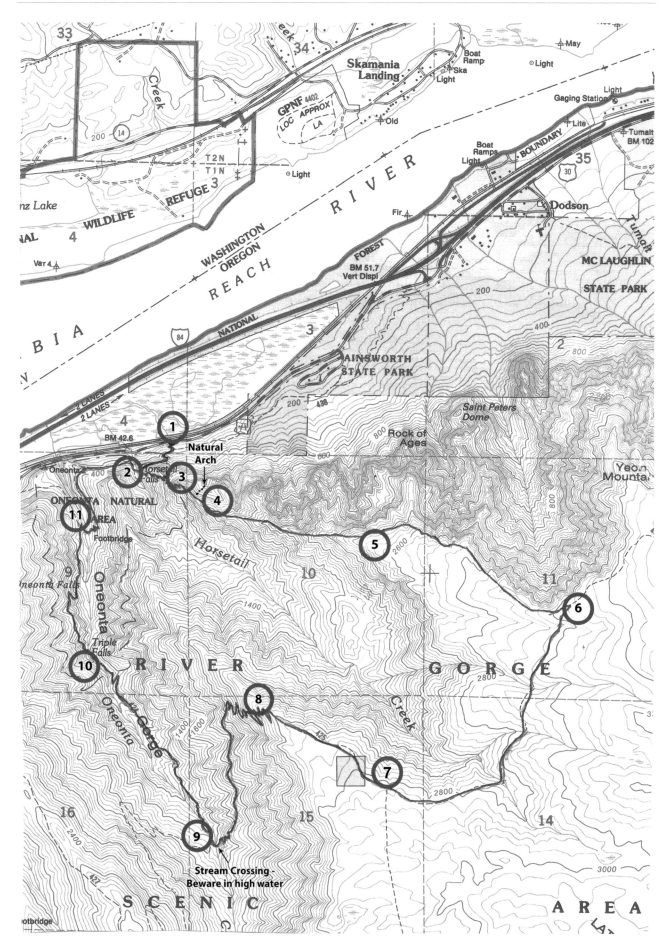

Rock of Ages/Oneonta Loop

The dictionary definition of "steep" should include as an example the Rock of Ages Trail but, if you don't mind mud, and a bit of scrambling, this "trail" may be a way for you to enjoy a varied and interesting flower show while at the same time fulfilling your exercise quotient for the week. This path is not advised for young children or as the first hike of the season.

Following the initial switchbacks of the Horsetail Creek Trail, Rock of Ages heads off to the left just after you curve into the canyon fronting the torrent of Ponytail Falls. If you go behind the falls, you've gone too far. A sign ("This trail is not maintained") 25 yards up this small offshoot of a path is neither reassuring nor close enough to the main trail to be seen at first. Never mind – the way is always clear and the adventure will satisfy your soul, even if it wrecks your knees.

Many hikers forego the pain of descent by looping back to their vehicles on the Horsetail/Oneonta Creek Trails (numbers 425 and 424, respectively). However, this loop trades the steep descent of Rock of Ages for the un-spanned crossing of Oneonta Creek. Folks not adept at rock hopping or log walking should carefully weigh the options, especially on an early spring afternoon, when the creek seems as quick and angry as a bunch of wasps defending their nest. Most people manage the creek well while still cursing the lack of a bridge. If creek crossings aren't your favorite part of the wilderness experience, you can still complete a loop by hanging a left on Trail 425, ascending to Nesmith Point, then heading down the Nesmith Point Trail (number 428), although this will require a short car shuttle. Either way, loop trips are fun – they provide a way to experience diversity without hassle, a special bonus to flower lovers everywhere.

CAUTION: The cut-off from the Horsetail Creek Trail all the way to the plateau at 2,480' is not well-maintained, rugged, and steep, though easily viewed and followed. The cutoff to the natural arch at **Point 3** and the traverse across the Devil's Backbone at **Point 4**, while not requiring technical climbing skills, should best be attempted by those with some experience using hands for balance; this should not be the first hike ever undertaken for anyone and might best be reserved for those experienced in cross-country skills. In addition, as mentioned above, the crossing of the main branch of Oneonta Creek at **Point 9** may be very difficult in all the winter and spring months, up until June. Only those accustomed to well-balanced rock hopping or wet log balancing should attempt this part of the loop. In these heavy run-off months, it may prove prudent to simply complete the first leg of this trip, up to the plateau at **Point 5**, then call it a day and return the way you came. This will have provided you most of the work-out anyway and avoided an unceremonious, or worse, traumatic dunking in those swollen rushing waters.

February and March

Joining the other cars at the Horsetail Falls parking lot, **Point 1,** early season hikers can expect to see some weeds abutting the concrete, but these often seem to change. During many years, Common Groundsel and Hairy Western Bittercress predominate, but Common Chickweed has also been present, though never abundant, apparently preferring lawns to sidewalks. These weeds may not look pretty but they are probably not as harmful economically as agricultural weeds can be. But heading up the switchbacks, this story takes a darker turn: Masses of Robert Geranium, also called Herb Robert (*Geranium robertianum*) cluster under tall Douglas Firs, thriving in the unnatural soil and shade. Pleasingly pink-to-purple blossoms innocently nod above lacy downy leaves, seemingly oblivious to a controversy raging in the cities: Should this decorative plant, an invader from Eurasia, be more aggressively controlled? Critics cite case after case in which alien plants have destroyed native ecosystems, yet more permissive proponents believe that drastic elimination measures would be too costly absent proof that this small flower is really causing harm. While it is common to conclude that only more research can inform these debates, funding for such studies is often as hard to find as a sunny day in February.

Robert Geranium

> Most botanists believe that many plant immigrants spread so widely and are so destructive because their adopted country lacks the bacteria, viruses, fungi, animals, birds and other destroyers which naturally evolved at home to parasitize or consume, and hence control, these aliens. When such a plant is accidentally or purposely transported to a new world, it can thrive unhindered by these organisms. A new study underscores the importance of these controlling factors. Biologists examined 473 European plant species introduced into this country, either on purpose or by accident. On average, they found that foreign plants in the U.S. were infected by 84% fewer fungal species and by 24% fewer viral species than the identical plants growing in their natural European ecosystems. This means that there are many fewer infectious agents to attack these plants in their adopted country. Of additional interest, the lighter the burden of pathogens, the more states in this country officially listed the plant as a "noxious weed". It appears that not only are there fewer controls on these plants in their new homes, but that they may be resistant to any novel pathogens they may encounter in their new world.

Around the bend heading into the gorge holding Pony Tail Falls (just before **Point 2**), some Plume-flowered Solomon's Seal is beginning to thrust its broad pointy leaves and spike of showy white flowers out by early March, seemingly in a greeting to the new spring, while a few specimens of Large-leaf Avens and Fringe Cup are just starting to appear. Starting up the steep side trail at **Point 2**, however, not much is happening botanically at this early time of year. Nonetheless, be on the lookout for the broad three-parted leaves of Vanilla Leaf, just beginning to send up their fuzzy stalks above the forest floor. Sweet Cicely is also beginning to bloom, adding to the sometimes-subtle romance of early spring in the Gorge. As an early springtime bonus, be sure to note the two specimens of Columbia Kittentails, an early favorite, with gorgeous purple flowers on the right at 950' elevation.

April

There is a certain satisfaction in seeing the progression of plants budding and blooming with the advance of the seasons in the Northwest. By April here on the switchbacks of the Horsetail Creek trail, masses of pink-purple Robert Geranium carpet the forest understory between **Points 1** and **2**, while Candy Flower and Plume Solomon's Seal provide bright points of white. Bleeding Heart is frequent as well. Spots of yellow indicate Large-leaf Avens, and pink here means Bleeding Heart. The Avens also frequents Portland and Seattle landscapes, growing in ditches, along roadsides, and in vacant lots and other blighted urban ecosystems. A Rose Family plant, the Avens is closely related to the cinquefoils. It can grow up to 3' tall, and is distinguished by its basal (lowest) leaves, which sport a large terminal leaflet, below which are progressively smaller leaflets down the petiole, the opposite of common plant architecture.

To further confuse things, this Avens grows in close association with Creeping Buttercup, another yellow-petalled flower. Cinquefoils and avens, however, often have notched petals which are flat and dish-like, while buttercups *never* have notches and, true to their name, most often have waxy shiny petals. Plant families are often distinguished by their seeds but, although in different families, both these genera, the cinquefoils and buttercups, often also appear to have similar seeds, with aggregations of bristly hooks which catch on fur and clothing, thus aiding dispersal to new locations.

As you head into the canyon holding Pony Tail Falls, look for the strange frilly flowers of Fringe Cup and the similar Youth-on-age. Many examples of Vanilla Leaf, Candy Flower, and Hooker's Fairy Bells are present here as well, but don't miss the sharp left turn onto the Rock of Ages trail at **Point 2**. Heading up this steep trail, Big-leaf Sandwort begins to appear, with Star-flowered Solomon's Seal and, between 400' and 800', Baneberry. Vanilla Leaf sends up its spiky plumes to about 500' elevation, while Sweet Cicely also provides spots of brightness against the green maze of spring vegetation. Be certain to take the rough but worthwhile side trail, at **Point 3,** to the natural arch at about 1,000', where a small yellow flower is just beginning to bloom. This plant, Martindale's Desert Parsley (*Lomatium martindaleii*), belongs to the Carrot Family (*Umbelliferae* or *Apiaceae*) and has several carrot-like characteristics, including finely dissected leaves, blue-green in this plant, and small flowers bunched together in umbrella-like clusters. However, it is an un-carrot-like ground-hugging plant, as opposed to most Carrot Family members such as the stately Queen Anne's Lace and Poison Hemlock. At any rate, most hikers at this spectacular viewpoint are more in awe of the view than the flora: steep headlands fronting an arch of vertical broken basalt above a canopy of Western Hemlock and Douglas Fir, one of the more extraordinary vistas in the Gorge.

Martindale's Desert Parsley

Back on the main trail, you tiptoe across Devil's Backbone at **Point 4** and, here and beyond, encounter several open rocky areas between 1,400' and 1,500'. In these openings, masses of the Large- flowered Blue-eyed Mary and its smaller sibling, Small-flowered Blue-eyed Mary, provide acres of blue, punctuated here and there by dots of crimson from Harsh Indian Paintbrush and the pink glow of Rosy Plectritis. The mouse-eared white petals of Field Chickweed wave in the frequent breezes, accompanied by the yellows of Nine-leaf Desert Parsley and the two common Oregon Grapes: the dull Cascade species found in woodland shade and the glistening-leaved taller Shining species, which flourishes in more open sunny areas.

In the woods beyond the Backbone, Star-flowered Solomon's Seal pokes its bright white blossoms up above its pointy strap-shaped leaves, while the occasional Trillium and the intricate and intriguing Calypso Orchid provide spots of color amidst the green maze. The shy, drooping, but strangely honest lavender four-petalled flowers of Oaks Toothwort dot the green carpet as well, particularly between 2,350' and 2,400'.

On the plateau, between **Points 5** and **6**, the yellow Evergreen Violet and the occasional Trillium stand out amid the tangle of Slide Alders and Vine Maples thriving in these flat wet environments. The jumble of these thin hardwoods, random and entwining, remind me of a Jackson Pollack painting, entrancing but difficult to navigate. At **Point 6**, 3,040', be sure to take a right down the Oneonta Creek Trail if making the loop, and prepare for the stream crossing below.

Just as in human families, two famous and colorful plants, the Tall Larkspur and the Corydalis on the Angels/Devils/Wahkeena loop, have relatives both lesser-known and somewhat more shy. The White Rock Larkspur (*Delphinium leucophaeum*), a shining white delphinium with central blue spots, grows only on cliffs and ledges in four counties in northwestern Oregon; Cold Water Corydalis (*Corydalis aquagelidae*), with more finely dissected leaves than its more common sibling and with paler flowers, is known from just three populations in the drainage of the Clackamas River and from a sole isolated spot high in the Columbia Gorge. Both plants are narrow regional endemics, considered to be threatened throughout their limited ranges. It is

> far from uncommon to learn about rare species closely related to better-known ones. They may differ only in minute technical details, but they are distinct species with their own habits and habitats. Often isolated and unappreciated (perhaps, again, just like in some of our own families), they remind us once more of the slow and special magic of evolution.

Heading down this loop on Trail 425, Stream Violet and Candy Flower punctuate the forest landscape with points of yellow and white against the mossy green at old stream courses, while Little-leaf Candy Flower can be found growing right off the rocks on the left just up from these frequent crossings. At this time of year also expect masses of Trilliums, mostly still sporting a gorgeous sheen of white, with a few just showing some burgundy streaks in the open woods and around streams between **Points 6** and **7**. As the hiker heads across a plateau at around 2,850', not many other flowers are yet in bloom, but, after crossing the west fork of Oneonta Creek and passing the junction with the Bell Creek Trail (**Point 7**), expect to be dazzled by two colorful denizens of deep and damp northwestern forests, as uncommon as they are spectacular: the Tall Larkspur, displaying the deepest purple the eye can discern; and the mysteriously beautiful Western Corydalis, in masses at streams between 2,100' and 1,900', its oddly-shaped pink blossoms stacked one on top of another like a wobbly deck of cards.

While Corydalis can be grown in the garden, particularly a yellow variety, it is only rarely seen, perhaps because of its finicky nature – it requires a rich, acidy soil and constant moisture. However, it can certainly be enjoyed here and near or in streams at low elevations in the western Gorge. Due to its scarcity, you should never attempt to transplant it from its natural settings. Opportunities to view this gorgeous flower occur at Multnomah Creek above the famous falls and at the big stream crossing on the Angel's Rest Trail. Look closely at the marvelous architecture of this plant: Between 10 and 25 horizontally-angled flowers perched one above the other are balanced on a stem up to 4' tall, each flower resembling a bird in the act of flight (corydalis means "crested lark"). Look below as well to appreciate the blue-green lacy foliage hovering over the damp environments this plant inhabits. These leaves betray their family resemblance to Bleeding Heart as both are members of the Fumitory Family. These magical plants flourish here as nowhere else, and cast a strong and mystical spell on these enchanted forests.

Other plants on this section of trail are also beginning to open their buds in April. As one descends the switchbacks between **Points 7** and **8**, Fringe Cup and its close floral associate, Youth-on-age, make their first appearance, the latter always crowding out the former the wetter the site. These plants are particularly common now below 1,800' as one descends toward the river crossing at **Point 9**. The silky tassels of Western Meadowrue and the drooping lanterns of Hooker's Fairy Bells can also be seen all along this stretch of trail as well. Lending a particularly quaint feeling to this area is the constant understory of Wood Sorrel between 2,500' and 1,700', the flowers just beginning to bloom.

After (carefully) crossing Oneonta Creek at **Point 9** (1,400'), look for Bleeding Heart, thriving in these moist and shaded woods. At this early time of year, Fringe Cup and Youth-on-age are just beginning to bloom, while Meadowrue and Hooker's Fairy Bells are sprinkled liberally along the trail as it parallels Oneonta Creek after crossing back to the east side on a bridge. (Where was a span upstream, when you needed one?) Just above and below Triple Falls, **Point 10**, masses of Scouler's Valerian are beginning to brighten the open woods, especially in rocky open areas by the side of the trail, while Merten's Saxifrage sends up its 10-12" spikes topped by dull white panicles of flowers on the wet rocks below the now-bridged crossing.

Woods Strawberry is frequent in openings below the Falls, as are the emerging popcorn-like flowers of Common Cryptantha. Openings, both natural and man-made, seem to provide handy habitats for several of our common species. Candy Flower is an excellent example of a plant which has flourished in altered landscapes. A denizen of open woods, it is far more abundant today in second- and third- growth forests than it was two centuries ago under the dense shade of the old growth.

The mossy rocks on the right just past Triple Falls also hold a special treat for lovers of arcane plants. Look closely on these rocks to see a miniature version of a monkey flower poking its yellow head out from the grey surroundings. *Mimulus alsinoides*, or the Chickweed Monkey Flower, prefers wet cliffs but is sometimes found in drier areas as well, as for example, growing right on the steep rocky trail up to Ruckel Ridge. Most often, however, as here and on the Hamilton Mountain Trail just before Rodney Falls, it thrives in the constant drip and spray so common to these woods in early spring. While its larger sibling, the Common Monkey Flower, is certainly more prominent and plentiful, a close inspection of this small plant reveals the delightful structure typical of all the flowers in this genus. In Chickweed Monkey Flower, the snap- dragon-like two-lipped yellow blossom sports a reddish-brown splotch in the middle of the lower lip, whereas in Common Monkey Flower this splotch is actually a series of dots. Combined with the stamens and pistil just above, these blooms give the appearance of a tiny funny face, although any simian resemblance is questionable at best.

> As a protective measure, many plants, such as Candy Flower and Wood Sorrel, fold up their leaves and/or petals under adverse conditions. By closing in on itself, a plant can conserve heat and prevent damage from splashes of heavy rain or even hail. This phenomenon, however, is not limited to mature specimens. Juvenile plants in bud begin in a curled up, almost fetal position, and then unfold as they mature – just think of tasty young fiddleheads on Bracken Ferns – in order to protect the emerging plant from early harm when it is in its most susceptible state. After all, plants don't have the benefit of a warm womb. Unfortunately, nature is not always amenable to facile explanation. Wood Sorrel, for example, also folds up its leaves when it encounters a patch of strong sunlight, perhaps to avoid scorching and excessive water loss from evaporation. It would appear that a single theory, however elegant, may be insufficient to explain the totality of nature's secrets.

Members of the Snapdragon Family (the Figworts, or *Scrophulariaceae*), monkey flowers (there are seven species in the Northwest) brighten our wetter environments with yellow blossoms throughout the spring and summer. They were originally named for the clown-like appearance of the flowers: *Mimulus* means "little actor" in Latin. Perhaps the Romans were reminded of a grinning ape when viewing this plant, but, look hard as I may, monkeys do not appear.

Heading down the trail, now on the west side of the creek between **Points 10** and **11**, many Plume-flowered Solomon's Seals arch out from the banks on the left. These plants seem to thrive in the well-drained soil of hillsides, as opposed to their close relatives, the Star-flowered Solomon's Seals, which do well on the flats. Starting up the brief switchbacks after crossing yet another bridge at **Point 11**, many Meadowrue plants can be seen, here consisting of mostly the male variety, with their silky tassels dangling like tiny chandeliers in the breeze. Woods Strawberry inhabits the drier open bluffs to the right above the trail.

Chickweed Monkey Flower

Proceed behind Ponytail Falls (yes, you will probably get wet from the spray) and complete the loop back to your car. Loops are such great fun, and on this one in April, the hiker/novice-botanist can truly appreciate the uncommon intensity of early springtime flora in the Columbia Gorge. It is long, occasionally arduous, and fraught with a tricky stream crossing – but what more could the Northwest adventurer wish for?

May

If April is pretty in the Gorge, May is spectacular, bringing on the greatest annual abundance and diversity of blooms and adding the bonus of warmer, sunnier weather, at least in most years. Here, the plants of April persist and join the emerging colorful flora of May to create a scene of unembellished eloquence. Masses of Robert Geranium greet the hiker heading up the switchbacks on the Horsetail Creek trail between **Points 1** and **2**, while Candy Flower smiles up at you, at least when the sun shines – this plant often folds its petals and pouts when the weather turns foul. Plume-flowered Solomon's Seal brushes up against your shoulder as it arches out from the banks on your left, infusing the forest with its pleasant perfume. Scouler's Valerian, Bleeding Heart, Pacific Waterleaf and Vanilla Leaf predominate but as you take the left turn into Ponytail Falls' canyon, Fringe Cup, Youth-on-age, Star-flowered Solomon's Seal, and Large-leaf Avens take over the show; all of these plants can be found on both sides of the trail.

Tackling the steep side trail to the left up Rock of Ages at **Point 2**, Hooker's Fairy Bells is still displaying its drooping white lanterns, especially on the left between 500' and 600', and both Solomon's Seals are present, although as one proceeds higher, especially above 500', the Star-flowered species predominates. An occasional Baneberry is out between 600' and 800', its spiky fluff of white stamens held at mid-height surrounded by its three leaves divided again into three leaflets, while Vanilla Leaf and Foam Flower rule these wooded slopes between 700' and 900'. The way is steep, often muddy at this time of year, but rewards exist, especially if you take the side trail at **Point 3** (1,000'), to the natural arch. Here, Martindale's Desert Parsley and a few Harsh Indian Paintbrush specimens decorate the basalt with yellow and red hues to complement the view of air, space and vertical cliffs.

Back on the main trail, make your way up to Devil's Backbone, **Point 4**, at 1,400–1,500' and tiptoe across its rocky spine surrounded by glorious color: meadows of blue Large- and Small-Flowered Blue-Eyed Mary's, then pink Rosy Plectritis, yellow Martindale's and Nine-Leaved Desert Parsleys, red Harsh Paintbrush, and of course, white Field Chickweed plus a few specimens of Northwestern Saxifrage. Even an occasional Chocolate Lily embellishes this scene. This kaleidoscope of varied hues may be enough to distract the casual hiker from the task at hand – balancing across the delicate traverse. It's nothing technical, and, although demanding, it's never desperate. Still, don't pay so much attention to the flora that you lose concentration here. This is like trying to identify flowers while you're driving – don't!

In the woods above the Backbone between **Points 4** and **5** (1,400–2,480'), the steep pace continues. Appreciate the exercise and the many examples at this time of year of Star-flowered Solomon's Seals carpeting the forest floor between 2,450' and 2,600'. This Persian rug of silky green leaves surmounted by twinkling starry blossoms of white seems to be almost part of a landscaper's design for these deep woods, each flower resembling a burst of light brightening the shaded trail. Lonely Trilliums occasionally pop up above 1,400', mostly dressed now in their late spring gowns of burgundy, while infrequent Fairy Slipper Orchids and Oaks Toothworts cheer the forest traveler between 2,300' and 2,600'. On the plateau (2,500'-2,900') between **Points 5** and **6**, not as much is happening botanically, although be on the lookout for the Evergreen Violet and an occasional Trillium – these love the semi-open woods and up here are still dressed in wedding-day white. If you opt to turn right and head down Trail 425 on the Oneonta Loop, be prepared for a continuing floral bonanza. The Evergreen Violet continues to punctuate the woods with its cheery yellow blooms, while gleaming masses of Candy Flowers decorate the wetter areas as one hikes the ups and downs on the Oneonta plateau. These particular specimens show off a pink candy stripe against their white petals quite dramatically, especially at the frequent stream crossings between 2,900' and 2,800', **Points 6** and **7.** Also, look for the Little-leaf version of Candy Flower, occasionally sprouting from the mossy rocks between 1,900' and 1,500'. Trilliums are still out around the streams and in the open woods on the plateau and many are still white, although lower down they are changing to their later-spring purple attire.

Descending from **Points 8** to **9** (2,600' -1,400'), the amazing purple blossoms of the Tall Larkspur are still in full and glorious bloom, at least early in May; these, our tallest delphiniums, reach 5' in height in areas of consistent spring moisture and it is here and at the stream crossing on the Angel's Rest Trail that they realize this potential. You will see taller larkspurs only in the most carefully tended of gardens or in a professional florist's shop.

Fortunately for the May hiker, the intriguing Western Corydalis is still in full bloom during much of this month and nowhere can it be better appreciated than between **Points 8** and **9**, where, especially at the streams at 2,100' and 1,900', then again between 1,800' and 1,200', it's strange, beak-like blossoms teeter on 3' to 4' tall stems, lending impressionist pink hues to the green backdrop of these west-side woodlands. Some Bleeding Heart persists on this section of trail, resisting the inevitable push to go to seed, but even those in seed are of interest. Look closely to see the expanding spur-like seed capsule as it emerges from the deteriorating blossom; it looks like it's splitting the flower as it emerges. These capsules contain an appendage rich in an oil favored by the ants which pollinate the plant. While we think of bees and moths as prime insect pollinators, creeping crawling insects, birds, and even animals, including us humans, can be just as important in seed dispersal.

As one descends toward the crossing at **Point 9** (1,400'), the familiar Fringe Cup and Youth-on-age co-mingle with Wood Sorrel, but the flowers of a new plant are beginning to emerge here as well – the colorful blossoms of the well-known Salmonberry (*Rubus spectabilis*), particularly common in wet areas between 2,500' and 1,400'. This familiar west-side shrub can grow up to 9' tall and often crowds wetter areas at low-to-mid-alpine elevations, particularly along stream courses and avalanche tracks. It can impede progress for those intent on bushwhacking, with its moderately prickly arching stems but, for stick-to-the-trail hikers, it can also supply tasty treats later in summer when in fruit. The pointy, toothed leaves look a bit like those from a maple tree, while the 1" to 2" flowers are imbued with a hue hard to describe – authors have tried "hot pink" or "magenta" – but to me they resemble raspberry sherbet – whatever, the blooms are hard to miss on this section of trail. The raspberry-like berries have been called everything from "insipid" to "delicious". This variability may be partly due to inherent differences in microclimate, soil conditions, or to differences in clones, as stands of these bushes may all stem from the same parent plant.

After the tricky crossing of Oneonta Creek at **Point 9**, deep woods inhibit the spectacular floral displays of the last several miles, but the amateur botanist will still appreciate the tenacity of the shade-tolerant plants able to bloom in the gloom, including Fringe Cup, Youth-on-age, and the occasional Evergreen Violet. However, as one progresses down Trail 424, things begin to open up. Masses of Scouler's Valerian begin to appear, especially on rocks and at the side of the trail at 1,300'. In addition, just above and below Triple Falls, at around **Point 10** (600'), Broad-leaf Arnica provides welcome splotches of yellow, particularly below the trail on the right. Big-leaf Sandwort is fully out in the woods, while Merten's Saxifrage peppers the wet seeping rocks on the left between 600' and 500'. These weeping walls, so common on the west side of the Gorge, form their own type of ecosystem and often harbor plants seriously specialized for survival in this soggy environment.

Another, even more localized specialist, Suksdorf's Mist Maidens, joins the show on these wet walls, especially at 500'. Woods Strawberry frequents the drier slopes above and below the falls, while Salal is plentiful in the forest below **Point 10**. Also, look for the cheerful orange-eyed Common Cryptantha, its small white flowers decorating sunny banks between 450' and 400' on the left. Related to the popcorn flowers, these Forget-me-not Family members *do* indeed look a bit like popcorn

Salmonberry

exploding out on dry slopes above the trail. Proceeding down the track between **Points 10** and **11**, look for frequent Scouler's Valerian in open areas, and for the two Solomon's Seals: Plume- and Star–flowered, brightening the woods. Here, especially on the left, also find the unusual American Wintercress (*Barbarea orthoceras*). Never abundant, this Mustard Family plant sports erect stems usually 1½-3' tall, topped by a cluster of small yellow four-petalled blossoms, often redolent with a pungent smell. This flower can also be found in low elevations in semi-open areas on the Ruckel Ridge Trail but in few other places in the Gorge. Note particularly the unusual shape of its lowermost (basal) leaves. These leaves have been described as "lyrate", that is, shaped like a lyre, with paired leaflets which get progressively larger up the leaflet's stem (petiole).

American Wintercress

Other plants brighten the path as you progress back to the Horsetail Creek Trail at this time of year. Little Buttercup can be found hiding along the trail just after the second bridge at **Point 10**, and up the switchbacks thereafter, while Meadowrue is abundant in this area as well. Baneberry is just making its appearance here and one large specimen of Red Elderberry can be seen about 100' downhill on the level area after swooshing and swimming behind Ponytail Falls (be prepared for the spray). Woods Strawberry, Candy Flower, Pacific Waterleaf, and Sweet Cicely complete the flower show as you descend back to your vehicle at the Horsetail Falls parking lot, well rewarded botanically for your considerable efforts. The good news is that this western section of the Gorge in May is abloom with a profusion of colorful flowers sufficient to satisfy the appetites of even the most ardent plant detectives. The bad news is…well, for the month of May in the Columbia Gorge, there really is no bad news.

Do plants which thrive in deep woods really enjoy the shade, or are they simply seeking refuge in less desirable areas while their light-loving relatives flourish in the open fields and meadows where they can worship the sun? In fact, many forest plants need shade because their leaves would burn easily in bright sunlight. Moreover, woodland plants may appreciate the respite tall trees provide as they may derive additional nutrients from the rich soil found beneath the forested canopy. In addition, such woodland dwellers have little need to invest the large amounts of energy necessary to produce colorful flowers; most under-story plants have purely white flowers. Their brethren in the open meadows face much stiffer competition and hence must shout for attention with pigment to attract potential pollinators.

June

Although the month of June may not feature as many blooms as May, there are still floral discoveries to be made. The woods going up the switchbacks from the parking lot at Horsetail Falls, **Point 1**, still teem with Robert Geranium and some Large- leaf Avens. Candy Flower, Scouler's Valerian and Bleeding Heart all persist in the areas between **Points 1** and **2**. Up the steep side trail after the left turn toward Ponytail Falls, **Point 2**, Big-leaf Sandwort rules this republic of plants in the dense undergrowth, seeming to favor patches of ground immediately beneath Douglas Firs and Western Hemlocks. While this plant may simply be seeking out shade, it is also possible that it thrives from nutrients derived from the roots of these trees. Other shade-loving plants which flourish here, between **Points 2** and **3** (the junction with the side trail to the arch) are also in bloom this month, including Vanilla Leaf, Foam Flower and the strange and spiky blooms of Baneberry, more fully described in the Starvation Ridge Chapter.

Out on the open ridges at the arch (around 1,000') and on Devil's Backbone between **Points 3** and **4**, the blues are fading as the Blue-eyed Mary's, both Large and Small, go to seed. Still persisting however are yellow Martindale's Desert Parsley, white Field Chickweed, yellow Mountain Oregon Grape, yellow

Nine-leaf Desert Parsley and the red of Harsh Indian Paintbrush. In the woods above the Backbone, look for carpets of Star-flowered Solomon's Seal sparkling like points of light in the green understory, especially between 2,300' and 2,500'. An occasional pink Calypso Orchid, or Fairy Slipper, graces the forest like the shy but beautiful nymph for which it was named. Even lavender Oaks Toothwort, 6-10" tall with four cross-like petals, perseveres here in June, although mainly at elevations above 1,500'.

In the woods above the Backbone, the twinkling stars of Star-flowered Solomon's Seal still brighten the understory, while Big-leaf Sandwort and Foam Flower add bits and pieces of white sparkle against the otherwise green monochrome. Note the absence of lofty plants here. Small ground dwellers like Vanilla Leaf and Foam Flower predominate low down, thriving under the considerable shade of lofty conifers. Where are the mid-size shrubs? It seems that common west-side shrubs like Elderberry or Red-flowering Current require more sunlight than can possibly reach the forest floor here, although exceptions exist. Cascade Oregon Grape and Indian-plum can be found growing in the shade, but thrive with more vigor in open woods.

Such woods are found on the plateau, between **Points 5** and **6** (at 3,040'), the trail junction with the Oneonta Trail, #425. However at this time of year at these relatively low elevations, only a few Evergreen Violets and the white strawberry-like flowers of Dwarf Bramble provide floral points of interest amidst the tangle of alder and vine maple branches sprawling across this upland plateau.

Making the loop by heading south at the junction with the Oneonta Trail, many summer flowers are in bloom between **Points 6** and **7**, including Foam Flower, Dwarf Bramble, Vanilla Leaf, and Candy Flower. Some springtime plants, however, overlap here, such as Fringe Cup and Meadowrue in areas wet in spring and Stream Violet near the several stream crossings on the branches of Horsetail Creek at 2,900'. Proceeding down the plateau, then down the switchbacks at **Point 8**, one can still see some tinges of purple on the dwindling blossoms of the Tall Larkspur, but the Western Corydalis has hidden her pink gems, perhaps fearing the heat of summer. No matter - the seeds of this Bleeding Heart relative are equally fascinating, with their pear shaped capsules bursting open enthusiastically at this time of year like Fourth of July firecrackers, ejecting their black lustrous seeds in celebration of the coming summer. In fact, June at these elevations is a bit like January, looking backwards and forwards, reflecting some of the best of spring and offering a foretaste of the warmer months ahead.

Heading down to the stream crossing at **Point 9**, some pink Bleeding Heart, dull-white Fringe Cup, and Youth-on-age remain in bloom. Western Meadowrue predominates between 2,000' and 1,600', while the magenta blooms of Salmonberry color the understory in more open, wetter areas, promising its tasty berries (well, to some tongues) later in the summer heat. A new arrival at this time of year, however, is False Lily-of-the-Valley, in places between 2,200' and 1,600'. This plant lays a bright green rug of wavy velvet leaves across the forest floor, each pair topped by a short spike of delicately-perfumed starry white blossoms. Look especially for these blooms underfoot at 2,150' and 1,700'. Even a casual inspection reveals that although they are both in the Lily Family, this "false" plant is not closely related to the "real" Lily of-the-Valley. Nonetheless, it's a pleasant enough name for an exceedingly cordial plant, one that the forest traveler will delight in finding repeatedly in our Northwest woodlands both at low and mid- range late spring elevations.

At this time of year, the Creek crossing at **Point 9** is mercifully easier, although care must always be exercised here. Thereafter, Western Meadowrue, and Fringe Cup can be found in the wetter areas all along the trail to the second bridge at **Point 10**, while Foam Flower and Big-leaf Sandwort populate drier areas of the forest. Just above and below Triple Falls in the welcome openings to the right and left of the trail, masses of Scouler's Valerian predominate while down the hill to the right, Broad- leaf Arnica brightens the hillside with its yellow daisy-like blooms. The Woods Strawberry is still in frequent bloom on sunny open slopes around the Falls, as are the cheery popcorn blossoms of Common Cryptantha. Just past the Falls, one can find many Scouler's Valerians still raising their flat-topped white compound blossoms on 1-2' stems at around the 600' elevation level. Farther along, the yellow American Wintercress nods cheerfully along the trail, especially on the left, accompanied by an occasional tiny waxy yellow flower – the appropriately named Little Buttercup.

Starting up the switchbacks after the third bridge at **Point 11**, while Western Meadowrue predominates, Baneberry is now showing off its spiky white "flowers" (mostly composed of stamens, really), while that Red Elderberry on the left downhill is turning from white blossom to crimson berry. Past Ponytail Falls, Woods Strawberry rules the open slopes, with some Candy Flower and Stream Violet still frequenting the wetter areas. Between 400' and 100', an occasional Sweet Cicely is still out, but the predominant bloom here in June is Pacific Waterleaf, its fluffy white-to-purple balls of wool completing the choreographed sequence of flowering in which spring blooms will give way to the blossoms of the summer months.

July and August

Yes, the main flower show is past prime by the summer months but aspiring naturalists should not despair as there is still much to admire during July and August in the western Gorge. For example, proceeding up the switchbacks from the parking lot at Horsetail Falls, **Point 1,** Robert Geranium continues to poke its sweet pink blossoms above its deeply-cleft leaves. This geranium is one of three common species in this genus found throughout the Northwest; all three are foreign invaders, although not unpleasant ones. The other two are Filaree, thriving in more plentiful hordes east of the crest, and the lovely little Dovefoot Geranium, spreading pink carpets of color in low-lying fields around Portland during March and early April. This Eurasian immigrant can be seen at those times as a rug of pink in the meridian on I-84 between mileposts 19 and 21 heading east and on the left turn-off for Multnomah Falls in grassy areas heading back west off that same Freeway.

> The common garden and flowerpot plant growing in all our cities that we call a geranium is actually a member of a different genus altogether, although a related one in the *Geranium* Family: the *Pelargoniums*. These larger plants have lent themselves well to hybridization and cross-breeding. Indeed, most of the "geranium" flowers one encounters in city gardens and windowsills are hybrid crosses of far smaller plants once found in the wild. This confusion of names is not atypical: Many "lilies" actually belong to different genera, such as the *amaryllis* and *brodiaea* clans, while some "daisies" are actually asters, sunflowers, or goldenweeds. That is why scientific names, though cumbersome, allow for more accurate classification. Plants, like humans, may have many nick-names, and many of these may be alike from individual to individual, but a first and last name assures more accurate identification.

All these geraniums share a peculiar trait exemplified in some of their common names, such as Stork's Bill and Heron's Bill. The style, that is, the female part in the center of each flower, elongates significantly as the petals fade, becoming much longer than one would expect, up to several inches in some species. This beak-like protrusion breaks off and falls to the ground, where it begins to descend into the soil, aiding germination. In Filaree, the style is spirally twisted: As it is dampened in the spring rains, the twist eases; once it dries, the spiral tightens again. This alternate coiling and uncoiling actually drives the seed into the ground as efficiently as a screw bores into soft wood. Sometimes, evolution creates structures and mechanisms so magically effective that they defy belief.

Between **Points 1** and **2** some Candy Flowers and Stream Violets remain out where moisture has not yet completely dried in the summer heat. A few Large-leaf Avens remain as well. Heading up the steep Rock of Ages path at **Point 2,** the summer woods contain many examples of Foam Flower and Vanilla Leaf, especially above 500'. In these forest depths, precious little light reaches the forest floor and these woodland plants display thin, delicate leaves held horizontally in order to intercept as much light as possible. These species are mostly deciduous and thus are beginning to deteriorate during the summer as they cannot afford the energy necessary to maintain their leaves during the dark winter months. Evergreen species on the other hand, such as Rattlesnake Plantain and White-vein Pyrola, with thicker leaves and more robust stems, can predominate and even flower later in the year. Look for some specimens of the related Pipsisssewa (also called Prince's Pine) on the Oneonta Plateau at around 3,000'.

Woods Strawberry

In the openings at the natural arch (the rough trail to it is our **Point 3**) and again on Devil's Backbone, **Point 4**, spots of color persist. Red indicates the Harsh Indian Paintbrush, while yellow can be the shrubby Shining Oregon Grape or the wispy Nine-leaf Desert Parsley. The colors are often set against the wavy white of Field Chickweed, still nodding in the wind at these magnificent rocky overlooks. As you reach the plateau, **Point 5** at 2,480', not many flowers are out, but this lack is compensated well by several overlooks to the left, where views to the north provide a sense of expansive space and light. Look down to the mighty Columbia and across that flood to distant peaks in Washington. One gains the feeling here of trespassing high in the sky and experiencing the uncommon intensity, yet strange closeness, of the Gorge.

In the deeper woods above the Backbone, not many flowers can brave the constant shade, but an occasional Foam Flower sends up a cheerful spray of white. On the plateau, between **Points 6**, the junction with the Oneonta Trail #425 and **7**, the junction with the Bell Creek Trail #549, look for the yellow of Stream Violet and its close accomplice, Candy Flower, at the many stream crossings near branches of Horsetail Creek. This pair of woodland plants flourish where water collects and both can persist throughout these drier months.

Descending from this plateau, down from **Point 7** to **Point 9**, the amateur botanist should take the opportunity to impress her or his less-knowledgeable companions by identifying summer plants by their leaves alone, minus any flowers. This is much easier than it sounds. Look for the huge deeply-cleft leaves of Tall Larkspur, especially on the right between 2,400' and 2,200', and the lacy blue-green fronds, resembling Bleeding Heart leaves, of Western Corydalis at streams between 2,100' and 1,900', then again along the trail between 1,800' and 1,200'. Just before and after the crossing of Oneonta Creek at **Point 9**, easier now in summer, watch for Meadowrue; some specimens may still be in bloom here.

Heading down the Oneonta Creek Trail after crossing the torrent, look for the strangely wonderful leaflets of the Inside-out Flower. Even though most of the oddly-inverted flowers have gone to seed, these webbed outward-pointing leaflets underscore this plant's other common name, Duckfoot. Although it is doubtful that this shape co-evolved with that of waterfowl to speed water travel, one can still sense the resemblance as both ducks and plant seem to thrive in these wet environs. Another plant to identify by its leaves here is the plentiful Western Meadowrue, with lacy blue-green leaflets floating in the forest, now largely bereft of its chandeliers of dangling stamens, at least at these low elevations.

Approaching the second bridge over Oneonta Creek at **Point 10** just above Triple Falls, color returns in the form of the yellow Broad-leaf Arnica, spread out in open fields, especially downhill on the right above the Falls. On these open sunny slopes, the Woods Strawberry (*Fragaria vesca*) can be found, its low-lying white crinkly five-petalled blossoms centered by a cluster of yellow stamens, all surrounded by sharply- toothed greenish-yellow leaves. (Compare this strawberry with the blue-green-leaved Broad-petal, or Virginia Strawberry described in the Dog Mountain Chapter.) The papery flowers and multiple yellow stamens in the centers are typical of the Rose Family, which provides so many of our Northwest berries and fruits, including the blackberries, raspberries, and even the cultivated apples and pears of the Hood River Valley. Indeed, this family is as important for its value in providing foodstuffs as it is in stocking flower shops and city gardens. Look closely at the Woods Strawberry beneath the cluster of its toothed leaves to hopefully find the scarce but delicious fruit of this small plant. Altercations among hikers have been known to occur in these woods over this luscious berry, so guard your cache carefully!

Heading down the Oneonta Creek Trail, an occasional Little Buttercup may still be in bloom in openings between 900' and 700', accompanied on meadowed slopes by the small popcorn-like white flowers of Common Cryptantha. In wooded areas, some Star-flowered Solomon's Seals may still be brightening the forest with their twinkly white blooms. Up the switchbacks past **Point 11**, the proliferation of Meadowrue plants cannot be missed, although it's the leaves which now comprise the show. Here and there throughout the forest as well, the spiky white flowers of Baneberry burst through the monotone of green. Never abundant, this 2-3' tall plant is nonetheless striking, with its blooms punctuating the woods like a starburst of white. The flowers do have petals and sepals, but most folks will notice only the spikes of stamens sticking out in all directions which give this plant the appearance of a fuzzy bottlebrush.

Past the switchbacks at **Point 11** after the last bridge over the creek, open areas become more prevalent, and on these sunny slopes the Woods Strawberry thrives. In the wetter areas not yet dry from the summer heat, some few specimens of Stream Violet and its frequent companion, Candy Flower, still survive. Turning right, the up- and-down trail holds few flowers, although the leaves of Sweet Cicely and Plume-flowered Solomon's Seal beg for identification here. Squeezing behind Ponytail Falls, watch for the spotted leaves of Western Waterleaf as you rejoin the Horsetail Creek Trail and complete the Loop back to your cars and SUV's at the parking lot. Although the hiker may dream of more color in the meadows at higher elevations in the North and South Cascades during these summer months, these trips in the Columbia Gorge confer twin benefits at this time of year: the proximity to town and the opportunity to identify much of the flora by their leaves alone. This last ensures the title of amateur botanist and perhaps more importantly, the opportunity to make a lasting impression upon one's fellow outdoor enthusiasts.

September and October

While it's true that, by the autumn months most Gorge flowers have gone to seed, there are still some blooms to be seen, and you can't forget the benefit of color-changing leaves as well. In this section of the Gorge, the Big-leaf Maples lower down, and the Douglas and Vine Maples at slightly higher elevations, begin to show some yellow and orange by mid-September, and paths are often littered with their leaves splotched with streaks and spots of yellow, orange, brown and red throughout the fall season. Eyeing the huge colorful leaves of the Big-leaf Maples, the derivation of its name becomes obvious; many are at least a foot long and wide.

Even some blooms can be found by the persistent and observant autumn hiker. Heading up the switchbacks on the Horsetail Creek Trail, a number of Robert Geraniums are still in flower, with their pretty pink blossoms carpeting the woods between **Points 1** and **2**. As you head up the steep path toward Rock of Ages, however, little in the way of flowers will appear. Nonetheless, look for the occasional Foam Flower still sending up its small white spikes of multiple flowers and seeds to cheer up the green forest gloom. Out at the rocky overlooks by the natural arch, after turning left onto its rude path, **Point 3**, and on Devil's Backbone, **Point 4**, the desert parsleys and paintbrushes are now gone, although of course the spectacular views remain in place.

Above these viewpoints, the woods hold few floral surprises, although an occasional Foam Flower still may be found between **Points 4** and **5**. The same may be said for the forested path up on the Oneonta Plateau, although again leaf identification is often fun here. Look for the leaves of Vanilla Leaf, Evergreen Violet, and Foam Flower in these woods. Heading down from this plateau, between **Points 7** and **9**, a few Candy Flowers and Stream Violets may still be found in flower at damp areas, while a lingering Fringe Cup or Youth-on-age may still be in bloom in the forest. The creek crossing at **Point 9** shouldn't pose too great a problem at this time of year. Look for the leaves of the frequent Meadowrues and Inside-out Flowers amongst the auburn colors of the maples lining the creek bottom along this stretch of trail.

Further on, at Triple Falls, just after **Point 10**, you may still catch a Broad-leaf Arnica out, and, at the meadowed slopes beyond, be sure to sample the fruits of the Woods Strawberry, still as luscious as earlier in the season. This berry signifies the values in hiking here in the autumn months: It is scarce yet precious, subtle but palpable, and, though you must bend and strain to partake of its sweetness, the rewards make the effort more than worthwhile.

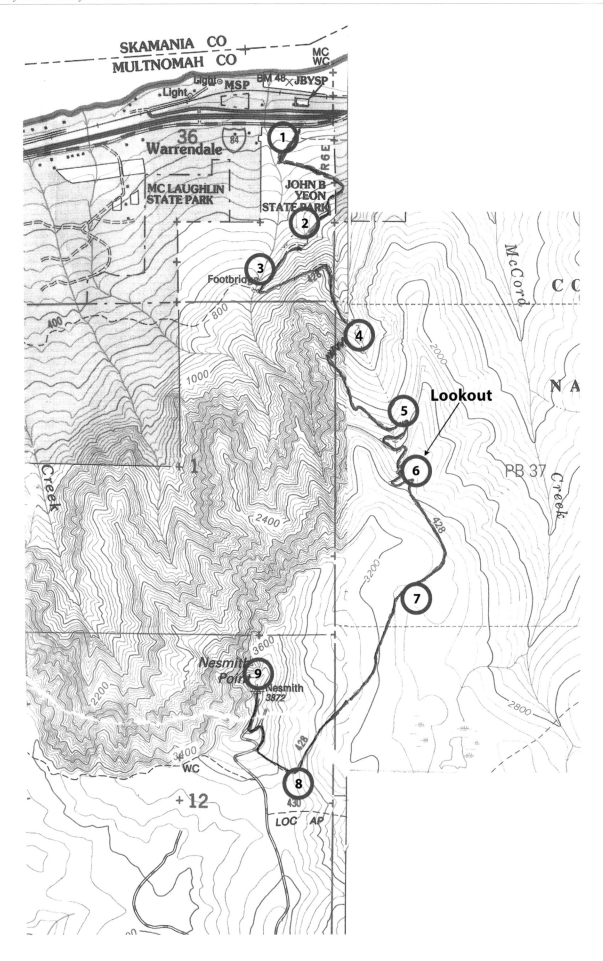

Nesmith Point Trail

For those interested in sampling a prime cross-section of woodland flowers in the spring months, the Nesmith Point Trail offers an ideal destination. Hikers can choose to ascend to the 2,780' rocky outlook to experience most of the prominent plants; the more adventurous traveler can continue through deeper woods to the very top for an unusual westerly view back toward suburban Portland and thus reap the cardiovascular benefits of an extra 1,200' of altitude gain. By proceeding about 200 further yards through woods at the top, a more northerly view can also be obtained. However, even the lower-elevation landscapes on this trail are noteworthy, with a spectacular viewpoint at 900' and a series of traversing switchbacks through bouldery basins rich with rock-loving flora between 2,000' and 2,600'. Here, a sweep of sky-scraping cliffs abuts meadows rife with red, yellow and blue plants floating on a sea of green. Here too, at lower elevations, this wet westside primeval forest features fallen nurse logs supporting a raft of life adrift on this sea. It is only in this westside soggy paradise that Maidenhair Ferns overwhelm the more common Sword and Bracken Ferns further east. Add to this botanical bounty the proximity to Portland and the popularity of the Nesmith Point Trail is easy to understand. Nonetheless, you'll encounter fewer hikers here than on Angel's Rest or Dog Mountain, an additional bonus, especially on weekdays or in early spring or mid-fall. Bring water in summer as the only reliable sources come within the first mile.

CAUTION: Those attempting this trail in winter or in years when heavy snow accumulations may persist through May, should be aware that it is easy to lose the trail just below **Point 6**, the rocky outlook at 2,780', where snow seems to persist later in the spring than one would guess. This is due to the trail's north-facing orientation and to the ravine in which a small stream meanders. Many hikers skilled in cross-country navigation kick steps in the snow up this ravine, then locate snow-free stretches above. However, post-holing is common here and if this is one of your first attempts at Nesmith Point, it's best here to simply turn around, content that you've experienced some steep hiking combined with discovering a host of spring wildflowers so typical of this western side of the Gorge.

February and March

While it may be misting and foggy on many early spring days and the western Gorge may resemble a cloud forest, the Nesmith Point Trail is still a good option at this time of year because it is relatively sheltered yet offers the beauty of floral blooms which hint at a sunnier springtime to come. Hiking up the green corridor at the beginning of the trail, between **Points 1,** the parking lot, and **2,** the stream crossing, gives one the feeling of being in an emerald alley, an experience actually enhanced by the occasional drip, drip, drip endemic to the Northwest. But green is not the only color encountered. Watch for the brilliant white of Trillium in late February to early March, not only scattered here and there in the open forest at the start of the trail, but especially concentrated in Trillium gardens at 450' and 600' between **Points 2,** the stream crossing, and **3,** the washout area. Again, you can encounter Trillium in open woods on the left at 700' to 900' elevation, then just following a right turn after the lookout at **4**, around 965'. These plants are just emerging in most years by the middle of March and will be at their peak, to the great joy of Trilliophiles, through mid-April.

> Most people interested in learning the names of flowers aren't as interested in leaf identification. Nonetheless, being able to identify a species by its leaves alone can prove to be a real advantage when flowers are absent (which is true for most of the year) and especially when a plant is an evergreen, as opposed to a deciduous species. Early in the season, even non-evergreen plants send up leaves in response to warming temperatures and lengthening daylight hours and especially briefer nighttime darkness. These often emerge in a compressed rolled up fashion as the plant tries to protect its young leaves and nascent flower structures from the early spring cold, and especially from the frigid February and March nights. Eventually, leaf and flower structures spread out, and in the dark forests, many leaves, such as those of Trillium and Vanilla Leaf, as well as many fern species and cone-bearing trees, spread their leaf structures horizontally so as to intercept as much light as possible filtering through the green canopy from above.

Another color to look for is pink, embodied in the peppermint stripes of Candy Flower (*Montia sibirica*). With its 6"-to-12"-high stems topped by dainty flowers, each with thin pink lines penciled in on a white background, Candy Flower does look good enough to eat. Adding to the culinary association is its other common name, Siberian Miner's Lettuce. Indeed, early explorers did use the paired opposite heart-shaped leaves midway up the stem as a salad green, but they must have been sadly short of other provisions as the taste is mostly sour rather than savory. No matter, as the flowers here, particularly just after the left turn at 100 yards, provide cheerful sprays of color and enjoyment even if you don't try to eat them (which you really shouldn't).

Candy Flower

This early in the year, a number of plants send up leaves from the forest understory several weeks to months in advance of any flowers. While most casual hikers probably aren't too interested in leaf identification, some of these plants are hard to ignore. In addition, learning the names of plants from the leaves alone can enhance your appreciation of them later in the spring, when they first come into bloom. One such species, easy to identify and plentiful from the first switchbacks at 300' and on up to 2,000' in shaded woods, is Vanilla Leaf (*Achlys triphylla*), a plant with 10" long three-parted leaves, each shaped like a wavy-edged fan. You must wait until late April or early May to see the showy spike of tiny white flowers, held 6" to 12" above the horizontally-oriented leaves, but as early as March you can stoop down to a leaf and inhale the faint vanilla odor. Old-timers hung bunches of these leaves in their homes as room fresheners and Native Americans used the plant as an insect repellent. Frankly, as a flavoring, you would be better served to buy the real thing from Madagascar or Mexico.

Vanilla Leaf

Also within the first 100' on this trail, just after the left turn at 100 yards, look for Robert Geranium or Herb-Robert as it is known in Europe, its continent of origin. This pretty little plant, with its 5-petalled ½" pinkish-purple blossoms atop a 1' tall leafy stem (the leaves are pinnately divided, giving them the appearance of miniature ferns), is equally notable for its seedpods, which develop following flowering. Typical of the Geranium Family, these pods resemble daggers, the fused styles (female parts) elongating into shafts almost 1" long. Once fallen from the stem, these seeds expand and contract as they are first moistened from intermittent rain showers, then dry out on sunny days. This action helps them actively dig into the ground, thus getting a jump-start on germination. Some members of this family have spirally twisted shafts which, as they are drenched by rain, then dry out, actually screw themselves into the ground. Plants are less passive than we presume.

While these Robert Geraniums don't look exactly like your garden geraniums (which don't even belong to the *Geranium* genus – they're *Pelargoniums*), they are all in the same family. In fact, these small geraniums are rather invasive and can easily be found in forested Portland parks, in wooded tracts along city streets, and in many urban gardens. Although they look attractive enough to be accorded the compliment of being cultivated plants, they almost always have crept their way into these plots in an adventitious fashion, although this does not in any way diminish their allure. Nonetheless, these attractive plants have created quite a stir within the botanic community.

Textbooks of the 1980s and 1990s insist that Robert Geranium is "rare in Western Oregon". If this was true 20 to 30 years ago, then this modest flower must have spread as rapidly as a computer virus because today, these plants seem to be ubiquitous in certain lowland forest habitats in western Oregon and southwestern Washington. They often carpet lowland forest floors and trail-sides, thus creating the not-unpleasant impression of a pink-purple haze above a field of leafy green, particularly attractive within the dark lowland woods in which these plants are so often found. This is especially true later in the season, as by May and June, their blooming is in full force (see especially the start of the Rock of Ages route in the Chapter above). However, some Robert Geranium flowers seem to blossom, as here on the Nesmith Point Trail, almost throughout the winter; indeed, it is not uncommon to see occasional flowers in bloom from November through February.

Unfortunately, not everyone views such scenes as rosy. Many ecologists fear that Robert Geranium is so rapidly replacing certain native plants as to push them in the direction of extinction, or at least, that its expansion into these habitats will eventually upset a natural balance. Ongoing studies are now trying to evaluate the extent, if any, of such damage. One reason Robert Geranium may be thriving in the Pacific Northwest, aside from a similarity to the soil type and climate of its native European home, may be the absence of natural predators such as the insects, bacteria, and fungi which consume parts of the plant in the areas where it first evolved. Absent the evolution of such natural predators here, Herb-Robert has taken hold with a ferocious, if beautiful, tenacity. Should efforts be made to eradicate this pleasant but invasive weed? Philosophy, aesthetics, and science will all need to converge on an answer.

Even as early as late February, several other flowers are beginning to bloom, and fortunately none of these are controversial. The gorgeous, deep purple of the narrow endemic (found only in the Gorge), Columbia Kittentails, can be found growing in its typical habitat, north-facing bluffs on the Oregon side of the Gorge (but see Table and Dog Mountain Chapters for exceptions). These occur at 1,550' between **Points 4** and **5**. While here only a few specimens of this locally-famous flower can be seen, each can perhaps be better appreciated than when one is overwhelmed by their abundance, as for example, on lower-elevation bluffs on the Starvation Creek Trail or at middle elevations on the Wygant Trail. (See the Wygant Chapter for a more complete description of this provincial treasure.)

By the way, note the many Maidenhair Ferns adorning this trailside, especially between the beginning of the trail and 1,500'. This is the largest concentration of these delightful plants alongside trails in the Gorge, probably because of how damp this far-west path always seems to be. Note their delicate fronds ("leaves") and thin black stems (called "stipes"), often hanging over the trail and dripping some of that west-side moisture onto your boots and pants.

Also at this elevation (1,550'), and in proximity to the Kittentails, one specimen of Western Saxifrage can be found in early March, in its typically wet habitat growing off a nearby weeping basalt cliff. Several other flowers make this early an appearance on the Nesmith Point Trail: Hairy Bittercress, the ubiquitous lowland weed which emerges by March, can be found in openings between 200' and 800' while several specimens of the shrub Indian Plum can be seen at about 1,400', with its drooping white-to-pink flowers just blossoming. These early spring months usually find snow above 2,000' but many hikers plow on; footsteps following the trail generally provide an accurate guide but also see the **Caution** at the beginning of this Chapter.

April

Exuberant plant life awaits the Nesmith Point hiker in April. Even at the parking lot (recently tidied up), **Point 1**, Robert Geranium, unaware of the controversy it has spawned (see above), cheerfully greets those preparing for the hike, while just up the trail to the water tower in the first 100', Candy Flower pokes its cheerful pink-striped blossoms up above its paired, fleshy leaves. By mid-month, several specimens of Plume-flowered Solomon's Seal send forth sprays of fragrant blooms within 100' after turning right onto the southerly extension of Trail 400 (which is now called the Nesmith Point Trail, #428 – the junction upstream at around 650' no longer exists as the bridge was washed out in the catastrophic 1996 floods). Also abundant here, in the deep woods both before and just after the left turn at around 240' elevation, are numerous examples of Western Meadowrue, the male flowers showing off drooping chandeliers of purplish stamens; the female plants, with their more modest star-like stigmas, are content to conserve their energy for more important tasks than display, such as perpetuating the species.

False Lily-of-the-Valley

Between **Points 1** and **2**, watch for the first emerging blooms of Star-flowered Solomon's Seal, announcing their presence first with tiny dots of white at the tips of arching, leafy branches just ½' to 1½' off the ground. Eventually, by the last of the month in lowland plants, but only by May to June at higher elevations, these buds will open up in true starbursts of white, creating, in places beneath the tall conifers, a constellation of flowers, each tiny on its own, but designing an overall effect quite beautiful in its entirety, like stars against a green night sky.

To add to this display, by late April, a relative of the Solomon's Seals, False Lily-of-the-Valley (*Maianthemum dilatatum*) is just coming into bloom. Watch for this plant amidst the tangled underbrush just after the left turn at 240' and continuing up to the stream crossing at **Point 2**. This Lily Family plant displays its 2"-to-4" cylinder of small white flowers on a 10-12" shaft in these lower elevation forests; as with all flowers of the mountains, bloom times progress to later months at higher altitudes.

The leaves of this plant deserve as much recognition as the flowers: broadly heart-shaped, often wavy, and up to 6" long, they can form a bright green mat lighting up dark forest floors from sea level to 5,000' in our mountains. False Lily-of-the-Valley is another humble plant maligned by taxonomists. I doubt the plant is attempting any type of deception – it simply resembles the true Lily-of-the-Valley, although it does not smell as deliciously as that common garden plant. Of all our lilies, however, False Lily-of-the Valley bears one distinction: Close inspection reveals *four* petals, rather than the six found in almost all other lilies (and the closely related irises and orchids for that matter). While technically, these are a combination of petals (the inner parts) and sepals (the outer, supporting parts), and thus are called tepals, this feature still makes False Lily-of-the-Valley unique in the Northwest.

All along this part of the trail between **Points 1** and **2**, fine examples of Hooker's Fairy Bells and Fringe Cups can be found. Also here and there on this portion of the trail, one cannot help but notice the small but voluptuous

Scouler's Valerian

blooms of the Calypso Orchid (*Calypso bulbosa*), or Fairy Slipper, the famous orchid of Northwest forests which has attracted interest from far-flung scientists and local mountain travelers alike. At casual glance, this 4-6" tall patch of pink on the forest floor, often growing off mossy substrates, presents a pleasant spot of color against a background of brown and green. Closer inspection is rewarded, however, for a full appreciation of its almost whimsical design: An exuberant flourish of five pink-to-purple tepals flare above a slipper-shaped pink-spotted lip sporting a pouch which seems truly designed to accept the slippered foot of a royal princess. In addition, this orchid bears a faint but pleasing aroma. While orchids grow larger in tropical climates and have attracted much attention from collectors and the media, few can match the ornate yet delicate beauty of the Calypso.

This orchid has but one leaf, egg-shaped, dark green and often persisting over the winter. Its stem usually lacks chlorophyll and some botanists believe this plant is on its way to becoming parasitic on the humus and fungi of the forest floor. Indeed the Calypso Orchid can frequently be found thriving best on moss (actually clubmoss). It grows from a bulblike corm and cannot be transplanted. In fact, although seemingly abundant here and on trails such as Nick Eaton Ridge and Starvation Creek, this flower is believed to be endangered by some botanists because of the threat of humans picking it. If you pick a Calypso, it will die because the life-supporting corm is so near the ground's surface that it too will be fractured. It's best to simply admire this flower and let it be, even if you want desperately to show it off to your non-hiking friends. In mythic lore, Calypso, the daughter of Atlas, was a bashful, but beautiful maiden, hidden in the Greek woods. How apt a name for this, our most extravagant, yet shy, orchid!

Calypso Orchid

> Orchids are generally associated with the tropics, but we have a fair number of plants within this family here in the Northwest. Unlike the Calypso Orchid described above, they typically grow in wet habitats, such as bogs. Orchids have been exquisitely shaped by evolution: Their lower lips and various appendages almost exactly fit the mouth parts of the insects which pollinate them. These designs have been spectacularly successful – the Orchid Family is estimated to contain between 23,700 to 30,000 members, even eclipsing the Aster (or Sunflower) Family in numbers of species. Besides the well-known florists' specimens within this family, orchids include Vanilla and the wild Ladies' Tresses of our higher Northwest Mountains.

Past the stream crossing at **Point 2** and through some deep Douglas Fir woods, one eventually makes a leftish series of turns in the trail and proceeds up past an open rocky slope (600' – 680') towards the washout at **Point 3** (680'). During the 1996 floods, a reservoir upstream broke its barriers, then overran the banks of this creek, creating a small area of deforestation. However, plants are rapidly re-colonizing this breach in the forest. Here, amongst the boulders on the left, you will discover the white-to-pink blossoms of Scouler's Valerian (*Valeriana scouleri*), rising on 1-2' tall stems with 2-3" forked leaves, usually appearing only on the lower third of the plant. This Valerian is not as commonly encountered as its high-meadowed counterpart, Sitka Valerian; it seems to favor open woods and rocky slopes chiefly in the Gorge, although textbooks award it a much broader range, from north of Seattle to northern California. This plant is part of the Valerian Family, one species of which has been used for centuries as a calming agent. The original sedative, *V. officinalis*, is from eastern Europe; *officinalis* is the scientific adjective denoting a medicinal quality in plants. While most people disavow any sedative effect from store-bought Valerian, some believe this herbal supplement helps them achieve a more restful night's sleep. Please desist from sampling this Scouler's species for your insomnia – it will do no good and picking plants in the Gorge is illegal.

The flowers of all three of our Valerians (Sitka, Scouler's, and Rosy Plectritis) occur in hemispherical to flat-topped clusters at the very top of their stems. Each individual flower is a tiny five-petalled tube and there can be 10 to 25 flowers in each cluster. A close look reveals their stunningly intricate structure and colors. While white predominates, shades of pink are almost always present on some petals and pink predominates in new blooms. In fact, a number of open-area plants in the Northwest bloom pink or red at first in their flowers or leaves, including members of the Carrot and Stonecrop Families, as well as the Valerians. These flowers and leaves eventually turn white and green but, as they first emerge, the red pigment protects the tender young plants by blocking the intense ultraviolet radiation in mountain environments; it also absorbs infrared radiation, thus warming the plants' tissues in the early spring months. Combined with the gradual unfolding of plants as they begin to emerge, these mechanisms have been selected by evolution to ensure survival of young blooms despite the capricious turns of unsettled March and April weather. Unfortunately, springtime human visitors have had inadequate time to evolve similar protective mechanisms and must make do with fleece, down and Gore-Tex.

Continuing up the trail, one can sense the magnificent tumult caused by the 1996 flood, which loosened huge boulders from upstream that came to rest here at 600' to 680', where the trail takes a left turn and climbs away from the noisy stream, our **Point 3**. Before you turn, be sure to look down-slope to your right at 640' to spot the fresh blooms of Red-flowering Current, a shrub common in these open woodlands.

Up the steepening slope, the hiker here is also rewarded with glimpses of many Calypso Orchids thriving off the mossy rocks on the left between **Points 3** and **4** (680' to 900'), as if scores of pretty princesses had stored their slippers in some forest shoe closet. Hairy

Oaks Toothwort

Western Bittercress, in contrast, looking quite weed-like, litters the trail in the open areas at 750' and 800'. A few Hooker's Fairy Bells grow on the right in the forest around 700' to 900', while Trilliums dot more open woods on the left at similar elevations, although, by mid-month they are turning a luscious deep burgundy.

Big-leaf Sandwort

By mid-to-late April, Big-Leaf Sandwort (*Arenaria macrophylla*) lines deep woods from 700' to 800' and is particularly prominent as one approaches a plateau at 850'. How did this miniscule forest dweller deserve a name with the word "big" in it? After all, the plant is barely 10" tall and its dull white flowers, just ¼-½" in diameter, hardly merit the attention of anyone other than a dedicated flower finder. However, relationships among species are oftentimes as important as obvious morphologic characteristics. In fact, among sandworts, this *Arenaria*, despite its apparently thin, pointy, elliptical leaves, actually possesses broader leaves than other *Arenaria's*, which generally grow at higher elevations and most often have grass-like leaves.

At the plateau, adventurous hikers can ascend on a side trail to the left 50' to an expansive lookout at **Point 4** (900') and see several more specimens of Calypso Orchid. Continuing up the main trail, past several rocky areas which afford views of the looming basalt cliffs to the east, many Oaks Toothwort (*Cardamine pulcherrima*) flowers can be seen growing off the slopes to the right, especially around 1,000'. Toothwort comes in several varieties (which I have never been able to distinguish), but our west-side flowers all seem about the same: dainty lavender ½" bells hanging their heads shyly in cloudy weather but opening up when the sky brightens to reveal the typical mustard flower pattern of four cross-like petals, these penciled in with darker lines. Indeed, the Mustard Family is called by some botanists the *Cruciferae*, underscoring the cross-like pattern seen in the petals of most of its members. Another name for this family, however, is *Brassicaceae*; many botanists prefer a system in which the family name matches the name of its most prominent genus. Thus, *Brassica*, the most common of mustards, is now used for this family, and *Asteraceae* is now preferred for the Sunflower Family, known for so many years as the *Compositae*.

> Many American plants bear names associated with their Old-World relatives. Cardamine stems from a root word referring to the heart; several members of this genus were thought to be helpful for heart conditions. Common names are often derived from purported health benefits as well: Toothwort leaves do look uncannily like huge molars. According to the Middle Age's Doctrine of Signs, a plant's appearance could reveal a medicinal significance. Thus, if a leaf or flower resembled a body part, it might be helpful to an individual who was afflicted with a corresponding illness. Kidney-shaped leaves were employed for kidney stones while heart-shaped flowers might be administered to someone heartbroken with the loss of a lover. In a similar fashion, the mottled leaves of some plants were thought to counteract the effects of a snakebite because they resembled the striations common on the skins of serpents. While we may be amused by what appear to be irrational superstitions of ages ago, some of these medieval plant preparations are now known to possess health benefits. At any rate, it is sobering to contemplate how advances in health care will undoubtedly make our use of medicinals seem equally feeble to Twenty-Second Century historians.

But enough of the obscurities within nomenclature. It's probably more important to simply be able to identify this Toothwort because it is everywhere on west-side trails in the early spring, even appearing in mild years as early as late February. It appreciates good drainage (its roots may be susceptible to root rot) and hence can most often be found growing on sloping banks, from where it often can be enjoyed at eye level. Frequently bypassed as just another small lilac flower, its delicate beauty can best be appreciated by a close look, especially in sunny weather when the open flowers, with their purple guide lines, reveal another wonder of evolution missed by more hurried travelers: These lines may steer pollinating insects to the sexual parts of the plant. Don't ignore the leaves of this plant as well. Their three-to-five-lobed or scalloped pattern is enhanced by their blue-green hue, a welcome contrast to the brighter monotone of yellow-green so prevalent in early spring leaves of most other plant life. Dividends accrue to those who take a bit of time to enjoy the smaller pleasures of our forests.

As you progress up the rocky switchbacks between 1,000' and 1,300', Glacier Lilies provide bright splotches of yellow color on the left at around 1,150' just before the trail takes a right turn. Also just before and after this turn, you will come across the tiny but delightful yellow Chickweed Monkey Flower, with a single burgundy splotch marking its lower lip. Meadowrue grows abundantly around the switchbacks up to 1,300', while a few Prairie Stars provide a welcome splash of pink, with their raggedy three-lobed petals batting about in the breezes frequenting meadowy locations at 1,200' and 1,500'. Yellow is continued with masses of Evergreen Violets growing frequently off mossy banks and are particularly plentiful at 1,800' on the left.

Proceeding up the switchbacks between **Points 4** and **5** (a mostly dry stream crossing), many Trilliums are at their peak, and can be seen on the right and left, especially in the open woods at 1,200', 1,450', and 1,700'. Trilliums are also just beginning to bloom higher up, in the wetter brushy areas associated with a small runoff stream one crosses repeatedly between 2,550' and 2,700'. Here at this time of year, the early Trilliums are most appealing and fresh, with their three spotless white petals just unfurling; indeed, such a large plant seems almost out of place in these frequently chilly woodlands.

Between **Points 5** and **6** this early in the year, one can expect fewer open blooms, but still the woods hold surprises. A particularly pleasant one is found between 1,400' to 1,500', where, on a north-facing mossy bank on the left, a number of specimens of Suksdorf's Mist Maiden poke out from the cliff. Its flowers sport five dainty white petals and its leaves are gracefully scalloped. A few yards further, spiky Columbia Kittentails display a magnificently deep purple hue. Further along, as one crosses and re-crosses the small stream between 2,550' and 2,700', Colts Foot pushes up its stout 2' tall stems and dull white stark flower heads. Unusual timing makes these blooms unique – they appear *before* the huge leaves, which are themselves notable not just for their size (1-1½' in diameter), but for their deeply-lobed and coarsely-toothed appearance. These Aster Family plants love moisture and can be found inhabiting streambanks throughout the Northwest; they are easily noticed on many roadsides and along I-84 as well, particularly between mileposts 49 and 50 heading east.

One notable burst of magenta at 1,900' on the right signals an early-season shrub, Red-Flowering Currant, common in open woods throughout our mountains all the way to their eastern slopes. Its maple-like leaves set off the intense hue of its dense, pendant clusters of bright flowers, often providing the only March and April color in these forests. While all our Currants, and the related Gooseberries, are edible, this simply means you won't die by ingesting them. Most, including Red-Flowering Currant, are insipid. Better to appreciate the vivid color of the flowers and contemplate their frequent use (as "improved" varieties) in city gardens and commercial plots, where their drooping clusters of brilliant pink or white flowers often brighten drab industrial landscapes. Also emerging, and emblematic of early-season shrubbery, are the many Indian Plums, their drooping white-to-pink blossoms dangling at the ends of branches bursting with bright yellow-green upright leaves. These 10-20' tall shrubs are plentiful at about 2,550' by the stream crossing, accompanied by a few Evergreen Violets.

At the spectacularly-situated rocky outlook at **Point 6** (2,780'), where most hikers take a well-deserved respite from the relentless switchbacks just negotiated, only a few additional flowers are encountered before May; indeed, patches of snow often persist throughout this month during most years. The truly intrepid off-trail explorer may wish, at this point, to head down the long-abandoned Nesmith Ridge Trail. Do so by all means if you are one of the two or three people in the world who enjoy bushwhacking through Salmonberry, Devil's Club, and Blackberry brush. A good sense of direction, a compass, a map, and a GPS have often proven useless in this green maze, but eventually most people who attempt this loop stumble out onto the Elowah Falls Trail and complete the circuit back to the Nesmith Point parking lot on Trail #400, although a bit battered by the experience. If your aim is to appreciate the flowers rather than the quintessential experience of Cascades bushwhacking, stay on the main trail here.

Despite the early season, by mid-April several species of flowers can usually be found on the higher reaches of this trail, especially between **Points 6** and **8**. Wood Violets are plentiful at the stream at around 3,160', our **Point 7**, along with their frequent companion, Candy Flower, while many Evergreen Violets line the trail between 2,700' and 3,000' and again above the right turn at **Point 8** (3,290') mixed here with Wood Sorrel. Precocious Trillium will be found, often just emerging from ground recently covered by snow. Look at the banks and woods on your right at 3,100' to see this icon of the Northwest, braving the chilly winds of April and the well-meaning but destructive hordes of hikers who want to pick it and bring it home. Plants near trails have more to fear than wind and cold.

Nesmith's summit, 3,872' at **Point 9**, is often sheathed in snow until late in April but the roads leading to it after the right turn are obvious, even absent any footprints. Proceed past the western lookouts on your left through dense woods to a northerly viewpoint 200 yards further down the trail and contemplate your foot-and-knee-sore descent to follow. Great exercise and keen flower observation will be your just rewards.

May and June

You may have noticed that, as opposed to other chapters, I have combined May and June for the Nesmith Point Trail. My reasoning is that, this far west, May's blossoms bleed over into June's to such an extent that separating each would entail too much duplication. Certainly, this is true for all our trails to some extent. However, I believe that the constant westerly winds bringing moisture to these most western portions of the Gorge may have merged both months' blooms to such an extent that they can be most easily considered together. Hopefully, this will not prove too confusing to you, the reader, or even to me, the author. I'll try to make the monthly distinctions clear as we progress up this steepening trail.

By May, you can even find flowers (well, actually weeds) in the parking lot, **Point 1** for the Nesmith Point Trail (which also hosts the Elowah Falls Trail, and is thus often crowded). The shiny yellow petals of Creeping Buttercup poke up just as you start up toward the water tower. This 1' tall Eurasian plant is commonly seen in urban environments, growing rampantly in roadside ditches and fields throughout the Northwest. It spreads aggressively by runners (technically, stolons), which are trailing stems which can send down roots at each node, thus often frustrating attempts to remove it where it is an uninvited guest. Despite its designation as a weed, Creeping Buttercup still has its charms, creating splotches of gleaming buttery yellow against the backdrop of its often-mottled green leaves. It brightens moist disturbed waste places around our towns, a cheerful ambassador from another continent. Where it overtakes native plants, however, or where it overruns landscaped gardens, it may not be viewed in so light-hearted a fashion.

Flower keys are guides to identification found in both technical texts and books for amateurs. They are of immense help in identifying a flower you are investigating in the field; however, they are not likely to be used by the typical hiker/climber because they require long pauses for study and reflection. Nonetheless, if you become serious about identifying plants, you will need a key, several of which are referenced in the bibliography. Keys begin with a description of a general floral characteristic, such as number of stamens or position of the ovary (yes, plants have ovaries), traits which can identify a plant family. They then progress to features of a genus within that family, such as whether the leaves are arranged in an alternate or opposite pattern on the stem, or whether the flowers all emerge from one or multiple points. The last branching point in the key is the distinction of a species (or, in technical keys, the differentiation between varieties within a species), usually by leaf shape or shape of sexual parts. Keys are thus algorithms which, if followed through their multiple nodes, should lead to positive identification in most cases.

A number of woodland plants, which first made their appearance in April, continue to bloom throughout May and June, and are especially prominent in the first ½ mile of trail, between **Points 1 and 2** (the crossing of an early-season stream at 550'). Hooker's Fairy Bells, False Lily-of-the-Valley (particularly just after the left turn at 240' elevation) and Plume-flowered and Star-flowered Solomon's Seals are all here, their white blossoms cheering up the shady forest. However, several new plants are now in evidence in these west-side lowland woods. A Lily Family flower first makes its gentle appearance in late May here: Queen's Cup (or Blue-bead Lily) (*Clintonia umiflora*), its low-lying parallel-veined leaves form a cup holding a lily-white six-petalled flower. Common throughout our range, it will grace low to mid-elevations in woodlands throughout the summer. Baneberry, with its jaggedy three-times-divided leaves and spiky blossoms, appears on the left and

Queen's Cup

right as soon as you enter the forest, while examples of the striking Tiger Lily adorn the left side of the trail just before and after the left turn after 200 yards.

A bit further on, the omnipresent Star Flower is just beginning to make its appearance by mid-May. Often carpeting swaths of forest floor, Star Flower sends up stems 6"-1' tall, with a whorl of egg-shaped leaves high on the stem surmounted by a spray of one to four star-like ½" flowers. Although keyed out as pink in most texts, the variably-numbered petals (usually six to nine) usually look more white than colored, though with a definite pinkish tint. They truly do look like stars, constellations twinkling beneath tall firs against a canopy of green. Learn this plant – it is everywhere in Northwest forests, blooming here in the lowlands by May, but appearing as a summer flower at higher elevations further north.

Several new flowers make their appearance by mid-month in these early woods. Between **Points 1** and **2**, the lovely Columbia Wind Flower (*Anemone deltoida*) can be found growing in even the deepest of woods as you ascend the long switchbacks (left, then right) between 240 and 400'. At first a globe, the flower, on 1' to 1½' stems, opens out into a 1" blossom with five petals above three egg-shaped and coarsely toothed leaves. It belongs to the anemone genus, a group of popular flowers within the Buttercup Family. They are represented in our mountains by several other white varieties at high elevations and by a gorgeous deep blue flower, the Oregon Anemone, particularly prevalent on Nick Eaton Ridge. Anemones are excellent examples of the obscurities inherent within the botanical kingdom and the differences between how we amateurs see and refer to plants when compared to true botanists:

Columbia Wind Flower

- Although they are in the Buttercup Family, anemones look nothing like what we consider buttercups. What makes them buttercups has to do with the structure of their sexual parts, not what they look like.
- Their three leaves are actually one. Technically, anything emanating from a main stem is a single leaf, which is then further divided into *leaflets*.
- Even their petals really aren't petals; they're *sepals*, the supporting petal-like structures behind petals, which, for reasons even botanists can only guess at, are absent in many Buttercup Family plants.

Despite these deceptions, all our wind flowers are beautiful; they are closely related to the early-blooming anemones so commonly sold at grocery stores in March. Fortunately, they can be appreciated without paying too much attention to the technical details noted above, although some small knowledge of plant anatomy doesn't hurt too much and may even enhance an understanding of the botanical world.

The second new plant appearing this low on the Nesmith Point Trail in May is less prominent than Columbia Wind Flower and certainly less frequently encountered. In fact, if you can find and identify this blossom, consider yourself an accomplished plant hunter. While not rare or endangered, Bronze Bells, also known as Mountainbells, are frustratingly uncommon in our parts. One specimen lives in the rocky opening on the left at about 600' elevation, about 150 feet from where the trail trends left just ahead of **Point 3**. This 2' tall plant bears strange, small, yet dramatic sprays of tubular flowers with six tepals (an anagram of *petals* and *sepals*) with re-curved, frilly tips which look disturbingly like Youth-on-age but are really in the Lily Family. The bizarre architecture of Bronze Bells is only infrequently encountered in Oregon and not much seen in Washington either.

Inside-out Flower

The color of the tepals is hard to describe, closer to an unholy combination of brown and purple than to metallic bronze. Nonetheless, the overall effect is pleasing, enhanced by a faint perfume not atypical of liliaceous plants. Stoop, sniff, look closely and you will be rewarded.

By late in the month, another newcomer to the lowland forest here, particularly within the first mile of trail, is the aptly named Inside-out-flower (*Vancouveria hexandra*), also known as Duckfoot because of the shape of its leaflets. These leaflets, as graceful in woodland breezes as those of the unrelated Meadowrues and Columbines, set off a spray of tiny white flowers 1½ -2½' off the ground which bear close inspection to appreciate their improbably whimsical structure: The stem of the flower (the petiole) seems to be attached to the wrong end, with the base of the flower unattached at all. In fact, however, the petiole actually does attach to the base of the flower but the six petals are reflexed and flare backwards to such an extent that the whole affair resembles a huge design error, until you realize that this arrangement, somewhat similar to a cyclamen or shooting star, promotes fertilization. This plant is serviced by ants, who can easily access the exerted stamens; thus the Inside-out Flower may gain some advantage from its oddly fascinating flowers, which indeed seem to turn themselves inside out, at least when compared to our common notions of what a flower should look like.

A large number of other blooms make their appearance in the flower-happy months of May and June; these are described more fully in other chapters. Three species of the Rose Family genus *Rubus* can be found by late May, including the fearsome, annoying weed, Himalayan Blackberry. It can be found all the way from the parking lot up to the water tower. The far more pleasant and equally tasty Thimbleberry displays papery-white 1-2" flowers which are particularly prominent in groves of 6' tall bushes between 250' and 600' of elevation, but whose yummy berries must await late July's hungry hikers. Finally, at 450' to 500', the tiny forest-floor-dwelling Five-leaf Bramble appears, whose low form contrasts with the shrubby nature of so many of its generic relatives. Often, morphology of size and shape is but a weak guide to family relationships, as can occur in our own species (or do you really think you look like cousin Ernie?).

A second genus with three members also thrives here on the Nesmith Point Trail, the *Montia's*, known variably as the Candy Flowers or the Miner's Lettuces. *M. sibirica*, Candy Flower, is the best known of these as it is the largest and showiest (there is some value in size). It sends up sprays of five-petalled 1" white blossoms with pink stripes 1' tall above a pair of heart-shaped, fleshy leaves in wet areas (are any Nesmith Point areas really dry in May?). It can be seen almost at the start of the trail, then again between **Points 4** and **6**, especially at the intermittent stream crossings. This Candy Flower's close sibling, *M. parvifolia*, Little-leaf Candy Flower, while slightly lower-growing and having fainter lines on its petals, still impresses with the sheer volume of its blooms, especially on the right in the semi-open rocky areas just after the viewpoint at **Point 4.**

The third Montia, *M. perfoliata,* most commonly called Miner's Lettuce, is the smallest yet the strangest of the three. It can be found blooming right on the trail itself and to the right at 590' to 650' elevation just before the trail reaches the large stream at **Point 3.** Its tiny white flowers beg indifference but its leaf structure merits attention because, instead of the typical Candy Flower pairing of two opposite leaves high on the stem, Miner's Lettuce stems appear to pierce a single round leaf seemingly too large for the tiny white flowers above. Botanists surmise that, in the course of evolution, the opposite twin leaves of this Montia fused to form a disk surrounding the stem at mid-height, although the advantage of this arrangement is less than clear. This green disk is often cupped and thus might hold water, although staying hydrated wouldn't seem to be a problem for these west-side plants. I suppose we cannot expect to

Candy Flower

comprehend all of evolution's historical convolutions in our time; indeed, some twists and turns may result from chance and quirk, rather than from system and logic, as we would like to believe. *Montias* are not named after mountains, but after the Eighteenth-Century Italian botanist Giuseppe Monti. American floral patriots need to recognize that there are also *Montias* in Europe.

A number of other flowers appear in late April and happily last through May to mid-June on the early reaches of the Nesmith Point Trail. These include Western Meadowrue, the broad fan-shaped leaves and white flower spikes of Vanilla Leaf, Bleeding Heart, Plume- and Star-Flowered Solomon's Seals, and Hooker's Fairy Bells. These blooms punctuate the deep coniferous forest with sparks of white set against the arpeggio of serrated greens cascading from treetops to the canopy's floor. White rules the forest's flowers while color dominates the republic of meadows, where light holds greater sway. This rule of thumb is reinforced just as the trail opens up a bit as you bear leftish into a wide rocky opening. Here, it's hard to miss the twin Tiger Lilies standing as sentries to this welcome, though steep, meadow.

Miner's Lettuce

Little-leaf Candy Flower

Continuing up the trail just below and above **Point 3**, by mid-May the fortunate hiker encounters on the right the beloved Scarlet Columbine, also less imaginatively called the Red Columbine. Its genus name, *formosa*, means "beautiful" in Latin; there is no mystery about this appellation. Distinctive in appearance, these exuberant blossoms send five red spurs tipped with yellow backward (these are the sepals), while the inner yellow petals hang straight down and cup a central tuft of dangling golden stamens. Turn the pendant flower upwards (gently!) to fully appreciate its magically symmetrical architecture and to view the five perfect yellow cups formed by the petals as they encircle the drooping stamens. "Columbine" means "like a dove" and indeed the lovely blossoms do appear to be winged scarlet birds about to take flight above the lacy blue-green leaves, twice divided into three's. Even when not in bloom, these leaves decorate the semi-open areas where columbines flourish, both here in the lowlands and, later in June and July, at higher mountain elevations.

To those who travel mountain ranges outside the Northwest, columbines should be familiar due to their prominence (they are usually 2-3' tall) and colorful flowers. The Wallowas and the Rockies of Idaho, Wyoming and Montana are awash in a yellow species, while Denver dwellers were so proud of their luscious sky-blue columbines that they made it their state flower. Even those who haven't traveled to distant ranges probably have appreciated the infinite varieties of columbines in city gardens. Many of these human-engineered plants owe their origins to our mountain columbines; indeed, the various color combinations seen in gardens have been produced by selectively breeding native columbines with desirable colors, shapes, and disease resistance. These experiments have not always been conducted by our own species: The ranges of native columbines often overlap and, where they do, cross-pollination can occur, frequently aided by hummingbirds. Thus in the Wallowas, one can find Yellow Columbines flushed with pink, indicating interspecies hanky-panky, hence bringing into question our narrow ideas about the definition of a species itself.

> Garden plants are, more often than we realize, simply oversized variations of native floral species. The roses, rhododendrons, and phlox planted in Portland and Seattle gardens are all derived from woodland progenitors, though for these examples, the original stock stemmed from Old World and Asian plants. Not so, however, for other common landscape favorites, including Mock Orange, Columbine, and the lupines, all of which were developed from American woodland and mountain species. Some native plant lovers may sneer a bit at the "monstrosities" in our gardens as unnatural mutants, but it's hard not to admire the work

of the plant breeders. Even so, it's equally important to recognize and appreciate the workings of nature in crafting structures of such delicate beauty as our native flowers, even if they sometimes dim in comparison to their lawn and garden counterparts. In our obsessive striving for gorgeous uniformity, however, we must be aware that selective breeding can narrow the gene pool and limit potentially productive diversity.

Just before and after **Point 3,** two other vanilla-white flowers punctuate the green of the forest and provide floral interest as you make your way up the steep trail. One of these, False Lily-of-the-Valley, pokes its 5-12" spike of tiny white blossoms above its glossy heart-shaped leaves by late May, and often persists throughout June and into early July at higher elevations. On the Nesmith Point Trail, this shady character can also be found in its usual habitat of damp woods between **Points 1** and **2**, often forming a lush groundcover under the tall firs and hemlocks within the first mile. Although almost all other members of the Lily Family have their flower parts in three's and sixes, False Lily-of-the-Valley attempts to confuse the botanist by perversely presenting four petals to the world, thus becoming a family outcast. It further confounds its derivative nomenclature by obstinately refusing to live up to its namesake, the common garden Lily-of-the Valley (*Convallaria majalis*), as it lacks graceful drooping bell-like flowers. However, it does possess a faint but delightful odor. Exactly how False Lily-of-the-Valley obtained its name may remain a mystery but the plant itself is hardly an enigma as it brightens dark forest floors with its brave spikes of white flowers throughout May and June on many low-lying Northwest trails.

The second white flower frequently encountered at this time of year here and inelegantly known as Field Chickweed (*Cerastium arvense*), commonly occupies rocky openings at low

Broad-leaf Arnica

elevations throughout our region. It can be found in profusion by mid-May, especially in a field just before **Point 3** and within earshot of the stream at around 590', particularly downslope. Growing just 6-10" high, these flowers attract attention with their grey-green lance shaped leaves and deeply-cleft petals, which look a bit frayed. While at first it appears as if each flower has 10 petals, close inspection reveals that each of the five petals is in fact composed of two lobes, and, regarded in this way, each looks a lot like Mickey Mouse's ears. Indeed, close relatives of this plant are often referred to as Mouse-eared Chickweed.

Field Chickweed

Continuing the Disneyland theme, you may recognize this flower as a relative of Candytuft (*C. candidissimum*), often seen tumbling over Northwest garden walls. All these chickweeds belong to the Pink Family of plants (*Caryophyllacaea*). "Pink" refers not to the color of these plants (ours are most often white), but to the cleft petals (think pinking shears). Why the "weed" designation? Well, Field Chickweed *is* related distantly to Common Chickweed (*Stellaria media*), a sprawling plant with tiny cleft flowers which occasionally terrorizes urban fields and lawns. It can also be found in the grassy areas in front of the vehicles in the parking lot for this trail; while they're both in the same family, the resemblance is fortunately remote.

A number of other plants appear as one heads up the slope to the expansive lookout at **Point 4**, around 900'. These have already been more completely described, or will be, in later chapters. These include the Inside-out Flower (see above), widely scattered in woods on the right; Plume-flowered Solomon's Seal, protruding from the slope on your right as the trail steepens at 700'; sprays of the tiny white-flowered Small-flowered Alumroot on the right in rocky but shady areas all along this section of trail; the white spiky Baneberry on the left at around 720'; two specimens of Harsh Indian Paintbrush in rocky openings on the left; several examples of yellow Broadleaf Stonecrop on rocks to the right, all at 750' to 800'; one specimen of yellow American Wintercress on the right at around 800'; the purple Cliff, or Menzies' Larkspur in openings on the right at around 750'; a cluster of yellow Wood Violets at around 930'; Western Meadowrue; Fringe Cup; Candy Flower; and, scattered but frequent, the tiny but shiny Little Buttercup. All these are sprinkled intermittently from **Points 3** to **4**. Don't neglect to accept the challenge of the spur trail to your left at 900'; it ascends steeply 50' to a marvelously revealing viewpoint, a nice reward after 1½ miles in the forest.

Common Chickweed

At this knoll in early June two yellow-flowered Aster Family plants (*Asteraceae*) reach the peak of their blooming season. Broad-leaf or Mountain Arnica (*Arnica latifolia*) is the more common of the two and, indeed, can be found throughout the Northwest, adding its welcoming yellow to the panoply of green in west-side forests from Alaska to northern California. In fact, this Arnica (there are at least six other arnica species in the Northwest) and the next plant to be described, Columbia Gorge Groundsel, or Senecio, may be (along with the low-growing yellow violets) the only colored flowers visible in many lowland forests, most others preferring various gradations of white. Broad-leaf Arnica typically stands 1½ -2' tall, with a pair of basal leaves, followed up the stem by two to four pairs of opposite egg-shaped and toothed leaves, all surmounted by one to five 1" yellow daisy heads swaying in the recurring breezes of the forest. Although you will meet other arnicas occasionally (Heart-leaf Arnica in east-side woods; Streambank Arnica by waterfalls and in high elevation damp meadows; and even the Rayless Arnica – see the Starvation Ridge Chapter – in rocky openings); it is Broad-leaf Arnica which will most commonly greet and cheer you in your Northwest travels. It appears not only in deep forest settings, but steals its way into openings from the lowlands to higher elevations, where it mingles with the blues of lupines and the blazing crimsons of paintbrushes to yield a prototypical scene of wildflowers in our higher mountain meadows. This vagabond Aster Family species is definitely a plant for all mountain travelers to know.

Common Groundsel

> Why are meadows so often rich in color while in deep woodlands a green monotone prevails? Sunlight can be a scarce commodity by the time it reaches the forest floor; therefore, plants in these habitats may not be able to afford the luxury of colorful hues, instead investing their resources in producing as much leaf surface as possible to intercept what light is available. In addition, many of the leaves on these deep forest dwellers are arrayed horizontally so as to catch the feeble rays which do reach ground-level. Splashy advertising in bright colors to attract pollinating insects and birds must be left to the wealthier denizens of light-rich meadows. Nonetheless, forest dwellers are often rich in devious ways – subtle shades of white and pink often grace the forest base, and these denizens of the deep green frequently shower the woodland air with sweet perfume, thus advertising their presence to potential pollinators and enriching our walks through the woods as well. Thus works the slow ballet of evolution.

A second daisy-like plant to brighten deep forests is the Columbia Gorge, or Bolander's, Groundsel or Senecio. Although its flowers superficially resemble those of Mountain Arnica, it is usually a shorter plant, with roundish, deeply-cleft, and lacy-appearing leaves, each growing in an alternate, rather than opposite pattern up the 1' tall stem. This arrangement of leaves, alternate rather than opposite, distinguishes the groundsels from their close relatives in

Hairy Rock Cress

the arnica genus. The wise amateur botanist will want to learn to identify several species in each genus; in fact, the senecios comprise the largest genus in the huge Aster Family. It is so large that the genus carries a multitude of common names, including Butterweed and Groundsel, and includes many urban weeds such as the Common Groundsel (*Senecio vulgaris*), with drooping yellow flowers and blackened bases. The main distinction of Columbia Gorge Senecio, however, is its limited distribution: True to its name, it thrives mainly in low-elevation forests on the wet west-side of the Columbia Gorge. While some texts describe its presence elsewhere, like a shy forest dweller it is exceedingly uncommon and uncomfortable outside of its home range. Look for it at this lookout at 900', and sporadically in woodlands from 300' to 2,000' throughout the Gorge.

As you wander up the main trail from the junction, passing rocky semi-open areas between **Points 4** and **5**, the views to the left encompass a spectacular basin rimmed by basaltic cliffs. Flowers appear here and there along this section by the trailside, including some lovely specimens of Trillium turning wine-purple on the right about 300 yards from the junction, and the rare, mysterious, and often misidentified dainty white Suksdorf's ' Mist Maiden, draped off dripping mossy rocks on the right at 400 yards. These pale, however, in comparison to the long-distance view to the left across the rocky span. There, mounds of soft blue-purple flowers cascading over the cliffs announce the presence of Spreading Phlox and Cliff (or Menzies') Larkspur, softening the stark grey basalt with lush color. Interspersed with the pillowy forms of the Phlox and Larkspur, yellow splotches of Stonecrop add to the colorful scene. Binoculars are handy here as the individual flowers are hundreds of yards in the distance, but nonetheless the mass of color is as well appreciated by the unaided eye.

While enjoying this massive floral display, be sure to also look down as well as out, for a more modest view of a humble but frequent Mustard Family plant, Hairy Rock Cress (*Arabis hirsuta*). Several specimens of this small pinkish-white flower, growing about 1½' high, appear on the right just after the lookout junction, at an elevation of 1,000'. Typical of the Mustard Family, the flowers have four petals instead of the usual five, and these are often arranged in the form of a cross, hence the old name of the Family as the *Crucifereae*. In accord with more recent botanical practice, the family name has been changed to the *Brassicaceae* to correspond to the major genus in the family. The Mustard Family counts among its most well-known members *Brassica campestris* and *B. nigra* (both the source of mustard seeds and the resulting condiment), *B. oleracea* (cabbage, broccoli, Brussels sprouts, cauliflower, and kale), and *B. rapa* (turnips). Radish (*Raphanis sativus*) and horseradish (*Armoracia rusticana*) also belong to this fiery but useful family. In fact, Hairy Rock Cress is a close relative of the garden rock cresses and the more common *Iberis* species, also known as Candytuft seen sprouting in bountiful abundance within many a Portland and Seattle garden. ("Candytuft" is also a name applied to another garden plant in the Pink Family and related to the wildflower, Field Chickweed.) Comparing this austere wildland rock cress with its garden relatives provides some notion about the alchemy modern breeders have practiced to produce manufactured, but beautiful, urban plants.

Between **Points** 4 and **5** the trail switchbacks up and across the rocky basin, encountering a number of flowers new to May but common at lower elevations during April. These include Cascade Oregon Grape, also known as Mountain or Dull Oregon Grape – again a confusion of common names. The taller Shining Oregon Grape (Oregon's State Flower) is a different plant. Many examples of Hooker's Fairy Bells are present on the wooded slopes above, along with both Plume- and Star-flowered Solomon's Seals; Meadowrue; a surprising clump of Fairy Slippers on the right at about 1,400'; clusters of Candy Flowers gracing stream crossings and wet areas; a few purple specimens of Cliff Larkspur coloring an opening at around 1,450'; and masses of the deceptively named Large-leaf Sandwort (its leaves are rather small and it does not grow in sandy areas) thriving around the roots of large conifers in the deep shade. In the frequent openings between 1,300' to 1,400', the strangely wonderful fuzzy purple spheres of Ball-head Waterleaf almost lend a bit of floral whimsy to the scene.

Other flowers have now been in bloom between these points throughout late April, and will continue to freshen the often steep sections of the trail between 1,000' and 2,500'. These include two shy white saxifrages – Merten's (see **July** and **August**, below) on the wet cliff faces between 1,300' and 1,500'; Fringe Cup, particularly when circling the basin around 1,200' to 1,300'; Youth-on-age, especially prominent around streams at 1,900' and 2,300'; Baneberry, with large specimens at 1,500' to 1,700'; and masses of two flowers carpeting the forest floor: Pacific Waterleaf, whose dull grayish fuzzy flowers and deeply cleft 5-7" leaves inhabit the shaded forest floors in luxuriant vegetation between 1,100' and 2,200'; and Scouler's Valerian, with its brighter 1' tall white flower clusters brightening partially wooded areas, particularly at 1,200', and again at 1,300' to 1,400'.

Here too, one sees several shrubs beginning to bloom at higher elevations in wetter areas, including Red-flowering Currant at 1,550' and Red Elderberry, both at 1,550' and 1,800'. To underscore this elevational gradation of blooming times, view the now-flowering Indian Plum on the left at 1,500'; in lower sites, such as in wooded areas of Portland, this plant comes into flower as early as mid-February, when only an errant Camellia, misdirected by genetic variation, brightens the rainy landscape.

The streams between **Points 5** and **6** are now just damp mossy memories of the robust flowing creeks of March and April, but at their crossings (2,080' and 2,550'), the broad but intricately-cleft jagged fronds of Colts Foot push up through the seepage, accompanied by their bristly white to pink sprays of multiple flower heads. This precocious perennial sends its flowers up *before* its leaves, rebelling against our notions of how a plant should behave. Accompanying the Colts Foot in these areas are several large examples of Common Monkey Flower and the well-loved Stream (Wood) Violet, gracing wetter areas in low-to-mid elevations throughout the Northwest. The variety of common names attached to this small yellow violet (not all violets are purple – in fact, most are yellow or white in our woods) attests to its popularity: Stream Violet, Wood Violet, Smooth Violet, Yellow Violet, even Johnny-Jump-Up. This nomenclatural Tower of Babel underscores the value of having one scientific name for any biological entity.

A second yellow violet, the Evergreen Violet, also appears sporadically along this stretch of trail, although in drier locations and often growing off mossy embankments. Scouler's Valerian reappears around 1,500', and the ubiquitous Big-leaf Sandwort seems to follow the trail as it veers back to the west between 1,700' and 2,100'. Vanilla Leaf is a frequent ground cover here, but, as the trail heads southeast and ascends switchbacks along the crest of a ridge, then deep forest, Cascade Oregon Grape begins to dominate. Hopefully you will spot one interesting specimen of Twisted Stalk on the left at 2,100'. Fringe Cup frequents the shady forest up to 2,200' and the related and similar-appearing Youth-on-age can be found in its preferred habitat, near seepages and drying streams at 1,700' and 2,080'. Look for Candy Flower in these wetter areas as well. As you ascend the multiple switchbacks, note the 4-5' tall evil-looking Devil's Club in moist areas, particularly on your right, just waiting for the unwary or clumsy hiker to take a tumble and suffer the medieval fang-like thorns protecting this common shrub.

You may wish to take a well-deserved break at the talus-strewn lookout at **Point 6** (2,780'), where some of the boulders seem conveniently shaped like a hiker's backside. While the view is mostly of distant conifers, the pause is welcome. Looking about, the dominant shrub, both at this viewpoint and from here up the trail as it ascends, at first steeply, then more gradually toward the plateau at **Point 7**, is the well-known Salal (*Gaultheria shallon*). While at the

Salal

Coast, Salal can grow to enormous proportions (up to 15' tall) and form impenetrable tangles of thick vegetation, in most areas frequented by mountain travelers it is a 2-5' tall shrub bearing tiny pinkish-white urn-shaped flowers similar to those of other Heath Family plants, such as the huckleberries and heathers of country gardens and mountain slopes. Salal is among the most common of understory plants in the Northwest, frequenting dry and wetter locations with seeming disregard to soil conditions, a vagabond tramp, although an attractive one, unconcerned with where she sleeps. Useful too – natives dried copious quantities of the bland berries into cakes for winter consumption. David Douglas was so taken with the plant that he shipped its seeds back to England so that, "It would make a valuable addition to our gardens". One wonders how the British now regard that sentiment; Salal is presently an invasive weed on that island – fit revenge for the invasion of the English Daisy in our lawns.

Above this rocky lookout, deeper woods rule the landscape and fewer flowers flourish. However, several Rhododendrons just before the lookout and again at 3,100' on your right, may be in bloom by mid-June. Their large pink flowers are a worthy rival to the bushes so common in our cultivated gardens in town.

Vanilla Leaf

Care is still needed on this gentle trail however as Devil's Club continues to pose a threat, especially in its first 300 yards. In addition, two familiar shade-tolerant friends appear to dot the green and brown milieu with some color: Trillium, now with some specimens still glowing white until mid-May (masses appear at about 2,990' on the right); and the welcoming yellow of the Evergreen Violet. A few Stream Violets hold up their more erect stems at wetter areas, accompanied by their frequent neighbors, Candy Flowers, while in the drier understory, the white stalks of Vanilla Leaf (*Achlys triphylla*) poke straight up like fuzzy radio antennae from the green forest floor. The species name *triphylla* refers to the three bluntly-toothed leaflets, usually arrayed vertically in young shoots but spreading horizontally as the plant matures, in order to capture as much light as possible filtering through the tangled forest awning above.

In some places, Vanilla Leaf forms an uninterrupted ground cover on the forest floor, a kind of living-room carpet of bright green leaves sending up, here and there, spikes of its bristly petal-less flowers (the spikes are all stamens). Its ubiquity in such locations is explained by underground rhizomes which spread widely and send up new buds from nodes along their not inconsiderable length. As a bonus, Native Americans proposed that the dried leaves protected you against mosquitoes and biting black flies. In my experience, the only way any leaf is going to help under those circumstances is to use it to swat away the pests.

> Just as in human families, species within floral families can have quite different appearances. Generally, species within a genus, however, share some similarities. The various candy flowers and the numerous violets all look pretty much the same. It gets a bit more complicated with plants either higher up the evolutionary chain or those within a genus but widely separated in locations or in time since they've diverged from a common ancestor. By looks alone, no one would link meadowrues with buttercups, though they're in the same family; buttercups and cinquefoils look a lot alike, although they're not even distantly related. Three of our most common understory plants in west-side forests, the Inside-out Flower, Vanilla Leaf, and the Oregon Grapes, are all really members of the same Barberry Family, but the casual hiker could never relate them one to another based on appearance alone. It's pretty fortunate that more careful observers, called botanists, figured out a long time ago that it's the sexual parts and seed constituents which determine relationships in the floral world, not the way a flower, or even a leaf, appears.

Unfortunately, between **Points 7**, (the marshy area at 3,160') and **8** (at 4.6 miles – the old trail junction where the main trail turns right and **Point 9**, the top of Nesmith Point, diversity is as rare as a cloudless day in December. A few Trillium (still white) and Evergreen Violets punctuate the deep woods, with some Oxalis leaves (no flowers left) littering the forest floor. Continue to the top, across the old logging road, however, for different rewards – a view to the west up the mighty Columbia and a feeling of accomplishment for gaining almost 4,000' of altitude. You can continue another 200 yards downhill, past an old outhouse, to another view east and north. Nesmith Point remains a great conditioner with typical west-side floral displays, not a bad combination for close city-side access.

Goat's Beard

July and August

While many Gorge hikers believe the main flower show has migrated to higher elevations by the summer months, there are still plenty of plants in bloom to intrigue the amateur botanist during the summer. Few are as delightful as False-lily-of-the-Valley, somewhat more pleasantly known as the May Lily. Although this plant may bloom as early as late April (see above) to early May, it is especially plentiful in early July, when it is found just after the left turn at 200', 1/4 mile in and on the left, and on up to 500'. These lilies are unique within their family in having flower parts in fours, rather than three's and sixes – not that most hikers would inspect the tiny individual flowers. Fortunately, the blooms are arrayed in a short white spike, usually 10-15" tall, composed of up to 20 flowers, all held above the plant's most attractive feature: two deeply pleated, vein-streaked and wavy thick leaves shaped like hearts and emanating horizontally from the stem. Native Americans used the mottled bronze berries in a variety of decoctions for various ailments, but, as anyone who has tried them can testify, they ate them only in emergencies.

As the summer hiker progresses up the wooded trail, Candy Flower accompanies her or him pretty much all the way to the by-now dry streambed at 2,600' and these will persist throughout the hot summer months. A few openings occur between **Points 1** and **2**, allowing sufficient light into the forest for many people's favorite flower, the Tiger Lily, to bloom. Several fine specimens can be seen, both on the left and the right, between 200' and 300' of elevation. (See the North Lake Chapter for a more complete description.) Still out as one progresses up the trail between **Points 2** (the early stream crossing) and **3** (the left turn at the washout), masses of the Inside-out Flower appear, although by early to mid-July, its blackish curved seed-heads are most in evidence. Look to your left as you enter this rocky clearing to see platoons of Little-leaf Candy Flower and, arching over the trail and also on your left, Turtle Head, with its tubular purple flowers resembling a penstemon, to which it is closely related.

Leafy Pea

Just making an appearance, though an odd one, at around **Point 3**, is the well-named Goat's Beard (*Aruncus sylvester*). The common roadside shrub, Ocean Spray, is often confused with Goat's Beard, but the leaves of Goat's Beard are much more pointy and usually divided into six-to-nine-toothed leaflets, while Ocean Spray's leaves, while also toothed, are rounder (really, egg shaped), and its flowers fuller; overall, Ocean Spray is a more robust plant.

Many people are familiar with sweet pea-shaped flowers from garden plants or plants from the wild, such as the lupines and Scotch Broom. Three fairly frequent Pea Family flowers in the Northwest are in the *Lathyrus* genus: The most apparent and showiest of this tribe is the Broad-leaved Pea (*L. latifolius*), also called the Everlasting Pea. It spreads its bright pink stalks of flowers over waste fields and rough urban areas throughout the summer months and well into fall. Not quite as

Nevada Pea

prominent, but brightening Cascade woods at low to mid-elevations is Leafy Pea (*L. polyphyllus*), a plant to be found just before and after the left turn at **Point 3**, called by some Purple Pea-vine and by others (in error), the Nevada Pea (*L. nevadensis*). Both these pea-vines save the energy required for stalk strength and place it into producing colorful pink-to-purple flowers which brighten deeper forests. Often, within the monochromatic maze of green, pea-vines provide delightful bursts of color. If you really want to distinguish Leafy from Nevada Pea, count the leaflets: The Leafy species has 10 to 16 leaflets arrayed like a row of soldiers in regular formation along the leaf stalk (petiole), while the Nevada Pea has 4 to 10. Look for Nevada Pea as well further up the trail throughout the summer, especially between **Points 4** and **5**.

If you really want to get technical, consider the distinction between a pea-vine and a vetch: The pea has hairs on the tip of its tiny style (the female part) on one side only, like a toothbrush, while the vetch has hairs that encircle the style, like a bottlebrush. To tell the difference, you can get out your magnifying lens or do what I do – call any flower near an inch long a pea and smaller flowers a vetch. You'd usually be right, as vetches are smaller plants with tiny leaflets, although they also cling to the underbrush. This habit has embittered many Northwest gardeners and farmers because some of the imported vetches, such as Cow Vetch and Bird Vetch, now cover vast tracts of fields, gardens, and roadsides. It's not that they're homely, what with their deep blue, purple, and red flowers; it's that they're devilishly difficult to eradicate, with intermeshing and tough underground root systems along with their habit of clinging to other vegetation. Although there are some native vetches (see the Dog Mountain Chapter), it's the foreign invaders that seem to have caused the most trouble, conquering portions of our terrain, at least near urban areas and on land already worked for other purposes, such as old fields and roadsides.

> No stranger to Northwest travelers, the Pea Family contributes not only weeds such as vetches and the infamous Scotch Broom to our landscapes, but also a number of attractive and useful native and garden plants, such as lupines, lotuses, alfalfa, and clovers. While some of these plants are considered weeds, almost all of them enrich the soil in which they grow by harboring certain bacteria on their roots which absorb the nitrogen in the air and deposit it into the soil. Nitrogen is an essential component for many plants, sort of like a plant vitamin, yet it is an element that is most often deficient in acid and humus-rich forest and mountain soils. Pea Family plants, by thus "fixing" nitrogen into these soils, help many other plants thrive which, without this assistance, would be sadly absent from our environments. One famous and fortunate example is the coexistence of lupines and paintbrushes which carpet our alpine meadows blue and red – thank the blue – and its nitrogen-fixing properties – for the splashes of scarlet.

Two quite common Aster Family plants live almost side by side at the lookout above **Point 4** at 900' and bloom throughout the early parts of summer: Broad-leaf (or Mountain) Arnica and the narrow near-endemic, Columbia Gorge Groundsel. While this Arnica, described more fully above in the May-June Section, is among the most widespread of our composites, in contrast the Groundsel (also called a senecio or butterweed) is only rarely found outside the Gorge. Both brighten deep forests with cheery yellow daisy flowers, but you can distinguish them by their size – Broad-leaf Arnica is almost always the larger and more robust of the pair, growing up to 2½' tall – and more definitively by their leaves: Arnicas have opposite leaves, and this arnica's leaves are egg-shaped. The groundsels (senecios) have alternate leaves and, in the Columbia Gorge species, these leaves are fairly distinctive: Round in outline, they are tipped by 5 to 7 broad teeth which themselves have 2 to 3 indentations. This groundsel is a fairly dainty plant, growing around 1' tall or even less, with flowers ½-1" broad; it can often be found on the margins of trails, and, although common, it is rarely abundant. It appears to thrive at low to medium altitudes in the Gorge and provides a welcome relief of color to break the monopoly of green in our Douglas Fir and Western Hemlock strongholds.

Arnicas and groundsels form one huge group (or "tribe") within the Aster Family; there are thousands of groundsels, for example, and many are common garden and roadside urban weeds. Fortunately, the casual botanist who wishes to learn the names of Gorge groundsels and senecios needs to be familiar with perhaps just three or four of these, as many others are found in alpine meadows. The same can be said about the arnicas. Just two are commonly found in the western Gorge: Broad-leaf and Steambank Arnica, while three or four others will be encountered only in higher mountain meadows. These Aster Family plants all carry the same yellow daisy-type flowers, so, as the beginning naturalist will discover, their true identities can best be revealed by studying their leaf structure – and remember: Arnicas have opposite leaves, groundsels are alternate.

As the hiker progresses up the often-rocky trail between **Points 4** and **5** at a stream crossing, now dry, at 2,080', many spring plants are showing their age with waning flowers or simply seeds to mark their senescence. Western Meadowrue, Baneberry, and Little Buttercup, among others, test one's ability to identify plants by their leaf patterns alone, a sign that the casual mountain explorer has become a true amateur botanist.

One plant still in bloom, however, deserves special mention here: Merten's Saxifrage (*Saxifraga mertensiana*). Like a medieval maiden secretly bathing herself in a hidden shower, this shy denizen of the forest thrives in the spray of waterfalls and rocky seeps. Although its loose panicle of dainty white flowers is short of spectacular, its nicely-scalloped leaves are distinctive, with a circular outline and roundish lobes, each itself sharply and uniformly toothed. These leaves help this bashful plant stand out amongst the arpeggio of graded greens blurring the forest floor. Merten's Saxifrage is one of a handful of Northwest plants inhabiting areas where spray from waterfalls or boisterous streams provides almost constant moisture. Others in this select group include narrow endemics such as the Oregon Sullivantia (found mostly growing at Multnomah Falls) and the Columbia Gorge Arnica (see the Eagle Creek Chapter), a variety of Streambank Arnica.

Merten's Saxifrage

> Why are there so many different species of flowers and how do they persist in seemingly hostile and narrow environments? One explanation may have to do with *polyploidy*, a condition in which plants often have what appears to be an excess of genes. Many of these extra genes don't seem to have much function in terms of creating useful proteins for the plant's survival. For example, fruits and vegetables, as well as the flowering plants, often have double or triple the number of chromosomes, and hence genes, that humans and other mammals possess. This doesn't automatically mean they represent more complex organisms but it does endow them with a certain flexibility we may lack. For example, it may have allowed them the ability to employ extra genes to survive and diversify more readily with changes in the environment because these extra genes could help when the usual compliment of a plant's genes might have been unable to cope with ecological shifts. For example, the end of the Cretaceous Period 65 million years ago signaled the end of the dinosaurs because of a general cooling effect from an asteroid impact (although it is more likely in my opinion that the asteroid was the proximate, but not the ultimate, cause of their extinction – it took five million years for these beasts to go extinct, a sign that it was the out-pouring of volcanism caused by the impact that was the actual source of the gradual cooling effect). Yet it was exactly at that period of time that the flowering plants began to diversify and populate the entire planet. Recent studies have shown that it was the extra, previously "useless" genes that empowered the explosion of flowering plants in the aftermath of the asteroidal "big bang" and the resulting outburst of volcanism thereafter. This deadly combination, cooling the atmosphere, ended the Cretaceous Period and resulted, over thousands of millennia, in the extinction of the subclass of reptiles we now call the dinosaurs.

Above **Point 5**, at stream areas mostly dry now, Stream Violet still pokes its cheery yellow blossoms up above the green forest floor, while its close counterpart, the Evergreen Violet, can still be found on drier sites scattered up to the rocky lookout at **Point 6**, 2,780'. The aptly named Devil's Club is beginning to bloom in damp areas between 2,500' and 2,650', where it can be seen in spiny thickets straight ahead as the trail makes a left switchback. Its pyramid of small greenish-white flowers are rapidly replaced by shiny red berries by late July. Don't be tempted to taste them; they are acerbic at best and wandering to close to this evil plant puts you in jeopardy of being pierced by its thick and wicked thorns.

Wandering the woods beyond **Point 6**, both violets, Stream and Evergreen, can still be found, although they have, in the main, retired. Salal has begun to droop its blackish berries and only bristly spikes of Vanilla Leaf remain above their tattered shards of sagging leaves. Here and there, Foam Flower throws up sprays of tiny white plumes of bell-shaped flowers, though these are not seen here in the masses more common on other summer trails. About 15 minutes up from the lookout, however, Tiger Lily specimens abound, especially on your right, and these will continue to appear in semi-open woods to the junction at **Point 8**. The stream area, **Point 7**, about 1/3 mile from the lookout, holds a summer surprise: False Bugbane, its spiky white "flowers" made noticeable not by petals but by its stamens protruding from its center. It is on your right, followed by pink Cardwell's Penstemon and a host of Tiger Lilies.

Bear-grass remains in bloom before and after **Point 7** (the first opening on the level plateau at a marshy area, 3,160') while the sharp-eyed budding botanist can still discover a few specimens of Broad-leaf Arnica in shaded areas. The 1-2' tall strangely-curved Sickle-top Lousewort is present, along with the waning pink blossoms of the Cascade Penstemon in open areas as one advances up the trail toward **Points 8**, a trail junction at 3,290', just at a right-hand turn at this old junction, and at **9**, the top at 3,872'.

Also look for the strawberry-like Dwarf Bramble, with its low white flowers on either side of the trial/road after it turns right at the sign at 3,750'. Right at a sandy lookout west toward Portland a single Tiger Lily stands as a floral exclamation point marking, for most (at least for me) the "top" of Nesmith Point, also designated by a sign affixed to a tree on your right. Wildflower junkies can take in the view from here and anticipate next spring's bursts of rich floral displays. For a more expansive view east and north, continue slightly down the trail 200 yards to another opening and remember that floral hues still remain in the meadows around our tall volcanoes, Hood, Adams, and Rainier. These still support lush displays of color at this time of year, as do the higher grounds of the Cascades further north; they will continue to do so even through the early fall.

September and October

A limited sampling of blooms presents itself for the autumn hiker in the western Gorge, although fall colors compensate, with Big-leaf Maple, Douglas Maple, and especially Vine Maple punctuating the largely green monotony. The Trail is littered with the multi-colored red, orange and brown splotches of Big-leaf Maple leaves, many more than a foot broad and long. A few Robert Geraniums poke their cheery pink-purple blossoms above their gracefully-dissected leaves between **Points 2** and **4**, while just before and after the stream, **Point 2** at around 600', the Little Wild (Baldhip) Rose may still display its simple yet elegant flowers above its prickly stems.

As one progresses higher, Foam Flower can still be found in wooded areas, especially at the higher elevations between **Points 5** and **7**. However, in place of blooms, mostly leaves persist, and even some of these are beginning to wither. Yes, the flower show is not as spectacular as in spring or early summer but, for plants, there is no unimportant time of year. Below winter's snows, these trails will rest until spring.

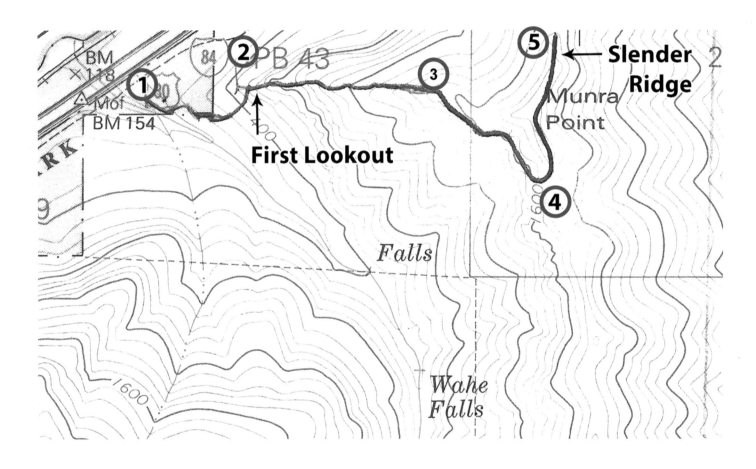

Munra Point

CAUTION: The first mile up to Munra Point is the steepest "trail" in the Gorge; the second mile is even steeper. This route, really more of a scramble than an easily-negotiated trail, was built by someone in a hurry. Sometimes it requires you to use your hands for balance; as a consequence, it should not be attempted in wet weather due to several sections of blank near-vertical sand (at least they seem that way) where it may be difficult to find adequate purchase, especially on the descent. Parties have sometimes brought a rope to this fray, although most hikers and outdoor adventurers will handle this route without undue trauma, especially in good weather. Young children and the inexperienced Gorge hiker are best advised to stick with better-defined paths. In addition, those with apprehension about the heights should chose prudence and pick a different trail. Even seasoned hikers should think twice (or more) about venturing beyond **Point 4** to **Point 5**, as this spine of a route, while brief, can be treacherous for anyone not experienced in balance and exposure to dizzying drop-offs on either side.

Then why should anyone go? Because of the feeling of adventure, the awesome views, and, of course the spectacular flower show. Because much of this route is in the open, one gets a feeling of space unlike any other trail in the Gorge, except perhaps from Indian Point or Table Mountain. And, because so much of the way to Munra Point is soaked with sun, the amateur botanist will have, in this short 2 miles, ample opportunity to identify more meadow wildflowers than on most other trails twice or three times its length. Add in proximity to Portland and a shorter-than-average trip time, and you have an ideal morning or afternoon excursion.

Two options exist when contemplating this drama: You can follow the description in Don and Roberta Lowe's *35 Hiking Trails – Columbia River Gorge*, which starts from the Bonneville Dam exit off I-84, or you can save about 1½ miles of a rolling forest walk by exiting the freeway at Cascades Locks and heading back west on I-84, then parking at the Moffat Creek bridge on the left, just after milepost 39.

A small parking area can be found here which affords ample space for a number of vehicles and, as long as you are off the pavement, the state police will not ticket you. Then cross under the freeway bridge as you head south to pick up a connector trail which intersects the Gorge Trail (Number 400); turn left here for 100 yards, then pick up the unsigned, but obvious Munra Point Trail on your left and begin the adventure. Munra may not be as tall as other targets in the Gorge but it cedes nothing to its loftier neighbors in spectacle and drama.

February and March

Some weeds may greet the early-season traveler plowing through the vegetation under the freeway bridge, though often these are quite variable. The all-season Common Groundsel (Senecio) can bloom at any time of the year, and, in accord with its Latin species name, *vulgaris*, it is a straggly looking thing, with its ferny drooping leaves and tiny yellow rayless plugs masquerading as flowers. It may not look like much, but this scrawny member of the Senecio Tribe in the Aster Family may be one of the most widespread and constantly-flowering plants in the world. Just like many Casanova's, its success lies not in its good looks, but in its ability to persist under adverse conditions and to sow its seeds perpetually.

Joining this groundsel in the thicket as you approach the woods and Trail #400, another weed may pop up as early as late February – Red Clover (*Trifolium pratense*). Most folks recognize the clovers but it pays to look closely at this common garden and lawn weed and at its close relative, White Clover (*T. repens*). The floral heads of all clovers are not just a single flower but are composed of multiple small, pea-like flowers packed into a compressed ball. Scores of individual flowers can thus be held aloft on a single stem, a device employed by many plants to advertise their presence without investing in huge individual flower heads. *Trifolium* means "three-leaved"; if you find one with four leaves, consider yourself lucky.

White Clover

Heading up Trail 400 between **Points 1** (the parking area) and **2**, the first really open viewpoint at 800' (spectacular!), a few Trilliums may be out in open woods by mid-March, and, as you turn left up the Munra Point Trail proper, an occasional Candy Flower can be seen. Given the relative absence of flowering plants, however, this is the time of year to practice your leaf-identification skills. Vanilla Leaf is just beginning to emerge here, its early leaves unfurling from a vertical orientation to their more familiar horizontal one later in the year. At the first left switchback, the thick evergreen leaves of the lovely Twin Flower (see a more complete description in the May section in this Chapter) can be seen spilling down banks and over the roots of ancient Douglas Firs. Further along, as the path grows ever steeper, the leaves of Plume-flowered Solomon's Seal and Hooker's Fairy Bells are beginning to appear. All is not lost for the color seeker, however, as an occasional yellow Evergreen Violet brightens the woods in March and, even at this early time of year, one or two Fairy Slipper Orchids can be found in the forest just before emerging into the open below **Point 2** after March 1st.

> Why are some plants deciduous and shed their leaves each autumn, while others, called evergreens, retain their foliage the year round? Plants living in tropical climes may have gained some advantage in remaining green all year in order to capture as much light as possible in dense jungles, while species further north or south would benefit by dropping their leaves and saving the energy required to maintain them throughout the low light of winter. With shifting continents, however, and changing climates, some tropical plants found themselves far from their points of origin and evolved mechanisms to survive in new lands. Hence, evergreen species in our seasonal environment often have thicker leaves than their counterparts further south, and are often darker from the extra chlorophyll needed to ensure their survival during the dim winter months. This is not, however, a complete explanation for the evergreen versus deciduous dichotomy. Many evergreen plants, such as the mighty conifers crowding our Pacific forests or the tiny Twin Flowers thriving beneath them, have all evolved in northern climes. Convergent evolution may have favored these mechanisms in entirely different ways. There is thus more than one way to survive a Northwest winter.

Many of the plants further up this rough precipitous trail are deciduous and their leaves are not yet out, although the first inklings of Wooly Sunflower leaves, toothed and grey, are just beginning to poke up by late March in openings at 800', and then in meadows between that elevation and **Point 3**, the second large opening. The woods may contain some of the same leaves as noted above, but the open meadows from here to the top usually host deciduous plants which can thrive only in the abundant light during the spring and summer. Thus these beautiful areas, so filled with blooms later in the spring, now hold only grasses and the promise of color to come.

April

By early to mid-month in April, Munra Point presents some of the finest flower shows in the Gorge, in part because the entire trip occurs at relatively low elevations. Even the parking area showcases a pretty flower, American Speedwell or Veronica, its small blue-to-purple blooms lining ditches here and there in the Gorge. Masquerading as a weed, it is actually a native plant.

Heading up into the woods, Plume-flowered Solomon's **S**eal, Hooker's Fairy Bells, and Vanilla Leaf are conspicuous along the early route between **Points 1** and **2**. A striking discovery awaits at about 450' elevation, where a single Chickweed Monkey Flower pokes its small but pretty head up from the middle of the dirt trail. This small yellow flower is most often found in wetter environments, but occasionally, as here, it seems to do well in the absence of constant moisture. Indeed, it can also be found on the Ruckel Ridge Trail, again at low elevation and again growing right in the middle of a sloping and sandy trail, a place where moisture would rarely collect. The inherent flexibility of this plant may help it survive during periods when damp soil is not abundant.

Racemose Pussytoes

As you enter the beautiful open areas around 750', just below **Point 2**, swaths of Woolly Sunflower begin to appear, just emerging from their long winter's rest. This, among our most admired of Aster Family plants, populates these open meadows all the way from this point to just below the top. Woolly Sunflower, also called "Oregon Sunshine", paints these meadows a gorgeous deep yellow, its ray flowers spreading out from a golden orange button of disk flowers at its center, thus comprising the composite flower typical of the Sunflower Tribe in the Aster Family. The "woolly" part of the name derives from the deeply cleft leaves, grey-green in color because of their hairiness. Some botanists believe that this feature helps such plants survive periods of drought by retaining moisture present in the air, while others think the wooliness helps reduce exposure to harmful ultraviolet rays in these plants, so often growing without the benefit of any shade. Both opinions may be true, although many other plants growing under these conditions lack hairs. We honestly don't know all the reasons for these many differences but that shouldn't stop us from admiring nature's multitude of variations nor from continuing to explore potential answers.

Between **Points 2** and **3**, at elevations varying from 800' to 1,250', note the Racemose, or Slender, Pussytoes (*Antennaria racemosa*) in open woods and particularly on wind-swept ridges. Of the five species of pussytoes found (sometimes with difficulty) in our mountains, this and Field Pussytoes (with the unflattering Latin name of *A. neglecta*) are by far the most common ones at low elevations. Both these pussytoes, usually 12-18" tall, share a common plant design: an unbranched stem with linear to lance-shaped simple alternate leaves, all surmounted by a cluster of soft wooly off-white flowers which, for all the world, look like an upturned feline's paw.

This flower show, however, conceals several mysteries within the Aster Family (*Asteraceae* or *Compositae*). For one, many asters do not look anything like daisies or asters at all. Consider such diverse plants as the thistles, Trail Plant (Pathfinder), Yarrow, Pearly Everlasting, Goldenrod, Colts Foot, or the Wormwoods. These bona fide members of the Aster Family present themselves as agglomerations of tiny flowers, each of which itself is a composite flower. Moreover, many of these plants can reproduce asexually, the ovules maturing into seeds without fertilization by male pollen. Thus the offspring may all be clones identical to their mother. This adaptation may have evolved in alpine plants growing in short-season environments where insect pollinators are unreliable. Mammals, including humans, have fortunately not adopted this mechanism.

Racemose Pussytoes

> Several plant species, particularly in the Aster Family, can create a bluish-ultraviolet light; invisible to humans but quite a sight for potential pollinators, who are attracted to the flower's sexual parts by these nanoscale ridges on the flower's petals, sort of like airport's blue guidelines for plane landings, though quite a bit smaller. This ultraviolet halo is common in certain species which manufacture their ultraviolet properties with certain irregularities, which have been shown to be even more attractive to insect pollinators than steady lines. The ultraviolet halo created almost always is crated with ridges that contain irregularities much like diffraction gratings in materials, giving them different colors with differing angles of approach. Like an airplane aiming for a landing, bees have been shown to prefer these irregular gratings to artificial flowers which present a regular pattern of ultraviolet hues. Such floral disorder actually helps potential pollinators hone in where it matters most: to reach the nectar in the center of the petals and thus to unwittingly collect pollen, then deliver it to the next flower's stigma and style.

Entering the open areas at and above **Point 2** by mid- to-late-April, the fortunate hiker will come across two stunning meadows at 1,150' and 1,250', which contain gorgeous displays of Large- flowered Blue-eyed Mary (*Collinsia grandiflora*). This Snapdragon Family member carpets these meadows in a lush blue at this time of year. These larger versions of Small Flowered Blue-eyed Mary are actually members of a different species: Not only is the plant much larger, growing up to 18" tall, but its corolla is bent more at right angles and the entire plant is more robust, with more blooms clustered around its top.

Although you can admire these plants from a distance for their riot of color, a close look rewards as well. Each flower is a miniature snapdragon, with an upper and lower lip, varying in shades of blue to white to purple. These hues combine with the punctuated red of Harsh Paintbrush and the yellows of Western Groundsel to render a pointillist scene worthy of Seurat – an extravagant flower show almost profligate with color.

Yet another Snapdragon Family species begins to bloom in mid-April – Fine-tooth Penstemon. This blue-purple flower is most abundant here, poking its multi-flowered heads 1-2' above its whorled leaves in brief openings at 750' and again in almost all sun-filled locations between **Points 2** and **4**. It is particularly frequent near the top as one scampers along the crest below the summit between **Points 3** and **4**. Count the stamens (without destroying the flower) to see that the fifth one is often hairy and sticks to the throat of the flower tube (the corolla). This sterile stamen (a "staminode") may serve to attract pollinating insects, which then crawl further into the tube seeking its nectar, located deep within the flower. The style, or female tube extending from the ovary, bends down after the flower opens so that the stigma, or tip of the tube, can receive any pollen carried in by hungry insects. This occurs only for several hours after each blossom opens, but by this time most insects have already entered. Thus the flower ensures that it will be cross-pollinated, rather than self-fertilized, contributing to the healthy perpetuation of the species.

Between 1,200' and 1,600', scattered examples of Cliff (or Menzies') Larkspur can be found sprinkling its purple hues in rocky meadows, while an occasional daisy-like Broad-leaf Arnica colors the woods at these heights with yellow. In the openings as well, between 750' and 1,300', the Nodding Onion (*Allium cernuum*) nods its pink, multi-flowered head and lends its distinct onion smell if you get close enough. Native Americans ate the bulbs, but, because of its restricted range (just here and in Puget Sound and the southern part of Vancouver Island), you should not emulate this habit. This curious plant, about 1' tall, has grass-like leaves which, strangely, wither before the flower blooms. The flower head is composite, with each pink-colored bloom sporting six petals. Onions are in the Lily Family; look closely to see

Large-flowered Blue-Eyed Mary

their remarkable resemblance to garden brodiaeas and hyacinths.

Several woodland plants are just now emerging: Plume-flowered Solomon's Seal is out in the lower woods, while the distressingly named Bastard Toadflax (see May, below) is just beginning to appear in open woods between 1,100' and 1,250'. Nearer the top, Wooly Sunflower is at its peak and, between **Points 4** and **5**, some Field Chickweed, Harsh Paintbrush, and Fine-tooth Penstemon can be found. Take care if traversing out to the viewpoint beyond **Point 4**; good balance is needed as you are on a ridge minus handholds.

Although the major flower show at these view-filled heights awaits May's blooming, there is so much to see and appreciate in April that the winter-weary hiker will rejoice in the spectacular climb up Munra Point at this time of year, as much for the color as for the exercise.

May

Nodding Onion

Every panting breath and aching step up the Munra Point Trail will be worth the effort during this premier month for blooms in the western Gorge. Climbers and hikers alike may approach this "trail" for the exercise, but most often leave it breathless as much from the flower show as for its thigh-burning steepness.

In the forest just after **Point 1**, one immediately notices the shrub-like Hooker's Fairy Bells, their white lanterns hanging from the undersides of pointy green leaves. This Lily Family plant is considered prototypical of many lilies growing in our wet forests as having "drip-tip" leaves; supposedly, the profusion of rain and the constant dripping from the trees could saturate these plants, but the pointy tips of their leaves act as natural gutters, directing the flow off the leaf and to the ground. Sounds nice – a grand evolutionary solution – except that these leaves already contain oils which prevent saturation. Moreover, most species in this drip-drip-drip environment don't have leaves which end in a tippy point, as for example Vanilla Leaf, Foam Flower and the waterleafs. It seems more likely that the drip-tip of these plants is a shared characteristic of this clan within the Lily Family rather than an adaptation specifically designed to accomplish water shedding. One problem with such theorizing is our inability to design experiments to prove or refute clues to distant evolutionary events.

Small-flowered Alumroot

Hooker's Fairy Bells accompanies the hiker up to about 600' between **Points 1** and **2**, along with other familiar woodland plants such as Big-leaf Sandwort (between 100' and 1,000'), Columbia Wind Flower (100' to 450'), Vanilla Leaf (below 600'), and the Inside-out Flower (below 900'). Masses of the Common Monkey Flower color a dripping mossy wall with welcoming bright yellow on the left at about 400'. In areas which retain some moisture, not easy in this inclined environment, Candy Flower (look especially between 300' and 500') and Fringe Cup (on the left at 300') thrive. Little-leaf Candy Flower springs up at this time of year from mossy rocks between 200' and 900', amazing us with its apparent lack of regard for the absence of sustaining soil, while its close cousin, Siberian Miner's-lettuce, appears occasionally in the woods below 900'.

One new plant appears here as well in these forests, the unfortunately-named Bastard Toadflax (*Commandra umbellata*), a species which rarely attracts attention except for its name. Perhaps the "bastard" part refers to the fact that this Sandalwood Family member has few close relatives in the Northwest, although that's hardly sufficient reason to denigrate it so. Nonetheless, this plant is attractive if viewed closely. Its foot-long spikes of

Bastard Toadflax

feathery-leaved stems are topped by compact, ball-headed clusters of five-petalled white flowers, giving the overall impression of a woodland alyssum. This plant, although often found under shade, does require some sun and can frequently be seen in the open woods on this trail between 700' and 1,500'.

Another plant of these low elevation forests, although one requiring more light, is Small-flowered Alumroot (*Heuchera micrantha*), sending up its 1-2' sprays of dainty white flowers above hairy maple-shaped leaves. This species thrives in semi-open forests amid rock crevices, on talus slopes, and in situations where mossy rocks predominate. Perhaps, as in human relationships, this alumroot benefits from the warmth which can accumulate from the stones which it seeks out for companionship.

On the Munra Point Trail, it can be found between 150' and 600', especially at the boundaries between deeper woods and open rocky areas. Don't forget to inspect the tiny individual flowers of this too-often neglected plant. Viewed up close, these magically change from a panicle of amorphous white spots to wee dainty bells dangling in the western breeze. This alumroot, and another common northwestern one, Smooth Alumroot (*H. glabra*), which grows generally at higher elevations but is less hairy and has leaves as broad as long, are both related to the garden alumroots, most of which derive from *H. sanguinea* and have sprays of reddish flowers, a red stem, and burgundy leaves. Native Americans ground up the roots of these plants to cure swollen testicles, a condition fortunately uncommon these days among American males.

> How simple a leaf appears to be. It merely gathers carbon dioxide from the atmosphere and converts it via photosynthesis into oxygen, which it fortunately delivers to our atmosphere, allowing us to live on this planet. The exact chemistry is still being elucidated but the generalities are well-known. What if we could manufacture an artificial leaf to mimic its natural counterpart? Could this help reduce the carbon overload heating up Earth? In fact, attempts have been made. One problem is that, despite our naive preconceptions of perfect nature, a leaf is not very efficient in reducing carbon. Several researchers, however, are pursuing accelerating this very process and enhancing its efficiency. Scientists are using the hydrogen in this system, utilizing only water and sunlight in a closed system in which the CO_2 would be transformed into hydrocarbons, thus manufacturing fuel without polluting the atmosphere. Thus far, this process, while more efficient than natural photosynthesis, is still not sufficiently economical to contest the alternative energy production of sunshine, wind, and tides. At present, however, scientists are experimenting with this system through the use of catalysts and genetically altered bacteria which may produce realistic energy amounts within reasonable costs. Such an achievement could enable the creation in which the carbon dioxide emitted by combustion could be transformed into fuels, fertilizers, plastics and even medications. All this from those leaves we pass by every day without a thought of how they may transform our ideas of providing energy.
>
> Innovations include seeds covered in fungi or bacteria which are resistant to drought, can deter plant pests, and expedite growth. It has just been in the last decade that biologists have recognized the complex web which is necessary for plant growth to occur. But if no changes are made in the production of food crops (especially cereals), the world could face a food shortage of dire extent, making the droughts of the past minor bumps on the road from farm to market. Some people are opposed to genetically altered crops but they have proven to present little harm thus far; indeed, humans have been altering the DNA of plants for millennia.

By mid-May, two Rose Family shrubs appear in these lowland forests, Little Wild (Baldhip) Rose, with simple rose-colored blooms and weak but plentiful spines, and the Himalayan Blackberry, it's pretty white-to-pink rose-like flowers belying its considerably stouter and often painful thorns. While the rose is a native, the blackberry, of course, is not, although we do have one native blackberry, the Trailing Blackberry (*Rubus ursinus*). Fortunately, on the Munra Point Trail, Himalayan Blackberry is not prominent, and is mainly confined to openings in the early woods between 100' and 300'. Little Wild Rose can be easily found in this same area, but in deeper woods. Look for it especially on the right between 200' and 300', where it is plentiful.

> Until the early 18th Century, plants were given increasingly cumbersome Latin names as more and more species were discovered. Thus, a new species of wild rose could only be distinguished from a close relative by a single word within a long string of descriptors, such as "the rose with white petals which grows in northern forests with stout prickles" versus "the rose with white petals which grows in northern forests with soft prickles". Needless to say, this system rapidly became unworkable as species from an expanding world became available for scientific study. Linnaeus arrived at an ingenious solution modeled in part after people's common names: He grouped plants based not on their appearance, but on their sexual characteristics, and dropped the large family name (as in a human clan, where many last names may exist), and instead focused on the genus name first, analogous to the last name within a first-degree family, then the specific species name second, analogous to the first name of any particular individual. Thus we have, for example, the *Montia's*, or Candy Flowers in the Purslane Family, and among these, are the individual species, Miner's-lettuce (*M. perfoliata*), Candy Flower (*M. sibirica*), and Little-leaf Candy Flower (*M. parvifolia*). Despite recent challenges to this system, it has served science well for the past 300 years, and, remarkably, has been, in the main, confirmed by modern DNA evidence, a development which could never have been foreseen by this 18th Century Swedish naturalist.

Another Rose Family shrub, Thimbleberry, can be found here as well, with its many white tissue-paper flowers lining the trail between 150' and 400'. However, hikers will have to wait until late July to August to gobble the tasty red fruits. Other plants common to these woods in May include Pacific Waterleaf (100' to 350'); Sweet Cicely (below 500'); the yellow Columbia Gorge (or Bolander's) Groundsel (on the left at around 500'); Star Flower (300' to 600'); and the hard-to-see, but delightful-when-you-do, Woods, or Small-flowered, Nemophila (see **June**, below). You must look with a small eye to appreciate its rare beauty.

Here too, at 300', is a small, yet better-appreciated flower, admired across the northern hemispheres in this country and abroad, the lovely Twin Flower (*Linnaea borealis*). In fact, it is rare to find its name, Twin Flower, not preceded by the adjectives "lovely" or "beautiful". The reasons are many: This sub-shrub (it has woody sprawling stems) has glossy deep-green elliptic leaves which form a perfect ground cover over the wooded banks it prefers to inhabit (it needs some drainage and thus does not tolerate the flats nearly as well). Its blooms, though small, are perfect little drooping pinkish-white twin dunces' caps, each mirroring the other atop 6" high stalks. Although these blooms each by itself may not emit much fragrance, if you bend to smell the entire plant, its perfume is both enigmatic and enchanting. Finally, this little shrub magnifies its charms, spreading across the forest floor by dispatching runners which, at regular intervals, put down roots so that the plant can extend its reach. All these characteristics have endeared it to outdoor travelers and gardeners alike, and it is as common as it is illegal to see some illicit green thumbs trying to uproot Twin Flower for transplantation to their city gardens. Don't be tempted by this plant's charms – it takes a mighty skill to get this flower to bloom in urban landscapes. Twin Flower was said to be the favorite flower of Linnaeus, the father of scientific taxonomy, and its genus was named after him; *borealis* means "northern".

Twin Flower

As you approach the viewpoint at **Point 2**, the forest gives way to more open vistas and the glorious meadows which have made this trip so famous. Cliff (Menzies') Larkspur has already shown its deep purple-blue face early this month and here it can be happily found dotting these upper meadows between 800' and 1,700' all the way to **Point 4**. A smaller denizen of these openings, though equally attractive to those who pursue close inspection, is the 1' tall

Varied-leaf Collomia, with its bright pink phlox-like blossoms trying to compensate for its diminutive size through color alone. Perhaps its flashy hue also makes up for its habit of usually standing alone; although widespread in distribution, this Collomia, a member of the Phlox Family, rarely grows in profusion.

The Borage family is allied in many ways to the Phloxes, and a familiar member of this family, Common Cryptantha, can also be seen as you enter the open areas just below 1,000', especially on your right. Often mistakenly called the "Popcorn Flower", this relative of the true popcorn flowers does look a bit like a kernel of the movie snack, with small white blossoms centered by a yellow-to-orange button of stamens and a pistil. Please don't try to munch on this flower – even loads of butter and salt will not render it palatable.

Between **Points 2** and **3** many inhabitants of these meadows thrive, including Field Chickweed, especially prominent in the clearings at 1,100' and again at 1,650', and the gorgeous, well-known, and often-fragrant Spreading Phlox, its low mats of pastels: pink-to-blue-to-purple-to-white five-petalled flowers carpeting these sunny fields, particularly at 1,200' and again just above **Point 3** near the summit trail junction. At this time of year, with the breeze just right, you don't even have to stoop to smell the sweet licorice perfume of these massed blossoms, so strong that it whets the appetite.

In counterpoint to the cool colors of the Phlox and Cryptantha, spikes of Harsh Indian Paintbrush punctuate these meadows in shades of orange and red, the combinations and permutations much like that of a Bach fugue, but in color. It is not all meadow to the top; indeed at 1,600', some of the steepest scrambling occurs. In these woods, as you cling off roots and scramble second-to-third class rock up the boulders, don't be too distracted by the yellow daisy-like Broadleaf Arnicas thriving in the shade.

If you missed the incredible flower show of Large-flowered Blue-eyed Mary's populating the meadows at 1,150' and 1,250' in April, do not despair, for this plant, mixed with its sibling, Small-flowered Blue-eyed Mary, continues in bloom through at least the middle of May before retiring for the summer months. Nine-leaf Desert Parsley provides yellow accents to the floral display in these upper meadows all the way between **Points 2** and **4**, as does Western Groundsel (Senecio) and, in a different hue, Rosy Plectritis, which appears in masses at 1,300' and 1,600'. The groundsel, a prominent member of lowland openings in the Gorge, comes in two different hues – yellow here on the west side and white on the east (see the Dog Mountain Chapter). It can grow here up to 2½' tall, with thick dark-green elliptic leaves and daisy-like flowers with yellow rays and a yellow disk as well. Here too the Fine-tooth Penstemon remains in bloom at these higher elevations throughout the month, as does the Wooly Sunflower, happily found in masses between 700' and 1,700'. In fact, at the very top of this climb, among **Points 3, 4** and **5**, Wooly Sunflower in yellow, Harsh Indian Paintbrush in red, and Fine-tooth Penstemon in purple put on quite a color show, a fitting climax and reward for the effort.

Perhaps the combination of plentiful rain in the winter months combined with abundant sunshine in the spring and ideal soil provide just the right conditions to promote such a profuse outpouring of colorful blooms in these happy meadows. So startling and brilliant is this flower show, one almost wishes for a return to the shaded silence of the forest – almost but not quite. At the very tip of Munra Point, and at the very end of the thin walkway out to **Point 5** (use care here!), Harsh Indian Paintbrush, Martindale's Desert Parsley, and Fine-tooth Penstemon can be found.

Both the effort required to reach these spectacular viewpoints, and the colorful floral display which accompanies it, makes the haul up Munra Point one of the most spectacular in this premier month of May in the Columbia Gorge.

June

There may be fewer hikers on the Munra Point Trail during the summer months of June, July, and August – all the more reason to stretch muscles and tendons and climb this route now, especially since many of the colorful flowers described above are still joyfully blooming at this time of year. While the Lily Family woodland plants, such as Hooker's Fairy Bells and the Solomon's Seals, are in the Assisted Care phase of their lives (gone to seed), other forest species continue to flourish despite the summer heat, protected by the arching shade of conifers.

Small-flowered Nemophila

For starters, the deep pink hues of Robert Geranium poke through the green understory on the forest floor in the first ¼ mile, while another pink plant, the Little Wild, or Baldhip, Rose is a prominent feature of the shrub layer on this early forest walk. It is gratifying to see such color in the woods, as most often, forest flowers are white. These understory species have little need to invest expensive energy in color when their white attire can more easily be noticed by potential pollinators within the greenery of these majestic corridors. Still, a number of deep forest plants do sport some color, proving again that living organisms follow rules about as religiously as your average 14-year-old.

Thus two other plants show some color in the low woods between **Points 1**, the parking lot, and **2**, the first lookout, at this time of year. Twin Flower, with its paired white-to-pink trumpets perfuming the air, continues to charm passers-by in the woods on the left between 200' to 400', and a new, but unjustly-ignored species is making its appearance: the Small-flowered (or Woods) Nemophila (*Nemophila parviflora*) in the forest on both sides of the trail between 250' and 450'. Here, one truly must look closely to appreciate this exquisite gem, which barely reaches 4-5" above the forest floor. Its leaves are nicely scalloped and almost club-shaped, generally with five lobes, and the flowers, which seem to bloom a bit later here than farther east, are a delicate bluish white, their five tiny petals no more than several millimeters wide. *Nemos* means "grove" in Greek while *philos* of course refers to "love of", hence Nemophila is an apt name for this shade dweller. As you stoop to examine this small denizen of our lowland forests, you can appreciate the relationship it bears to its larger cousins in the same genus – members of the California clan of the Waterleaf Family – including California Five-spot and Baby Blue Eyes, common garden plants. Perhaps these Californians grow larger in the southern sun, but *our* nemophila makes up in grace and delicacy what it lacks in size.

These low forests are brimming with other flowers as well during June. Star Flower twinkles forth here and can be seen particularly after the first ½ mile, between 300' and 600' elevation. Although keyed out as pink in most guides, many of this charming plant's specimens are more white than pink. The Columbia Gorge Groundsel (*Senecio*) lends its cheery yellow to the woods, especially on the right of the trail at 500' elevation. Also, the common weed-like Annual Bedstraw (*Galium aparine*), also known as Cleavers, sprawls across other vegetation within the first ¼ mile. Its tiny white flowers are almost unnoticeable, but its square sticky stems with hooked bristles enable it to cling ("cleave") to other vegetation like a poor relative seeking support from its neighbors.

Lest we be too critical, however, we should allow that a number of plants, such as the colorful sweet peas, have adopted this habit, which saves the leaning plant from having to invest too much energy in sturdier stems. Cleavers is weedy, growing rampantly in city lots and roadsides, but is apparently a native, giving the lie to the dictum that all weeds are immigrants. Indeed, many weeds are quite beautiful. Again, generalizations prove as inadequate in the plant world as they are in animal life.

Cleavers is related, though at a distance, to *G. verum*, the bedstraw supposedly lining the manger where baby Jesus lay. Bracken Ferns were mixed in to soften the flooring. The Bracken failed to acknowledge the child and lost its flowers, while the bedstraw welcomed the baby, and was rewarded with bountiful blooms. Some legends die hard.

About a ¼ mile before the lookout at **Point 2**, the scenery changes as you break out into more open woods and meadows, and the way becomes progressively steeper. Bastard Toadflax is plentiful in the open forest between 700' and 1,500', while Racemose Pussytoes remains out on the windswept ridges just above **Point 2**. Between this lookout and **Point 3**, many of the species described above in May are happily still in bloom throughout June, including the fields of Phlox, Blue-eyed Mary's, Harsh Paintbrushes, and Western Groundsels, which color the upper meadows between 800' and 1,600'.

However, several equally colorful and unusual plants can be found just beginning to bloom this month. The Ball-head Cluster Lily, also known by the Native American name of Ookow, is conspicuous in sunny openings at 800' and 1,250', with its tight knot of deep purple six-petalled blossoms standing above 2' to 3' stems amid the grassy swaths. Spectacular in their own way are the weird and wonderful blossoms of Mountain (or Bronze) Bells, their drooping brownish-purple bells spread up and down its 2' stem, with each petal sporting flared tips. These never-common Lily Family members can be found in the grassy parklands at 1,200' and 1,600'.

Between **Points 3** and **4**, a few Cliff (Menzies') Larkspurs offer their pleasing deep blue-purple hue to the rocky meadows, while Rough Wallflower lends its striking yellow-orange in counterpoint. As you make your careful way along the rough path beneath Munra Point proper, **Point 4**, note the Nine-leaf Desert Parsley in yellow, the Harsh Indian Paintbrush in delicious crimson, and the Spreading Phlox in tones of white to pink to purple. Tiptoe carefully along the trail's extension from **Points 4** to **5** if you are sure of your balance, and, even if you are a solitary hiker, you will be accompanied by many familiar friends, including Wooly Sunflower, Common and Harsh Paintbrushes, Small flowered Blue-eyed Mary, and Western Groundsel. Head up to the very top of the **Point** for an air-filled view of the Gorge or balance your way south along the lessening trail to breath in the glorious flowers along these exalted ridges. As you scale these lichen-encrusted rocks and take in the view, often as the sun feathers through a reef of clouds, you will be absorbing some of the best northwest Oregon has to offer. Just ignore your aching knees on the near-vertical way down to your vehicles.

Cleavers

July and August

Take plenty of water or sports drink up this steep path if you plan to scale it during the summer months as there is not a drop to be had on this mostly ridge-top route. While the heat can be draining, this half-day trip in summer still rewards the hiker with a combination of great conditioning and plentiful flowers. The parking lot and the areas under the Freeway bridge still contain weeds, but the fact that they're aliens doesn't mean we shouldn't try to identify them, especially since these here are so common in the city as well. Common Groundsel and both familiar clovers (White and Red) are present in abundance, along with Queen Anne's Lace, ubiquitous along roadsides and in waste fields by August.

Another summer weed almost as common is Hairy Cat's-ear (*Hypochaeris radicata*), an Aster Family plant masquerading as a Dandelion. Lining highways throughout the Northwest throughout the summer, this plant looks for all the world like the weed we call a Dandelion (*Taraxicum officinale*), and almost everybody except a botanist wouldn't notice much difference. After all, it is yellow, produces only ray flowers, has toothed basal leaves, and frequents weedy places. This is guilt by association, however, as Hairy Cat's-ear is not even in the same genus as the Dandelion. It can be distinguished from the common lawn weed by its slightly taller habit, its wiry stems, its hairier leaves, and, most importantly, its several heads at the ends of its branches; Dandelion has but a solitary head at the terminus of each branch. A general rule for distinguishing among these two look-alikes is that Dandelions besmirching your lawn are really Dandelions, while the "Dandelions" lining I-5, I-84, and US 26, are probably Hairy Cat's-Ears or Smooth Hawksbeard (*Crepis capillaris*). Don't tell any botanists I told you this.

Away from these Freeway fields, native plants take over the floral spotlight. In shady woods down low, between **Points 1** and **2**, look for Miner's-lettuce, the Candy Flower relative with perfoliate leaves, a fancy term that means the stem seems to pierce the stem leaf. In fact, this apparently single leaf is two leaves, originally opposite but now joined, as if to support the tiny white flower above.

Candy Flower itself is still in bloom at areas once damp earlier in the year, especially at 300' and 500' elevation. Where the forest gives way partially to more open woods, Small-flowered Alumroot may still be displaying its 1-2' spray of dainty white blooms, while Robert Geranium continues to grace these lowland woods with its pink flowers, especially prominent in the first ½ mile of trail. While these forest denizens were present in the spring months as well, a new shade-loving plant now appears in these woods, Foam Flower. This common woodland flower sends up 1' tall sprays of white flowers beginning in mid-July at these elevations (here at 500' to 1,200') and these can last through September at higher altitudes. At times, this plant provides an almost continuous ground cover beneath the conifers, thus very much appearing as ocean foam arising from an unseen and distant sea.

Hairy Cat's Ear

Throughout the summer heat, several other woodland plants continue to thrive here, including Bastard Toadflax, Small-flowered Nemophila, and the Little Wild Rose (see June above for descriptions and locations). Star Flower persists as well in the woods between 300' and 600', although by the end of July it has pretty much gone to seed. Fortunately, Twin Flower continues to charm visitors where it blooms on the left between 200' and 400', about ¾ of a mile from the trail's start. Its scent seems even stronger in the still summer air, lending a special enchantment to these perfumed woods. In the meadows on the right at 1,200' to 1,350', one can also find another splendidly fragrant plant, the yellow-orange Rough Wallflower, its spicy smell adding to the allure of these meadows.

Starting at about 750', and continuing between **Points 2** and **3**, the two yellow Aster Family species described above, Wooly Sunflower and Western Groundsel, brighten the open meadows all the way to the top, while Harsh Indian Paintbrush continues to dot these openings with its splashes of crimson-red. Broad-leaf Arnica will continue to bloom in the open woods between 700' and 1,300'. Some Cliff Larkspur remains in bloom at 1,450' and 1,600'; a few specimens can be found as you hike from **Points 3** to **4**, while at these uplands, Field Chickweed graces the rough path as well, then reappears as you tiptoe carefully out to **Point 5**. Thus, while the flower show may not be quite as spectacular during these summer months as in the spring, there is much to admire on Munra Point for the summer adventurer. Catch your breath on these high and rough stones and admire some of the most inspiring views in the Gorge, sights well worth the considerable effort you expended to reach these heights. Moreover, these primitive ridges highlight the jagged up-thrusts which gave rise to the Gorge itself.

> This seasonal succession of blooming times may relate in part to temperature differences at lower and higher altitudes as well to differences in light exposure, the life spans of pollinating insects, and the varied breeding strategies of different species. Early spring brings forth the most varied profusion of blossoms in these lower Gorge woods, while late spring and early summer are peak times for the flower lover visiting the higher elevations. Nonetheless, fall brings its own charms, evident in the crisper days absent the stifling heat of summer and in the subtle mix of orange, red, and yellow on the fallen maple leaves gracing the trail.

September and October

While it's true that the famous flowers of Munra Point are not blooming as profusely in these autumn months, the trail is just as steep so the conditioning just as healthy and the views of the Gorge just as spectacular. From the parking area, **Point 1**, through the grassy disturbed landscape under the Freeway bridge, the familiar list of suspects, weeds all, are still blooming. Queen Anne's Lace, Hairy Cat's-ear, the Common Groundsel, and both Red and White Clovers will remain in bloom throughout the fall. However, deeper in the woods, as you traverse the Gorge Trail, then head up the steepening Munra Point Trail proper between **Points 1** and **2**, a maze of green awaits, with few actual flowers to help in identifying the plants. Here, leaf study and identification will bring rewards.

Impress your friends by pointing out, in these early woods, the triplet of fan-shaped leaves of Vanilla Leaf, now turning from a light green to tan, or the hairy maple-like leaves of Fringe Cup on the left at about 300' elevation. Although the lovely blooms of Twin Flower may now have gone to seed, you should be able to identify this plant, classified as a shrub due to its woody stems, by its trailing habit and glossy egg-shaped leaves spilling down the slopes between 200' and 400'.

Star Flower's rosette of compact pointy leaves can be found between 300' and 600', while the dainty club-shaped leaves of the Small-flowered Nemophila grace the forest floor between 250' and 450'. Many of these plants, which thrive in the magical shade of the forest, such as Fringe Cup and Vanilla Leaf, are deciduous, their leaves appearing straggly as they wither and die. But appearances can be deceiving. Were it not for their sacrifice, these plants might not be able to conserve sufficient energy to survive the winter. Moreover, the decay of leaf detritus on the forest floor enriches the soil and thus prepares the ground for a flock of new plants in the coming spring.

Most people believe that the identification of seeds is a matter best left to professional botanists with magnifying glasses and microscopes. They are correct. One of the problems in identifying seeds is that, without proper equipment, which most amateur naturalists would never carry on a hike, most of these fruits of a flower's labors are too tiny to be distinctive to the unaided eye. In addition, most seeds appear as nondescript colorless afterthoughts to a flower's blooming, merely the dying detritus of the plant's former glory. Thus autumn may appear to be the geriatric season for all of plant-dom. But of course, it's quite the opposite from the scientist's point of view. Not only do these dried-up remnants of spring's and summer's colorful blooms contain the wellsprings of new life within their shriveled forms, they also display, under the proper magnification, an amazing array of species-specific shapes and forms (think of those electron microscope images of pollen or insects). They are so distinctive that their architecture is employed by taxonomists to help clarify the relationships among species and varieties of plants. In fact, before DNA analysis became possible, seed identification was a mainstay of species' confirmation. While we amateurs cannot be expected to identify species by their seeds alone, we can at least admire the ingenuity and efforts of those professionals who do.

Despite the relative absence of flowers, a few blooms grace the lowland forest in these autumn months. Robert Geranium may poke its charming pink flowers up to provide spots of color in the first ½ mile of trail, and specimens of Candy Flower continue to bloom in areas wet in spring, particularly between 300' and 500'. Otherwise, these woods are usually absent of blooms at this time of year, even though meadow flowers at higher elevations may persist.

Above **Point 2**, and in the openings all the way to **Point 3**, two yellow Aster Family flowers are often in bloom from September into early October: Wooly Sunflower and Hall's Goldenweed. Look for the Sunflower in the open meadows at 800', 1,250', and 1,600', as well as just under the top of Munra Point between **Points 3** and **4**. The Goldenweed, one of the latest blooming species in the Gorge, populates the ridges around 900' to 1,400'. Color aficionados will appreciate the crimson of some late blooming Harsh Indian Paintbrushes around **Point 4**, and even a few Common, or Tall, Indian Paintbrushes (yes, a different species) down-slope to the left as the rough path forces a balancing act to reach **Point 5**. (*Warning:* Again, do not attempt the exposed traverse to **Point 5** unless you are certain of your balance and comfortable with some degree of exposure.)

Despite the lack of blooms, seedpods abound from the many plants which were in flower earlier in the year. From the leaf and seed patterns, aspiring botanists could identify many of the meadow plants mentioned above, including Ookow in the openings between 800' and 1,200'; Wallflower at 1,200' to 1,350'; Large and Small-flowered Blue-eyed Mary's between 1,200' and the top; and the Western Groundsel, scattered in the meadows around **Points 3** and **4**. Be careful not to examine specimens and walk at the same time on some of these stretches, as care and balance are not always compatible with close inspection. Tumbles are not uncommon among curious scientists in these regions.

No matter what season you scale Munra Point, its severe beauty and gripping scenery will captivate you and reward your considerable effort. Go in spring for the woodland flowers, go in summer for the lush pigments of its blossom-choked meadows, or go in fall for the subtle hues of the forest and the crisp views from on high. Go before winter's snow and cold evict sane travelers from these trails. But by all means, go.

Barry Maletzky, M.D.

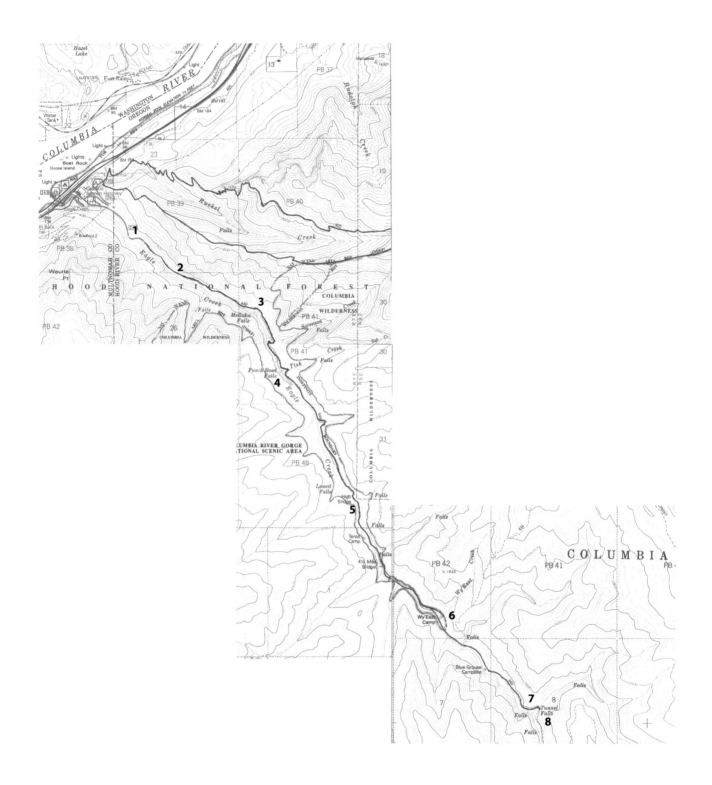

Eagle Creek

The prospect of a trip up Eagle Creek (Trail #440) creates conflicting feelings among Portland-area hikers. On the one hand, this is not usually an outing on which to cultivate solitude; Eagle Creek can be as crowded as a mall on the day after Thanksgiving. Moreover, you won't gain much elevation, no matter how far you travel, unless you tromp all the way past the 8½-mile point, so folks bent on conditioning must trudge long flat distances to accomplish much, and there are no grand scenic overlooks until after 9 miles. That said, however, this river walk can be a delight in so many ways that even the hardened climber anxious to rack up thousands of feet of elevation gain should not dismiss it. Its popularity stems from its many beguiling treasures. These include magical combinations of towering cliffs, mossy weeping walls, legendary waterfalls (including Metlako and the oft-photographed Punchbowl, plus Tunnel Falls), and of course enough species of wildflowers to keep even the most jaded botanists entertained.

To add to this joyful mix, you can hike this lowland trail in all seasons – the roar of the creek and its many tributaries and falls make this trip as pleasurable in December as in May, especially because snow will rarely block your access to this low-elevation wilderness. Even though this trail may provide fewer blooms and color in winter, you may experience this quintessential rain-forest walk in relative solitude during the darker months. However, even in spring or summer, the crowds often disappear in mid-week, so don't deprive yourself then of Eagle Creek's many charms.

Because this trail doesn't gain much elevation, marker points and items of interest will be denoted in the text mainly by mileage rather than altitude. In addition, the trail will be described up to the 6-mile point at Tunnel Falls for several reasons: Most outdoor travelers will probably only hike this far on any given day unless they are backpacking; a 12-mile day, even without much climbing, is enough for all but the most driven of adventurers. Moreover, much of the best this trail has to offer botanically comes in those first six miles; beyond, repetition is the rule and even the most dedicated of naturalists can get bored with recycled flowers.

Caution: There are several narrow spots on this trail in the first several miles or more that squeeze the hiker through a cliff on the left and a steep drop-off on the right. The narrow passage-way is wide enough for two people to walk side-by-side, but single file is easier on the nerves. Metal cables had been installed years ago but are not regularly maintained; still, they can be helpful if needed. Many children handle these sections with ease but for those uncomfortable with exposure, hand-holding for young folk might be wise.

February/March

Even by late February, and certainly by early March, a parade of blooms is beginning to appear on Eagle Creek, prompted by the twin cues of longer days and warmer temperatures, supported by the more-than-ample rainfall. In fact, Eagle Creek is the prime forest and cliff-side walk on which to experience the constant drip of seeping, weeping walls. Indeed, at ¾ miles and again at 1, 1¼, and 2¼ miles, hikers must tiptoe along a narrow tread hanging over steep slopes above the river roaring hundreds of feet below; steel cables (some are loose!) provide handholds for acrophobics but there is often no shelter from the miniature waterfalls constantly dripping from above. Even on warm days, most folks wear Gore-Tex here or at least a hat.

As you start on the trail during these early spring months, watch for Candy Flower on the left during the first ½ mile, and, later in March, yellow Stream (Wood) Violets – despite the nomenclature, most Northwest violets are yellow. These native plants, however, must often surrender domain to weedy species or foreign invaders on this popular trail, especially in its lower reaches. Thus, Hairy Western Bittercress, with its lyre-like leaves and tiny head of 4-petaled white flowers appears to have conquered swaths of real estate within the first mile of trail. Even at this early time of year a few specimens of the pink Robert Geranium may be in bloom starting from the trailhead and extending up to 3 miles. Here and there in the first several miles, tangles of the Himalayan Blackberry (see below for a more complete description if you really want to know) sprawl across openings and open woods alike with arching branches armed with vicious spines, while Common Chickweed (*Stellaria media*) lines the early reaches of the trail, right and left, with its weak and trailing stems, egg-shaped opposite leaves and tiny white mouse-eared flowers. And you thought this Chickweed only littered lawns. Do not confuse this alien with Field Chickweed, a much larger and more pleasant native plant.

Common Chickweed

Another early species to observe on this trail between **Points 1** and **2** is Bleeding Heart, especially prominent at ¾ mile in wet woods on the right. It appears again at 2½, 3 and 4 miles; look for it in March at the turnoff to the Metlako Falls viewpoint at **Point 3**, and also at the Punchbowl turnoff at **Point 4**, 2 miles from the trailhead. One other woodland plant first making its appearance in March here is the lovely Hooker's Fairy Bells, with its hanging white lantern-like flowers peeking out beneath tapered leaves. This shrub graces the forested parts of the trail at 1, then 1½ miles. At the wetter areas in the first 3 miles, also note the frequent Stream Violets, often accompanied by their moisture-loving friend, Candy Flower. Late in March, Fringe Cup begins to blossom; its 1' to 2½' tall stems support scores of white-to-pinkish cupped flowers with frilly thread-like filaments masquerading as petals. This common woodland plant occurs in masses at places between 1 and 3 miles, especially between **Points 3** and **4**, and will continue to bloom throughout April.

> Why are the first stretches of so many of our well-trod paths lined with Eurasian weeds? Perhaps it's because people unknowingly transport seeds on their boots, socks, or clothing. This explanation makes some sense for plants such as Common Chickweed and Cleavers, which produce roughened seedpods that can adhere to fur and clothing. However, the common Dandelion and Wall Lettuce, both troublesome weeds in the Gorge, have non-sticky seeds resembling downy parachutes and these are wind-dispersed. Moreover, one of the most common of westside "weeds", although a pretty one, is Robert Geranium, a plant whose seeds also would not appear to be spread by clinging to clothing. The answers are best thought of as diverse. For example, many weeds, such as the clovers and Pineapple Weed, thrive in openings where there is compacted soil, such as on the well-trodden earth of trails and roadsides. Our native plants often require looser, better drained soil, and so avoid these disturbed areas, leaving them open to invaders. Human-manufactured alterations thus can pave the way for immigrants, some less welcome than others.

There are five weeping walls along the first 4 miles of the Eagle Creek Trail, and each supports a unique and diverse flora throughout the spring and summer. While later in the hiking year more colorful plants will bloom here, two unusual early blooming members of the Saxifrage Family can be found flowering in March on these steep and rough walls. These deserve special attention as much for their tenacity as for their simple beauty. Western Saxifrage (*Saxifraga occidentalis*) and Northwestern (and also called Grassland) Saxifrage (*S. integrifolia*) paper these walls, especially the ones at 0.4, 0.7, our **Point**

2, and 1¾ miles. At these points Western Saxifrage is plastered against the seeping walls in an almost continuous layer. This plant grows about 4-12" tall; actually, on these walls it often grows horizontally seemingly right out of the rock. Its leaves are elliptic to egg shaped, with a smooth edge to their bases but with cogwheel-like coarse teeth around their margins. Its flower cluster consists of 5 to 10 white-to-pink blooms atop arching yet stout stems. Although here this plant prefers to cling to vertical walls like a rock climber, in other locations in the Gorge, such as on the open slopes near the power lines on Starvation Ridge or at the hanging meadows on the Ruckel Creek Trail, it acts more like an upright casual hiker, occupying grassy hillsides and wet seeps on gentler terrain.

Northwestern Saxifrage

Western Saxifrage's close sibling, Northwestern, or Grassland, Saxifrage, is less common on these vertical gardens, but can be found readily nonetheless. The easiest way to distinguish it from its relatives is by the lack of teeth on the margins of its leaves. In addition, this saxifrage is a slightly more robust plant, with taller stems (up to 1½' tall), and its flowers are arranged in a conical inflorescence. These two saxifrages are the earliest blooming of this family in the Northwest, but other species of this genus will be appearing, especially at higher elevations, throughout the spring and summer. Because the insects pollinating these plants are out and about soon after the winter rains abate, March and early April happily welcome these premature bloomers.

Two other early-season plants can also be found on the Eagle Creek Trail, sometimes even in late February. The first is the common wooded hillside flower, Oaks Toothwort. These 4" to 10" lilac four-petalled beauties are not as common here because of the terrain – they require forested slopes – but can be found scattered between **Points 2** and **4**. The other February-to-March flower to note is the charmingly-named Gold Stars, its small yellow daisy-like blooms gracing the openings at 1, 2¼, and 3½ miles on both sides of the trail. Look for it especially in masses downslope on the right in the open hanging meadow just after **Point 4**. Happily, not all these early-spring bloomers are plain vanilla white.

These flowers, along with spotless white Trilliums, in open woodlands betyond **Point 2** prove that, this early in the season, the budding naturalist will have plenty of plants to identify and will also appreciate the lack of crowds April and May will surely draw to Eagle Creek.

April

The great exhibition of blooms on the Eagle Creek Trail reaches its fullest expression in mid to late April, when scores of both lowland forest and meadow species are in riotous bloom. Indeed, many botanists choose this very trail to introduce novice flower hunters to the glories of Columbia Gorge wildflowers. Directly from the parking lot, Stream Violet and Candy Flower are blooming in abundance, especially on the right, and these blooms continue to grace the trail between **Points 1** and **2**. Miner's-lettuce is plentiful in the woods during the first mile, while Scouler's Valerian can be found at 1½ miles. A new flower now appears on the steep banks to the left within the first 3 miles, the Woods Strawberry, its white petals and three leaflets hugging the ground; they welcome the sun in these open sandy realms. Although common in many openings in the Gorge, this fruit-bearing member of the Rose Family seems to thrive especially well here, promising its small but luscious berries to mid-summer hikers. Look for it particularly on the sandy banks to your left at ½, ¾ and 1¾ miles.

The weeping walls between **Points 1** and **3** continue to display both Western and Northwestern (Grassland) Saxifrages in April, although by the end of the month these two species have largely gone to seed. Nonetheless, in a wonderfully colorful example of plant succession, these walls now seep not only with moisture, but are dripping with blossoms as well, primarily a mixture of two yellow Monkey

Flowers, the Common and the Chickweed species. The larger of the two, Common Monkey Flower, makes the greater show, but both put on a dazzling display here this month, with their bright yellow flowers and grinning faces splotched with bold maroon spots. They are especially plentiful on the walls at 0.7, our **Point 2**, and 1¾ miles. These are Snapdragon Family plants and, indeed, they can be seen to have a lower lip composed of two partly-fused petals and an upper one with three flaring petals, sort of similar to a penstemon or a garden snapdragon. The resemblance to a monkey's face, however, is suspect; whoever named this flower may never have actually seen a simian.

At 1¼ and again at 2½ miles, more color appears in the form of the gorgeous Few-flowered Shooting Star. A more flattering common name is Pretty Shooting Star, especially because there appear to be as many or more blooms on this species as on our other shooting stars. The fascinating architecture of these flower forms, with their swept-back pink petals, white bands at their base with an encircling filigree of wavy purple lines, and an orange to purple tube thrust outward like a bird's beak, belies their Primrose Family affiliation. There are seven shooting stars in the Northwest, all sharing this spectacular flower plan, and all but one of similar color. (One, found in wet areas usually further east, is called the White Shooting Star). These flowers are truly stars of our meadows and their resemblance to the common garden and house plants, the Cyclamens, can easily be seen. Indeed, many gardeners believe the blossoms of shooting stars hold greater allure, but they are devilishly difficult to grow outside of their natural habitat.

On the right, at about ½ mile from the start, look for several specimens of American Wintercress, with lyre-shaped basal leaves (the terminal leaflets larger than the proximal ones), and pretty yellow four-petalled flowers in clusters at the top of 1' to 2' stems. This Mustard Family plant has edible, but bitter, leaves best left on the stem. Also within the first mile, several Plume-flowered Solomon's Seals are, by March, sending their sprays of white fragrant flowers out from wooded banks on the right, while some specimens of Star-flowered Solomon's Seal are twinkling in the forest between **Points 2** and **4**. Hooker's Fairy Bells are very much in evidence in the forested areas within the first 2 miles. Fringe Cup is now pretty much continuous between the start of the trail up to High Bridge at **Point 5**, and occasionally beyond as well. It lines the trail in shaded places much like an army of individual soldiers standing at attention. Please inspect this seemingly weedy plant up close to appreciate its delicate architecture, with its pinkish petals cleft into multiple frilly lobes. Although textbooks claim this homely plant, found throughout waste areas in our city environments, has a pleasant odor, it is difficult to perceive this in individual specimens; however, when Fringe Cup presents itself in abundance, as here, its fragrance, combined with that of the Plume-flowered Solomon's Seals, lends an angelic perfume to these woods, a bonus for the April traveler on this part of the Eagle Creek Trail.

Just before and after High Bridge, **Point 5**, False Lily-of-the-Valley is just beginning to make its appearance by late March. These premature specimens should be regarded as special because this lovely plant doesn't usually bloom until May or later in many other locations, regardless of elevation. Perhaps the combination of abundant moisture and lower altitude encourages these particular flowers to show their pretty white spikes of blossoms above their wavy heart-shaped leaves in these wooded locations.

Also in this vicinity, at about 3 miles in, some equally early specimens of the gorgeous blue Oregon Anemone can be found on the left, accompanied by another deep woodland plant, Youth-on-age. These latter have not yet begun to produce the new buds which sprout out from above its mature leaves, thus giving the plant its name. However, they are in flower so examine these Fringe Cup relatives (and Saxifrage Family members) closely to see their mysterious brownish-bronze-to-purple little tubes ending in frilly threads which look like small afterthoughts to the bloom. Later in the year, the tiny buds of entirely new plants will appear in the leaf axils (the joints where the leaves meet the stem) and either fall from the stem to take root in the ground as a new plant, or throw down new roots to become a separate plant when the mother plant's stem droops and reclines in the autumn.

As you begin to cross the bridge over a subsidiary creek at 2¾ miles, a new Lily Family plant, Twisted Stalk (*Streptopus amplexifolius*), appears on the left just before the span. Never common nor plentiful, though well dispersed, this unusual shrub-like species also deserves close inspection. The stems of this plant reach 2' to 4' above the ground and produce oval leaves with parallel veins and the pointed "drip-tip" characteristic of the Lily Family. However, it's the flowers which give this plant its unusual appeal. Small, but unique in form, these little white bells hang from beneath the upper leaves like tiny drooping teardrops. Perhaps the plant is tearful because it's lonely, as Twisted Stalk rarely occurs in bunches. A closer inspection reveals the source of its common name: Each flower stalk (technically a pedicel) actually originates from the axil of the leaf below and climbs the stem to droop from the next higher leaf. In order to hang pendant, each pedicel must make an abrupt bend in its course halfway beneath the leaf, resulting in a unique twist in its stem. Like a shy woodland maiden discretely hiding her origins, Twisted Stalk thus delights the visitor who takes the time to discover her mysteries.

Twisted Stalk

A close relative, the Rosy Twisted Stalk (*S. roseus*), further described in the May Section of the North Lake Chapter, has single, rather than branched stems, lacks a twist to its pedicels, and, of course, has delightfully-colored rosy petals. In the Columbia Gorge, Rosy Twisted Stalk is less abundant than regular Twisted Stalk, but can be found along streams such as McCord Creek and Eagle Creek in the western part of the Gorge. It blooms slightly later than its sibling, generally by mid-May, and is usually not quite as tall. Brush and cliffs in these hard-to-reach places may hide this treasure from the view of most casual hikers, but these obstacles, like dragons guarding a treasure, simply enhance its charming mystique.

> Fortunately for our eyes, which appreciate diversity over monotony, the various species populating our mountain meadows distinguish themselves with different colors. These hues, in all likelihood, did not evolve to please our aesthetic sensibilities, but instead to distinguish themselves as species one from the other. While we humans may have developed our appreciation for color because our primate ancestors needed to identify palatable fruits, insect pollinators may also need to distinguish their plant food sources not only by shape, but by hue. Of course, some pollinators can also sense the infrared and ultraviolet parts of the spectrum, but many still rely on visible color as a means of identification, a fortunate turn of events for us. Thus each species growing in the same meadow will produce a different color than its neighbor in order to distinguish itself, while in the forest most flowers are white, having no need to invest energy in producing color as white will stand out against the ubiquitous green. In addition, most woodland plants are not massed, but stand relatively isolated from their peers. Yes, exceptions such as the Oregon Anemone (blue) and Broad-leaf Arnica (yellow) exist, but these are as uncommon as a blackberry without a thorn.

While the beguiling small yellow daisy flowers of Gold Stars are beginning to fade by the middle of this month, several new plants are taking their place to add color to the meadows at 1, 2¼ and 3½ miles. Cliff (Menzies') Larkspur (or Delphinium) is in full and glorious bloom, coloring the meadows especially between **Points 4** and **5** with deep purple blooms enhanced with its backwards-pointed spur. Rosy Plectritis (couldn't someone have bestowed an easier common name on this plant?) now carpets these same open meadows with its 6-18" stems bearing heads of fuzzy pink flower clusters in wonderful profusion. While these charming blooms can be found in openings throughout the Gorge, nowhere are they in greater abundance than down the steep slope to the right at 2¼ miles. Two Carrot Family plants in the same genus lend spots of yellow to these openings as well: Martindale's Desert Parsley, a low-lying species of rocky substrates with frilly fern-like leaves (technically called "dissected"); and Nine-leaf Desert Parsley, a 1' to 2½' high plant with distinctly un-carrot-like grassy leaves.

Trooping further up the Eagle Creek Trail, between **Points 5** and **7,** the April hiker must be content with the glory of these woods and the magic of the creek, always close at hand. While flowers are never plentiful in this forest, they do exist. The yellow Evergreen Violet plays its sprawling game over mossy banks particularly on the left at 3 and again at 4½ miles, and at this latter point, Scouler's Valerian can be found frequenting the open woods, its white to pinkish clusters of flowers appearing in fuzzy heads atop 12" tall stems.

Just after you cross the span of High Bridge, **Point 5**, look for a dense concentration of the yellow Martindale's Desert Parsley populating the cliff facing west (and on your left at this point). Otherwise, you will not find many blooms beyond here, although you will discover a few more sheer mossy cliffs, rushing tumbling creeks, and the amazing Tunnel Falls and its associated canyon, all reasons enough to tread further up the trail during this month. While the weather in April may not always be predictable or to your liking, you will always benefit from hiking Eagle Creek, not only for the spectacular show of blossoms, but from the display of so many justly famous waterfalls and the views of so many soaring cliffs - in the minds of many, the quintessential Oregon outdoor experience.

May

Does April surpass May in botanical abundance and spectacle in the western Gorge? Hardly. Although it barely seems possible, the floral tapestry woven on these seeping walls, mossy cliffs, and sunny meadows during this prime spring month equals that of April, both in depth of color and in diversity. Moreover, the predictable April rains ensure that the springs, creeks, and waterfalls are just as wonderful in May, and, as a bonus, your chances of a sunny day actually improve. This is one trail, however, on which you don't always need sunshine to enjoy the trip; many Northwesterners actually prefer a misty day here, the hovering fog caressing the cliffs and accentuating their mystery. Others among us, however, don't mind a bit of sun at this time of year, especially as we've solved most puzzles the ever-present fog has posed during countless days over the past winter.

> Actually, the Streambank Arnica here is a variety of this, an Aster Family plant, more properly, and somewhat chauvinistically, called Columbia Gorge Arnica *var. piperi*. More common at higher elevations in the Northwest is *var. amplexicaulis*. We amateur naturalists have trouble enough learning about genera and species, so specific varieties are rarely mentioned in this book (although see Western Groundsel in the Munra Point and Dog Mountain Chapters). You should know, however, that professional botanists further dissect species into varieties based on minute technical characteristics, such as hairiness of leaves or length of a calyx (the outermost tube of some flowers) or a corolla (the innermost circle of petals). Most often, these different varieties are found in different geographic locations and these subgroups are thought to have evolved slightly varying properties based on the different conditions they encounter in their disparate environments. As in animal and human populations, however, these subspecies and varieties can still interbreed. Perhaps an analogy would be the different races we encounter over the wide spans of distance across our continents. It may be that the plant world is trying to tell us something about interconnectedness. Then again, perhaps we're not listening very well.

Rain or shine, the Eagle Creek Trail holds many new plants to identify this month, although the already familiar ones cited in April are largely still in bloom. Thus, at the very start of the trail, the omnipresent Candy Flower and its frequent partner in grime (the wetter areas of ground-in soil), Stream Violet, persist, at least early in the month, while the low-seated, but pretty white

Naked Broomrape

Woods Strawberry, being untrue to its common name, papers the drier open slopes, all between **Points 1** and **3**. Unfortunately, the two Saxifrages noted above earlier in the year, the unimaginatively named Western and Northwestern (Grassland) species, have mostly gone to seed, but replacing these small but fascinating white plants at **Point 2** is an even more impressive species, Columbia Gorge, or Streambank, Arnica (*Arnica amplexicaulis*). This plant can be found in glorious yellow abundance lapping over the weeping cliffs at ½ and ¾ miles, and to a lesser extent, on the cliff at 1¼ miles. This is a large plant for the Gorge, actually garden size, with stems reaching 3' in length (although these here are usually reclining over the middle levels of these walls). These hold daisy-like yellow flowers up to 2" in diameter. *Arnica* means "lamb-skin" and the opposite pointy leaves of these specimens are particularly soft yet robust, attaining a length of 6-8". In contrast to the much more common Broad-leaf Arnica, which has two to four sets of stem leaves, Streambank Arnica has five or more sets, and, in this steamy wet environment, such lush foliage cascading over the mossy banks above the trail on the left lends an almost tropical feel to these first several miles.

Columbia Gorge Arnica

One most unusual plant occupies the second seeping cliff, at ¾ miles, Naked Broomrape (*Orobranche uniflora*). Never common or clustered, there are usually one or two 4-6"-tall specimens on this wall (and sometimes on the talus at around 700' on the Nesmith Point Trail). The pretty purple flower of Naked Broomrape, with its cute twin orange stamens, belies its evil intent: This plant is parasitic on other species. Its true nature is revealed by a close look at its stem, devoid of green chlorophyll, and its scale-like leaves, which have lost their purpose, as this invader feeds off its hosts and thus has little need for photosynthesis. Even its name betrays its greedy nature – *ancho* means "to strangle" in Latin and we all know the meaning of "rape". This wicked little plant feeds off saxifrage and stonecrop species, both of which are abundant on these walls. The two stonecrops include the Broad-leaf and Oregon species, and both plaster vertical rock walls throughout the gorge with their fleshy jade green (Broad-leaf) and bright green (Oregon) leaves and star-shaped brilliant yellow flowers. We probably should not be too critical of the manner in which Naked Broomrape makes its living, however, as both the saxifrages and the stonecrops here appear to be thriving in masses, while the lonely *Orobranche* remains rare and isolated. Moreover, evolution cannot be judged by human moral values – we should probably not attribute motives to our flora.

These weeping walls still contain the two Monkey Flowers mentioned above in April plus much purple Self-heal, and, at 1¼ and 2½ miles, the last stands of Few-flowered Shooting Star, its pink bird's-beak blossoms softening the stony cliffs and open meadows as a reminder of early spring. In the woods this month, between **Points 2** and **5**, the Evergreen Violet continues to bloom off mossy banks, particularly on the left at 3 miles, while Fringe Cup lines long reaches of the trail between **Points 2** and **4**. The Inside-out Flower sparkles in these deep forests, with its surrealistic flowers held above its lacy leaves, while Western Meadowrue can be found at the smaller stream crossing at 3¼ miles.

Just before and after the crossing of High Bridge, **Point 5**, look for the white spikes of False Lily-of-the-Valley popping up above its wavy pointed leaves, and don't miss the many examples of Scouler's Valerian in the open woods at 1½ miles and between 3 and 3½ miles. This 1-2' tall plant, with its fuzzy white flower clusters, is still in bloom until mid-to-late May; its close relative, Sitka Valerian, will only be found at much higher elevations, its seeds still germinating under the snow banks in our western mountains. Look too for the familiar Vanilla Leaf in these low-level woods all the way between **Points 3** and **7**, its spikes of dull white flowers held above its elephant-ear leaves, and for Star-flowered Solomon's Seal twinkling in the forest at 2½ and 4 miles in.

Beyond **Point 5**, Foam Flower can be found just beginning to emerge late in the month in the deeper woods, especially between 4 and 5 miles, while Twin Flower graces shady banks on the left just before and after High Bridge. Examples of Star Flower brighten this forest as well between **Points 3** and **6**, and occasional Western Meadowrue appears in the wetter areas up to Tunnel Falls at **Point 7**, especially at Wy'east Camp and at the many small stream crossings thereafter. While other woodland flowers must await the sunshine of June, the combination of vertical cliffs and rushing waterfalls makes this more-shaded farther part of the Eagle Creek Trail beyond High Bridge just as interesting, if less colorful, than the weeping walls and open meadows of the first 3 miles. In addition, you may well have this part of the trail to yourself in springtime, particularly on a weekday or even a rainy weekend, an added bonus to this already wonderful, and not too strenuous, excursion.

June

The waterfalls may be slightly less full, and the flowers less prolific in June, but just barely so. Indeed, the soaring cliffs and open vistas for which the Eagle Creek Trail is so well-appreciated are still just as wonderful throughout the summer months. Heading up from the parking lot, pink-streaked Candy Flower in the open woods, and the rose purple Robert Geranium in more densely shaded areas, both greet the June hiker. These plants, common between **Points 1** and **2**, will continue to bloom throughout the summer and both exemplify how man-made alterations to the landscape have changed what Native Americans might have seen here in these low-elevation woods centuries ago. This geranium, now so common in the western Gorge, would have been absent entirely before early white settlers inhabited the Americas. It was inadvertently imported by European travelers, whereas the Candy Flower, though present in scattered populations, would in all likelihood have been far less common. This Purslane Family plant thrives in lightly forested areas with ample light. Because the old-growth forests of the pre-logging era were fairly shady places, Candy Flower probably existed in marginal areas where enough light could have penetrated to the forest floor. With the opening up of trails, sufficient light could enable these two species to flourish while deep-shade plants, such as Foam Flower, Vanilla Leaf, and Pathfinder, were displaced.

Another import, Wall Lettuce, offers up a similar performance further up this trail. This Aster Family species is so common in shady areas of the Gorge between trailheads and 3,500' of elevation that it is difficult to convince historically naïve hikers that this European weed was introduced to this country only a century ago. In the woods between **Points 2** and **5**, Wall Lettuce appears frequently, sending up 1-3' tall stems with consistently five-petaled small yellow flowers; in fact, Wall Lettuce is the only Aster Family flower in our mountains with the same number of "petals" (actually ray flowers) on every single plant. All other asters, daisies, arnicas, groundsels, sunflowers, dandelions, and the like, of which I am aware, have a varying number of ray (and disk) flowers. If you find a Wall Lettuce with more or fewer petals, report this immediately to your local emergency botanist. Also, examine the strange leaf pattern of these plants. The irregular, deeply-lobed 4-8" leaves, with a terminal segment shaped like a psychotic maple leaf, have defied a technical description. If you devise one, again contact your local botanist.

> Most folks know that the main function of leaves is photosynthesis, the conversion of sunlight to carbohydrates and energy. However, this process is fairly inefficient; plants produce energy at barely 3% efficiency. This limits how plants may be used as energy generators. The U.S. Department of Energy is trying to rectify this through genetic engineering to enhance plants' efficiency by inducing pine trees to manufacture turpentine and grasses to produce vegetable oil, both as potential fuels. Other such projects are attempting to coax microbes to make oil and to engineer artificial leaves to fracture water into oxygen and hydrogen for use as fuel. Who could guess that some answers to our impending energy shortage might be growing all around us?

Another apparent weed appears here on the early reaches of the Eagle Creek Trail, even lining the trail right out of the parking lot: Self-heal, also called Heal-all (*Prunella vulgaris*). This 6-14" tall plant can be found in sunny positions on many early stretches of our Gorge trails, leading to the opinion of many botanists that this humble plant, with its spike of purple ½" long two-lipped flowers, is an introduced weed, thriving in the disturbed ground visited and trampled upon by countless Northwest hikers. This opinion is no doubt reinforced by the species designation, *vulgaris*, which really doesn't imply being crude or offensive, but only means "common".

Self-heal

However, recent research indicates that Self-heal, one of the few plants to occur on all continents, may have several varieties, one of which may be native to America. Regardless, its medicinal properties, revealed in its multiple common names, have been recognized since antiquity. As with other Mint Family plants, Self-heal has square stems (roll them between your fingers to appreciate this property) with opposite smooth-edged leaves and two-lipped flowers, the upper lip with a hood and the lower one with three lobes. As with many other Mint Family members, including the common mints and catnip, horehound, and pennyroyal, medicinal properties are said to include strengthening the heart and accelerating the healing of wounds. Please don't eat this plant if you have heart disease, however; it will prove safer to seek medical attention.

As you pass under several of the seeping walls at ½, ¾, 1¼ and 2½ miles, watch for the two species of stonecrop which flourish on rocks and seem to grow right off the stone: Broad-leaf (*Sedum spathulofolium*) and Oregon (*S. oreganum*) Stonecrops. Both of these low-growing plants have star-shaped flowers with yellow petals which look like little

Broad-leaf Stonecrop

sunbursts of color splayed across these near-vertical rock faces. Their leaves are just as delightful, appearing plump out of proportion to their diminutive size. In fact, these leaves function much as cactus stems do, storing water in anticipation of the long dry summer ahead.

Why would such water storage mechanisms be necessary in our rainy part of the world? Perhaps because of the particular environment in which these plants have evolved. Sedums thrive on rocky walls where little soil and few nutrients are available and where, in summer, the seepage can come to an abrupt halt. Like a prudent desert traveler, these clever little plants plan for their drier future by storing water just in case. Related to the common Chicks and Hens of city gardens, and the indoor giant, Jade Plant, six species of stonecrops decorate the rockeries of our mountains and lend their yellow cheer to the grey stones lining trails and routes at all elevations. You can distinguish Broad-leaf from Oregon Stonecrop by its sage-green leaves covered in a dusty white powder; these sometimes turn red when exposed to continuous sunlight, perhaps a mechanism to protect the leaves from harmful ultraviolet exposure. Oregon Stonecrop's leaves are a bright green. Also, Broad-leaf Stonecrop blooms somewhat earlier here in the Gorge, often by late May, while the Oregon species (which is found throughout Washington and British Columbia as well) may not be in full bloom until mid-July.

Further up the trail, Foam Flower is just beginning to emerge in the wooded areas between **Points 2** and **5**, but especially at 1, 2½, and 3 to 4 miles. Other woodland plants populate these woods as well, including Cascade Oregon Grape, with its woody stems and clusters of yellow flowers, and Bunchberry, a ground-hugging version of the flowering Dogwood Tree, carpeting the forest floor at 2 and 3 miles in. Some Sweet Cicely is still out between 1 and 2 miles, while further on in the forest, some specimens of Star Flower remain.

Just before and after High Bridge, at **Point 5**, a sharp eye will detect the shiny heart-shaped leaves of Wild Ginger hugging the ground. You will need an even sharper eye to detect its odd brownish-purple flowers, which hide under the leaves lying on the ground just beneath. These bell-shaped flowers have three flaring lobes and are pollinated by crawling insects; thus they have no need to advertise their presence atop erect stems. Other plants likely to be out beyond **Point 5** include patches of Twin Flower at 5 to 6 miles and more Foam Flower and Bunchberry, particularly before and after **Point 6**, the junction with Trail #434, and just before **Point 7**, near Tunnel Falls. Even though the forest here is never filled to profusion with blooms, the trip up to Tunnel Falls still rewards with spectacular views in the open areas and a peaceful respite from the heat now so often suffocating our cities.

July and August

Throughout the summer, many of the same species of wildflowers which first made their appearance in the spring persist to brighten the Eagle Creek Trail, particularly those in the open areas in its first 4 miles. While forest dwellers, such as the Solomon's Seals, Fairy Bells, and Bleeding Hearts have gone to seed, meadow and partial shade plants, such as Candy Flower, are scattered in the first ½ mile. Nine-leaf Desert Parsley, found in the openings at 2½ and 3 to 4 miles, and the purple spikes of Self-heal, all still lend color to both sides of the trail. Plant life is particularly abundant on the weeping walls at ¾, and 1¼ miles, where isolated stands of Western Saxifrage remain throughout the summer, clinging to the vertical cliffs where moisture persists at least into early July. The stubby-toothed leaves and 4-12" stems of this plant require a steady supply of water to produce its flat-topped cluster of tiny dull white flowers.

However, as summer progresses, these rocky walls become progressively drier as the rains diminish and snow disappears from the heights above. Replacing the moisture-loving plants here and there on the now drier parts of these cliffs between **Points 2** and **4** are legions of Broad-leaf Stonecrops, with their fleshy jade green leaves in rosettes festooning the rocks. These are topped by bright yellow star-shaped flowers. The overall effect in places is of green wallpaper blotched with yellow because these plants think nothing of growing on vertical surfaces; indeed, they seem to prefer it, perhaps because, while their succulent leaves store water like a cactus, they disdain standing water which might predispose their roots to the horror of fungal rot. Species which seem to thrive on slopes, such as Twin Flower or Woods Strawberry, may also be sensitive to root rot and hence require good drainage.

Bull Thistle

In summer, several new flowers make their initial appearance in the open areas within the first several miles of the Eagle Creek Trail. One of these, Scouler's Hawkweed, can be found on the left at 1½ miles. It looks like a multi-branched dandelion, although taller at 1½-3'. Another new plant here, although perhaps not as welcome, is the gangly weed, Bull Thistle (*Cirsium vulgare*). Just like a bull, this robust member of the Thistle Tribe within the Aster Family intrudes its way into lowland meadows and, although never found growing in abundance, can stamp out other native plants because of its invasive root system. Bull Thistle, a Eurasian import, can grow as tall as 4', and, with its spiny-winged stems and leaves, it does look not only "vulgar" but also monstrous and threatening. However, every plant contains redeeming features: Bull Thistle not only has pretty reddish-pink compound flower heads, but it is edible as well, the peeled and cooked stems having provided a tasty treat for early pioneers. Indeed, as most people know, the artichoke is simply the budding flower of another species of thistle. Several other thistles can be found in our mountains, although none are common except perhaps the Canada Thistle, which did not, confusingly, originate in Canada. See the Starvation Ridge Chapter for a more complete description of this less sinister-appearing Eurasian weed.

Although we may notice the plants in openings more readily, woodland species are also abundant at this time of year, particularly between **Points 3** and **5**. Pathfinder, also known as Trail Plant, can be found in the woods at ¾, 1½ and between 3 and 5 miles. This shade lover gets its names from its ability to mark a trail through the understory because of the two-toned nature of its leaves. Once wayward hikers tramp through a patch of these plants, they can look back and see exactly where they've been thanks to weak square stems which are easily overturned and remain that way. The path is thus marked by the overturned leaves, a contrasting white superimposed on a carpet otherwise a monotone of green. Present-day hikers should have no need to attempt this experiment and trample off-trail. Take my word for it and stick to the path.

Pacific Dogwood

Unfortunately, the flowers of this species are tiny and unremarkable, almost lacking color. They are pollinated by insects attracted to their smell. In these same forested areas, also look for Foam Flower, now fully in bloom, with its 4-12" high stems topped by sprays of dainty white flowers; the woods between 3 and 6 miles are often awash with these blossoms by mid-July.

Another summertime woodland plant frequent on the Eagle Creek Trail is Canadian Dogwood (*Cornus canadensis*), more commonly known as Bunchberry. This plant seems to thrive growing over tree roots and stumps, and thus is found in the deepest parts of our forests. Look for it especially at, and around, **Points 5** and **6**. It is difficult for many hikers to believe that this trailing plant, barely rising several inches above the forest floor, is actually a close relative of the flowering Pacific Dogwood tree (*C. nuttallii*). In fact, their white "flowers" are almost identical. Each has four white "petals" which are actually bracts (modified leaves); the real flowers, with their associated sexual parts, are tiny greenish-white specks held within the center of the more showy bracts. Such costuming is common in the floral world, where it is often important to advertise your presence to potential pollinators without actually investing in resources to produce flamboyant blossoms. Colored bracts (as in the dogwoods), flashy leaves (as in White-vein Pyrola) and compound flower heads (as in the Aster and Carrot Family plants) all market their wares with a bit of Madison Avenue-type deception, but with perhaps a nobler purpose than monetary reward. The "flowers" of Bunchberry are followed by bright red berries in the fall; while birds devour them, to most human travelers, they taste insipid.

Bunchberry

Recent research has amazed us with how plants obtain their energy and fend off the multitude of bugs, bacteria and viruses attacking them. We have also been astounded in recent years at the number and complexity of microorganisms dwelling in our guts and, indeed, in every organ of the human body and on our skin as well. Instead of making us feel creepy that bugs are crawling around inside and on us, we are beginning to understand that, without these bacteria, viruses and fungi, we could not digest our food or fight off multiple infections. Now, plant biologists are discovering the same is true of plants. Not only do all plant species require bacteria and fungi in the soil to thrive, they apparently rely as much as we do on those same types of organisms to survive inside them as well. Microbial communities, each somewhat different based upon its plant species host, for example, help mustard plants know when to flower at the best times in spring while antibiotic bacteria formulate compounds which protect them from more harmful bugs. Commercial companies have taken note and are now spending millions to investigate how to employ such microorganisms to thrive free of disease without the overuse of potentially harmful fertilizers. They are even investigating how such bacteria, viruses and fungi, once thought to be harmful to plants, may actually be employed to help crops grow healthier and be more productive. And all this without genetic intervention – a savings for companies but more importantly, a more natural and less harmful way to feed our growing and ever hungrier populations.

Further up the trail, beyond **Point 6**, not many woodland species are still out, although Foam Flower and Bunchberry can be found in patches right up to Tunnel Falls, our **Point 7**. One last flower to emerge however by late July is the Annual Fleabane or Annual Daisy, "fleabane" being the old name for daisies. It was thought that some of these plants repelled fleas, an idea not really borne out in the New World. This daisy, 2' to 5' high in some places but about 3' tall here, can be seen on the left at 5½ miles. As usual, Aster Family plants require a good deal of sunlight, and this specimen grows in a small opening here. Also, the Aster Tribe within this family often blooms later than most plants so, in the Gorge, we can look forward to daisy-like plants well into the fall. This particular species inhabits wet lowlands throughout the west and can be identified by its toothed leaves and its numerous loose heads of light-purple ray flowers surrounding a yellow-orange disk. It is somehow comforting to know that this daisy, along with many other Gorge flowers, will still be in bloom throughout the summer months. While the peak flowering season may have passed by August, the intrepid plant hunter will still find much to savor well into autumn on the Eagle Creek Trail.

Why do some plants seem to bloom over such a short period while others flower from spring to fall? The activities of pollinators may provide one clue because, if an insect which is the main pollinator of a species lives only a short time, valuable resources would be wasted by blooming longer than necessary. On the other hand, genetic variability may also play a role. Subtle mutations in a plant's DNA or epigenetic changes may provoke earlier or later blooming. Such variations could be based on environmental conditions which confer some advantage, such as the opportunity to attract new pollinators. Genetic variation also provides the opportunity to flower over a longer period. Our city gardens present many examples, what with some camellias blooming as early as February while different varieties of rhododendrons are in flower by early March and others just beginning to bloom in mid-May. Who knows, with global warming, what trends we may see in the distant future? Daffodils in December? Fritillaries in February? Waterlilies throughout the winter? We (or our children) shall see...

September and October

Fewer hikers and cooler temperatures highlight the Eagle Creek Trail during the fall months. Those looking for flowers won't be totally disappointed however. Along the early reaches of the trail, some Robert Geraniums persist along with Candy Flower, especially on the left in the first ½ mile. These two plants display a wide variation in times of blooming, so they may be enjoyed over the course of an entire hiking season, with early flowers appearing in March or April and later ones clinging to life into the autumn months, some as late as November. Also in the first few miles within the more forested regions, note that Foam Flower is still blooming, especially early in September. The sprays of this woodland denizen last well into the fall, although a close look may reveal that what appear to be flowers are really the seedpods of this omnipresent forest ground cover. Foam Flower may also be found in bloom into mid-September higher up the trail in wooded areas between miles 2 and 4, especially just before and after **Points 3** and **5.**

Common Snowberry

Despite the relative lack of showy flowers in the fall, one plant stands out in the lower reaches of the Eagle Creek Trail – Common Snowberry (*Symphoricarpus albus*). Although texts give the height of this plant as between 2' and 6', in the

Gorge it usually grows only to 2' to 3' tall. Its bell-shaped flowers, pinkish white and only ¼" long, hang inconspicuously below the elliptical wavy-toothed leaves. These rarely attract much attention when in bloom in June. It's the berries, white as ping-pong balls (though much smaller) that dominate the plant, especially in the fall and winter, when its stems are leafless. In fact, stands of this shrub may be the most notable vegetative feature of autumn's lowland woods, with their skeletons of bare branches reaching above the stark forest floor or even from above a cover of frozen ground or snow, seeming to be gasping for a last breath of frigid air. Look for Snowberry on the right within the first 100 yards of the trail's beginning and then scattered between ¾ and 1½ miles. Look for it also in your home gardens and abandoned city fields, either growing naturally or planted as a pleasant winter shrub. Avoid the berries however; they are not only tasteless but some aboriginals thought them poisonous and fit only for children to throw at each other in mock battles.

You may chance upon some Candy Flowers and Foam Flowers still out in September further up the Eagle Creek Trail but you can't count on it. Still, one flower may still be in bloom in forested areas in the first 3½ miles, the ubiquitous weed, Wall Lettuce. This Northern European plant has invaded many of our trail-sides in the Gorge, thriving in the wet west-side shade up to 3,500' in many areas. Its yellow five-petalled flowers can be found in bloom from June through October, while its deciduous leaves, weirdly lobed with flanges here and there, may appear as early as April and hang on through November. You may also find it along shaded city roadsides and abandoned fields, a reminder of how easily some foreign species can thrive in nonnative environments to their liking. Yes, this plant *is* related to common lettuce (*Lactuca sativa*) but the resemblance is marginal at best.

Eagle Creek offers an easier trip than most of the more strenuous hikes in this guide, no matter the season. Combine the flowers with the basaltic cliffs and the famous waterfalls (particularly The Punchbowl and Tunnel Falls) and you've got some of the best the Gorge has to offer. To boot, this lowland trail can be enjoyed throughout the winter, a feature not shared by many of the higher-elevation trips in the Gorge. Spring brings the color of most of the Northwest's best wildflowers, while fall offers more solitude and deciduous color. In addition, you won't sprain your back or your knees while enjoying Eagle Creek at any time of the year!

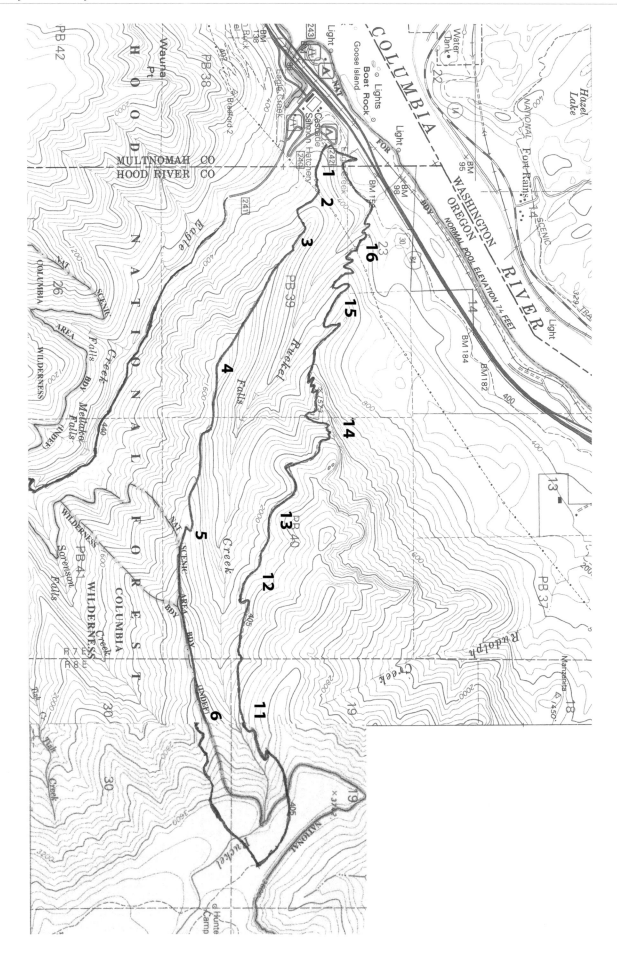

Ruckel Ridge and Ruckel Creek Loop

No finer display of west-side wildflowers can be found in the Gorge than on this famous loop, strenuous, challenging, yet more than worthwhile. More of a route than a trail in places, the climb up Ruckel Ridge requires some easy route finding, although by and large it is well marked. At times, you can even opt for a moderate class 3 scramble at the 2.6 mile mark at the infamous "catwalk" on the craggy crest of the ridge, although an easier route, preferred by many hikers, lies just below on the right. Steep in places, yet level in others, this classic ridge route offers the joyous surprise of many unexpected openings bursting with the color of wildflowers throughout the spring and summer months. Taking the faint trail across the plateau to reach Ruckel Creek brings you to the top of the loop and the beginning of the seemingly endless, knee-jarring descent back to the cars. The way down rarely comes close to the creek, but offers its own delights as it winds through a shawl of green which surrenders at times to magnificent open meadows and hanging gardens ablaze with flowers. A rite of passage for many Northwest hikers and climbers, and featured in many national publications, this spectacular loop is unforgettable.

Although most trail guides direct you to park at the Eagle Creek parking area at Exit 41 off I-84, a more convenient and safer place for the loop is actually just ¼ mile past this exit, at a green highway sign with a small parking area right by the sign. It is legal to park here as long as you are off the pavement; in addition to being closer to the start of the hike, the risk of vandalism is actually lower than in the official parking area. Head south across the weedy patch, easily surmount a concrete barrier, then hit the paved bike path, heading left. At the signed trail junction just beyond, take a right to ascend to the campground, our **Point 1**. Once there, keep to the left on the paved road to find the "Buck Point Trail" sign. Above the power lines, you actually need to head right (not left, as in the Lowe Guide) to continue onto the Ruckel Ridge route.

CAUTION: The ascent up Ruckel Ridge is on an unmaintained path, now easy to follow but difficult to negotiate for anyone inexperienced in Grade 3 rock climbing. I strongly recommend that you *ascend* Ruckel Ridge and *descend* Ruckel Creek, unless you wish to challenge your down-climbing skills on technical rock. Moreover there is a loose wall to climb at **Point 5,** around 2,200'. Even going up the Ridge route, you must encounter the famous "catwalk" at **Point 6**, 2,660' elevation, where one must delicately cross a brief, but exposed ridge. Again just after the catwalk, another scramble awaits. While technically, all these rocky points would be designated as Class 2 to 3 in the climbing literature, they all require an acute sense of balance and are definitely not for the novice hiker, those uneasy about exposure or the use of hands for balance or for children. Best here to simply tackle the Ruckel Creek Trail up and down, remembering that in the description below, "right" and "left" will be reversed.

I apologize for the absence of **Points 7 – 10**. The references in the text, however, are correct, ending with **Point 6** near the plateau on the top of the Ruckle Ridge Trail and beginning with **Point 11** at the beginning of the descent of the Ruckle Creek Trail.

Up Ruckel Ridge

February/March

Not much is blooming at the parking area or the campground at this time of year but at the power line throughout the winter, Pearly Everlasting continues to live up to its name, with its white berrylike blooms hanging on in low-elevation openings, although they do appear somewhat withered. An early surprise, however, are several specimens of intense purple Columbia Kittentails just beginning to bloom in early February at the 230' elevation on the right just before the power line (**Point 2**). By early to mid-March in this area, Snow Queen, a close relative of Kittentails, enlivens this stretch of the trail. Also bright purple, a single specimen of this early species of the Snapdragon Family might go unnoticed due to its small size; however, when blooming in masses, as here, Snow Queen's tiny but showy flowers attract attention as much for their color as for their happy announcement of spring's arrival.

By March, several wildflowers are already in bloom, especially on the lower reaches of this trail. Stream and Evergreen Violets grace the start of the trail up to the campground on both sides, while the Evergreen variety, with thicker darker leaves predominates in the woods above the campground between **Points 1** (the campground) and **2**, (the power line). A few Woods Strawberry are out at the power line and beyond, in open woods, Toothwort blossoms nod their lavender four-petalled heads in a welcome introduction to a flower-filled spring. On the talus slope at **Point 3** in late March, some Martindale's Desert Parsley creeps across the rocky substrate with its blue-green highly dissected leaves and fuzzy small yellow flower heads. The ubiquitous weedy Hairy Bitter Cress appears here as well, its ½' tall stem topped by tiny white four-petalled flowers that always seem closed against the elements, perhaps because it thrives best in wet westside environments.

Don't miss the left turn, marked by a cairn in the midst of the talus; the route now steeply ascends what appears to be overhanging sand. By the time you reach the top of this precipitous trail at approximately 950', take a breather and admire the Martindale's Desert Parsley and Western Saxifrage, the latter often growing off the rock faces. This saxifrage has small white flowers atop a 6-8" tall thick stem, all springing from a base of toothed green leaves. In the steep woods beyond, occasional Trilliums and many Toothworts beckon, particularly between **Points 3** and **4**. At the first opening above these woods, at about 1,200', some magnificently- colored purple Grass Widows grace the hillside, nodding in the breeze and accompanied by the low yellow of daisy-like Gold Stars.

In the openings at 1,400' and 1,600' a few Glacier Lilies (*Erythronium grandiflorum*) are just beginning to bloom by early March. The much-beloved Glacier Lily is a close relative of the Avalanche Lily, the main difference between them for us amateurs being color – Glaciers are yellow, Avalanches white. Both display their nodding re-curved petals atop 1' tall stems but, in the Gorge, you will find the yellow Glacier Lilies much more frequently than their white siblings, which prefer higher elevations.

Glacier Lily

Masses inhabit open woods on the Vista Ridge and Paradise Park Trails in the Mt. Hood National Forest. Both are appropriately named: The white variety does seem to avalanche across slopes and open woods in the southern Cascades and Olympics, while the yellow Glacier Lilies *do* bloom very near ice throughout the Northwest, frequently pushing their noble way bravely up through the snow lingering on open slopes. However we may like to attribute human virtues to plants, it appears to be the heat of the flower that melts the snow rather than any persistence on the part of the lily, thus allowing the flower to rise above its icy surrounds.

Other flowers are also making their early appearance by mid-March. Kittentails are prominent on the east- and north-facing bluffs just before and after the wooded knoll at **Point 4**, and also at 2,200' on the left. A few purple Cliff (Menzies') Larkspur (see the Ruckel Creek section in **May**, below) are also beginning to appear in openings at 1,650' and 1,750' along with some Small-flowered Blue-eyed Mary's. Toothwort continues in woods even by late February, especially on the sides of moderate slopes between **Points 4** and **5** up to 2,400'. Above this point, this early in the season, you may find more snow than greenery but that is no reason to turn back. The thrill of the catwalk and the prospect of the hanging gardens on the loop await, not to mention the apprehension, if you don't make the loop, of descending what you just climbed up!

April

April can be the best of months on Ruckel Ridge for the flowers or the worst, if rainy, because of the slippery mud and rocks, but April will never bore you here. Purple Dead Nettle, a common roadside weed, can be found at the parking area just off I-84. As you briefly head up the paved road from the parking area, then make a right turn at the sign to climb up to the campground (**Point 1**), many Robert Geraniums show their pretty pink smiles, while male Meadowrue plants hang their brownish-purple silky tassels (really stamens) above their lacy thrice-divided leaves on both sides of the trail. A few Plume-flowered Solomon's Seals and Fringe Cups are just beginning to bloom here by later in the month. Entering the woods after the campground at the Buck Point sign, a single Hooker's Fairy Bell is on the left, along with some Toothworts, here more white than lavender. Trillium persists in the open forest here, although by late March these are blushing purple. Along with the omnipresent yellow Evergreen Violets, below the power line are the delightful purple Snow Queens, mentioned above in March.

Just before the power line (**Point 2**), at about 550' elevation, many examples of both Nevada and Leafy Pea can be found, along with the Evergreen Violet. Here too, by mid-month, on the right there are three specimens of one of the strangest plants you will ever see in our mountains, Striped Coralroot (*Corallorhiza striata*). One of three coral roots in the Northwest, these plants are a reddish brown color and lack any green parts whatsoever. They are orchids, although no one would associate them with their flamboyant and colorful tropical cousins. One can see the tongue-shaped lip underneath the five other tepals (petal-sepal combinations) hovering above, and in this species, all these tepals are marked with three maroon stripes each. Since green signifies chlorophyll in plants, and it's the chlorophyll which provides the energy for plants to grow, how do such species live? They do so by deriving their nutrients from decaying organic matter within forest soils. In fact, such saprophytes don't even need sunlight; the soil provides all they require (although see the sidebar for a disclaimer) and in this way, they are very much like mushrooms, which also lack the color green: Given the right conditions of the soil in which they are placed, they could grow in your bedroom closet.

> Saprophytes are organisms which live on dead organic matter, as opposed to parasites, which get their nutrition from living organisms. They are therefore "nicer" to others in their neighborhood, in that they don't prey off them to benefit themselves. The situation is, as usual, more complicated in nature, where we cannot impose our anthropomorphic and naïve ideas of good versus bad to the botanic world. In fact, coralroots, and other saprophytes in our area, such as the Phantom Orchid and Pinedrops, exist in a mutually beneficial arrangement with tiny fungi populating all soils worldwide. These fungi send out small hairs called *hyphae* which then penetrate the developing plant embryo and begin to suck out important nutrients from it. Despite what appears like a hostile invasion, the new plant welcomes the hyphae as the fungus actually delivers water and minerals to the plant, while the plant, in return, transports sugars to the fungus which it cannot make itself. This symbiotic arrangement is needed for almost all orchids to grow, and, indeed, for a majority of all plants to thrive as well, a fact not well publicized. In most saprophytes, this mechanism has assumed control over *all* nutrition, while in many plants, it operates only as a boost to growth. Such cooperation between two different Kingdoms, the *Plants* and the *Fungi*, could serve as a model for us in the Animal Kingdom as well.

At the power line a number of Chocolate Lilies and Woods Strawberries are in bloom. Frequent here, and at all low-elevation openings in the Gorge from late March through April is the well-known Shining Oregon Grape (*Berberis aquifolium* – note that many texts use the genus name *Mahonia* for the Oregon Grapes). This ubiquitous plant, the state flower of Oregon, has been cultivated since David Douglas sent the seeds back to England in the early 19th Century. It stands from a few feet tall in sheltered areas to over 4' in the open, although cultivated forms in city gardens can grow to 7' and more.

Shining Oregon Grape

Even more common is the Cascade, or Dull, Oregon Grape (*B. nervosa*), a shrubby plant usually arching its 1-2' long stems as it carpets deep forest floors from sea level to mid-elevations.

Cascade Oregon Grape

This latter species seems to thrive in once-or twice-logged westside forests now planted with Douglas Fir, where it often forms a continuous ground cover. It is not an anxious plant, despite its Latin species name; the *nervosa* refers to the three prominent veins on the each leaf. Both species have evergreen holly-like leaves with pointy spurs and yellow flowers in erect clusters, followed by blue berries with a whitish bloom. Despite the "grape" sobriquet, these berries, while edible, are not very much like grapes. Still, mashed up and with generous dollops of sugar, they were often used in earlier times to make a jelly or wine. Modern-day hikers would probably be only practicing their pucker.

You can tell the difference between the two Oregon Grapes by their habitat – the shiny species prefers open sunny locations while its duller relative seeks the shade; by their heights – the shiny one is often taller or at least has a more erect structure; by the number of leaflets – the shiny one usually has five to nine while the more common dull type, even though it's often smaller, has more than nine, and sometimes up to 19; and of course by the luster of the leaves – shiny or dull. Perhaps the Cascade species needs more leaves than its larger cousin as it occupies shadier neighborhoods. It's good to get to know your Oregon Grapes as they are frequently encountered and it's not every plant that has a state named after it!

Between **Points 2** and **3**, the talus field, the forest holds several small delights. Fairy Slipper (the Calypso Orchid) is at its gorgeous peak here, as well as the small white Large-leaf Sandwort. As you descend slightly onto the talus area at **Point 3**, the yellow-headed flowers you see, mostly on the left, are either Martindale's or Nine-leaved Desert Parsleys. The flower heads, actually bunches of tiny flowers each, look the same in these two species, but the Nine-leaf variety is taller and has linear leaves, while Martindale's often hugs the rocky ground and its leaves are finely dissected in keeping with its Carrot Family tradition. As you next struggle up the steep loose and sandy "trail" from here to the overlooks at about 950', watch for Small-flowered Blue-eyed Mary's and the tiny but beautiful Chickweed Monkey Flower; several examples can usually be found growing at around 650' to 700', some right in the middle of the trail. This small monkey flower often goes unnoticed but a close look rewards with a view of a single burgundy splotch on its lower yellow petal.

At the top of the talus, around 950', along with Martindale's Desert Parsley, Western Saxifrage, and Broad-leaf Stonecrop, an example of a Serviceberry shrub in full flower graces the view. Also here, find the beginnings of what will be several fields of Glacier Lilies in the openings between this point and in openings to 1,500'. This yellow lily should be distinguished from its sibling, the white Avalanche Lily, which is not easily found near the trails in the Gorge. This white lily *is* frequent, however, in higher elevations on Mt. Hood, the other Northwest volcanoes, and also in the Olympics. The icy references in their names may derive from these lilies' habitat, but more likely, it relates to their uncanny ability to inch their way up through melting snow banks in the early spring. In our mountains, Glacier Lilies can be confused with nothing else, with their gracefully nodding yellow flowers with petals re-curved almost from their base. (But compare the rare and not re-curved Yellow Bells in the North Lake and Dog Mountain Chapters.) These "lilies of the field", admired by all mountain travelers, seem to embody almost human traits of innocence and persistence in the face of environmental adversity, but in fact they are well suited to their subalpine environments. They thrive in the breezes of April here in the Gorge and can be found at higher elevations in the Cascades poking their cheery heads through the snow throughout the early summer. In fact, their heat is sufficient to melt the snow and ice through which they emerge, smiling and laughing at the cold and wind and mocking our feeble attempts to comprehend beauty in the face of such odds.

The steep, muddy ups and occasional downs of the trail beyond **Point 3** and between 950' and 1,100' is made a bit more tolerable by the presence of a smattering of Calypso Orchids, Toothworts, and Trilliums, the latter turning purple by the end of the month. Most hikers look forward to the three or four openings to the west between 1,000' and 1,400', as much perhaps for an excuse to rest and admire the view as for the bountiful flowers just beginning to bloom. Harsh Paintbrush, Small-flowered Blue-eyed Mary, Glacier Lily, Chocolate Lily, Gold Stars, and fields of Spreading Phlox weave a rich tapestry of red, yellow, pink and blue across these meadows, all backed by the occasional white of shrubby Serviceberry. The openings at 1,200' and 1,400' add the deep purple of Cliff (Menzie's) Larkspur while the opening closest to 1,300' adds the light pink of raggedy Prairie Stars and the brighter pink of Rosy Plectritis to this colorful mix. Higher meadows along the ridge between 950' and **Point 4** display Phlox, Martindale's Desert Parsley, Gold Stars, Larkspur, and Rosy Plectritis as well.

In the woods between the openings, all is not simply brown and green. Cascade Oregon Grape is in bloom here, with its bright yellow flowers tucked below its arms of spiky leaves. Look closely beneath the tapered leaves of Hooker's Fairy Bells to see its dainty re-curved white blossoms. Here too find Large-leaf Sandwort, with its tiny white flowers in mats especially around the roots of Douglas Firs at 1,400' and just before and at the wooded knoll at 1,850' at **Point 4** (a not unpleasant resting spot). Not all flowers are white in these woods: Fairy Slipper delights the careful observer with its pink-to-purple petals and maroon spotted lip; it is common growing off the mossy understory at 1,175', 1,450', and 2,100'.

On the dry ridges at 1,700' and 2,300', Manzanita is just coming into bloom. The openings after the steep rocky climb back up to the ridge from the east at 2,120', just before **Point 5** (the scramble back up to the ridge at 2,250') display yet more Phlox and Martindale's Desert Parsley, while in the woods between **Points 5** and **6** (the catwalk), an occasional Trillium can be found. Although there are no flowers yet in bloom at the catwalk, just beyond it, at the bottom of a dip in the trail, two specimens of White-vein Pyrola appear; its name derives from the white stripes outlining the veins on their low-lying leaves. In fact, it's the leaves which mark these plants now; the drooping pinkish flowers must await summer to bloom.

As you labor up the steepening trail, the yellow Evergreen Violet appears up to 3,000'. Beyond the overlook at **Point 7**, the only plant in bloom will be Oregon (or Mountain) Boxwood, a shrub whose dull red flowers are so tiny as to defy notice by all but the most dedicated amateur naturalist. Nonetheless, the struggle up to the plateau at **Point 8** (often blanketed in snow until later this month) will probably preoccupy most hikers and flowers may not be as appreciated as on the lower reaches of this trail. Nonetheless, a number of plants will be in bloom even here by May and June.

May

Fortunately for the flower lover, many of April's blooms last well into May, and these are joined by scores of new plants, making the steep climb up Ruckel Ridge a worthwhile outing not only for conditioning but for plant identification as well. Several weeds greet the hiker parking in the area just off the Freeway as you make your way through the grass, including Common Groundsel (Senecio) and Dandelion, not to be confused with Hairy Cat's Ear, one of the most common dandelion-like weeds yellowing our summer roadsides. The real Dandelion prefers lawns to gravel, is usually shorter, has milky juice in its thicker stems, and a single flower head, while the Cat's Ear's stems are wiry, and often topped by multiple yellow heads. Clambering over the concrete barrier and proceeding up the paved road to the trailhead, yet another import, Robert Geranium, greets you with its cheerful pink blooms, especially on your left and appearing very much like a planned floral tapestry in the shade. This prolific invader, though replacing many native plants in the Gorge and in the city, is actually so pretty that many folks mistakenly think it's an on-purpose ornamental when they encounter it in urban yards and roadsides.

Taking a right turn at the sign to head up to the campground, Plume-flowered Solomon's Seal is in full bloom early in the month, its sweet-smelling spray of tiny white flowers appearing to reach out to grab passers-by. Hooker's Fairy Bells, Candy Flower, and several specimens of yellow Large-leaf Avens on the right provide floral interest right from the start. After a few feet of elevation, Meadowrue and Thimbleberry (the white flowers, not the berries yet) appear, and by late in the month, the Inside-out Flower as well. Walking through the campground (our **Point 1** at 200' elevation), Wood Violet can be seen poking its yellow head up here and there, accompanied by the ubiquitous Robert Geranium, growing so fast here that you'd think it might cover slower-moving campers even overnight! Above the campground, Snow Queen has become a geriatric plant, now setting seed, but in the wooded areas Plume-flowered and Star-flowered Solomon's Seals (the latter at the first left switchback), Star Flower, Meadowrue, Columbia Windflower, and Alumroot are all present, along with Little-leaf Candy Flower on the mossy rocks on the left at 350', about midway between the Buck Creek sign and the power line; what these woods lack in floral color during these months they make up for in variety.

With a bit of sunshine at the power line (**Point 2**, 550') however, even though they are human-induced, more colorful plants appear. Just before this artificial opening the strange saprophyte, Spotted Coral Root sends up several pinkish purple stalks with orchid-like flowers spotted purple on your left, while the shrubby Scotch Broom appears in splotches of yellow up and down the swath cleared for the power poles. Several white flowers are in bloom here as well, including Woods Strawberry and perhaps the most despised plant in the Northwest, the Himalayan Blackberry (*Rubus discolor*, also named *R. procerus*). Its flowers appear innocuous enough, with papery white to pink blooms trailing along the ground or at the ends of the stout arching branches. However, as everyone knows, this plant is armed with evil re-curved spines and, if allowed to grow in stands, it can form skin-shredding thickets. Here at the power line, however, it seems to control itself, so the hiker can admire the flowers in anticipation of the tasty berries to follow in August.

Himalayan Blackberry

We have three varieties of blackberries in our woods and towns: The Himalayan kind is the most common and, of course, the most vicious, but two others (the Evergreen Blackberry, *R. laciniatus*, and the Wild or Trailing Blackberry, *R. ursinus*) are occasionally found. One example of the latter, our only native blackberry, can be found just above the power line here. For those who want to know, this blackberry can be distinguished from the Himalayan kind by its prickles, which are round, not flattened. Folks who brush against either kind probably would not care, as they are both just as sharp! The Evergreen type, which has deeply incised leaves, is rare in the Gorge. It also originated in the Himalayan region and its thorns, as well as those of its relatives, may have provided protection from grazing mammals. Often, we say that a feature such as thorns evolved *in order to* deter predators. In the course of blind evolution, traits simply develop, then are selected if they confer some advantage. In August, mammals of the human species are not so easily deterred, however, and the thorns of spring are forgiven for the black feasts of summer.

In the woods above **Point 2**, and heading down to the talus at **Point 3**, many examples of Star Flower can be found, seemingly taking merciful respite from the blazing sun. In these woods, other shade lovers appear as well, such as Vanilla Leaf, Sweet Cicely, Star-flowered Solomon's Seal, and Columbia Wind Flower. Not all forest blooms are white, however, as here and there poking up in the shade is Woods (Small-flowered) Nemophila, with its tiny but pretty blue blossoms and club-shaped leaves. At the talus, two kinds of Desert Parsleys are in bloom. Martindale's, with its ferny blue-green leaves, sprawls on the ground, while the Nine-leaf species stands more proudly upright, with linear leaves masquerading as grass. The pink or lavender Varied-leaf Collomia and yellow Western Senecio add color here. This collomia is small but notable for providing a tiny but bright splash of color in more open areas of the Gorge, here, on the early switchbacks of Dog Mountain, and the meadows by the power lines on Starvation Ridge. Its five-petalled flowers sit atop 5-10" tall stems and the plant secretes a resinous substance which coats its upper parts. *Collomia* is derived from the Greek word for glue, but it's probably the sticky seeds which give this genus its name. The "varied-leaf" part comes from the way in which the upper leaves, rather simple in design, differ from those lower on the plant, which are pinnately divided, like the pinnae on the feather of a bird. Descriptions of leaves on other plants include "palmate", with the leaflets all radiating from a single point like the fingers on a hand (think lupines), and "whorled", with the leaves in rings around a stem (like in the Tiger Lilies).

Up the steep rocky trail beyond **Point 3**, the grunt is cheered by the jewel-like Chickweed Monkey Flower. This yellow-flowered tiny gem marked by a single maroon spot on its lower lip can be found growing in the midst of the trail on the stony substrate at 850', rather than in its usual moist habitat. While this monkey flower usually prefers wet seeping banks, such as on the Hamilton Mountain Trail just below Rodney Falls and the Pool of the Winds and also on the Horsetail Creek Trail above Triple Falls, here it appears to thrive on the hard-packed ground of this rough path. Here too are the white-to-pink blooms of Miner's Lettuce and the extraordinary, yet tiny, blossoms of Small-flowered Tonella in the Snapdragon Family. This often-overlooked plant deserves a closer view as its orchid-like flowers are a delicate reminder of the beauty in nature's ways. It sometimes seems that these flowers are more beautiful than they need to be.

At the top of the steep talus, at around 950', the persistent hiker is rewarded with splashes of bright color, for here in May the crimson of Harsh Paintbrush mixes with the yellows of the two Desert Parsleys described above. Yellow predominates here, with Broad-leaf Stonecrop growing right off the rocks and the beloved Glacier Lily now in full bloom. These lilies, with their drooping golden-yellow heads of six re-curved petals and a pair of yellow-green leaves at their base, are unmistakable members of the Lily Family, and, in keeping with the lily plan of parts in sixes, each flower sports six drooping stamens at its center, all staring straight down at the ground. Many folks get their lilies confused – are those white ones Avalanche Lilies or Glacier Lilies? Since both avalanches and glaciers are white, it's hard to keep these two species straight. Unfortunately, I can offer no convenient mnemonic. However, the white Avalanche Lilies prefer higher elevations and are more common around the Cascade volcanoes and in the Olympics, where they often carpet open woods and meadows, while yellow Glacier Lilies roam openings at all elevations and are more common in the Gorge. These latter will sometimes appear to poke their way up through melting snow banks, as if they were filled with the courage of spring. More likely, they actually play a role in melting the snow with their inherent heat and that of the sun reflecting off the plants.

Now the trail eases off just a bit and, between 1,000 and 1,200', hikers are rewarded with a series of glorious meadows alive with the reds of Paintbrush, the blues of Small-flowered Blue-eyed Mary, the orange-yellow of Wallflower, the pure white of Field Chickweed, and the pinks of Rosy Plectritis, along with gardens of Spreading Phlox. The steep root-filled trail in between these openings holds fewer treasures, although early in the month, Toothwort and Fairy Slipper persist here and there. Some of the best May has to offer in the Gorge comes next as the trail winds up, steeply at times, between forest and open meadow. In the openings between 1,200' and 1,800', Glacier Lilies sway in the breeze accompanied by deep purple Cliff (Menzies') Larkspur, the furry balls of Rosy Plectritis, and Phlox so thick it looks like it was planted by a garden designer. The air here on a hot afternoon can be redolent with the sweet licorice smell of these blooms. On the other hand, the woody intervals contain few blossoms, although the small white Large-leaf Sandwort, the pinkish Toothwort ("wort" is the old English word for plant), and some yellow Cascade Oregon Grapes are in bloom. These shaded interludes seem to hold time hostage while awaiting the more colorful meadows ahead, perhaps like the silent pauses between movements of a symphony – ripe with the promise of light to come.

Yet more openings appear between **Points 4** (the wooded knoll at 1,850') and **5** (the moderate cliff you will need to scramble up to regain the ridge), between 1,800' and 2,250'. Here, Martindale's Desert Parsley mixes with Phlox, Chickweed, and Larkspur. In these upper woods, shade and silence predominate, but cresting the ridge from the east after the steep rocky climb (use of hands advised), more Phlox and Martindale's Desert Parsley await. These woods still hold a few of April's Toothworts and Fairy Slippers, especially at 2,500' and 2,650', just before the infamous catwalk. Some Cascade Oregon Grape and Vanilla Leaf are in bloom here as well. You will not be distracted by flowers as you tread carefully over the catwalk in May (our **Point 6**, at 2,660'), although toward the end of the month, expect some whitish-pink Scouler's Valerian here. Beyond, you will find only the occasional yellow Evergreen Violet to accompany you to **Points 7** (a merciful resting spot with an overlook at 3,200') and the plateau at **8,** at about 3,640'. No matter though – the exercise and the flowers have made this trip in May well worthwhile.

June

For some reason, the blooms on Ruckel Ridge, far from withering away in the heat of summer, only seem to become more prolific in June. The omnipresent pink-purple Robert Geranium accompanies the summer hiker up the road, then, after a right turn, up the path and through the campground (**Point 1**).

These contrast well with the yellow Large-leaf Avens, Little Buttercup and the white Inside-out Flower and Candy Flowers present. Just as you start up this first bit of trail at the sign (and below the campground), an intriguing plant appears this month, just out on your right – Bittersweet Nightshade (*Solanum dulcamara*).

Bittersweet Nightshade

Look closely at the purple shooting-star-like flowers hanging beneath the egg-shaped dull green leaves. A yellow cone protrudes down from the center of each flower, the result of the five stamens being fused together, all ending in a pointed style that completes the picture of a miniature comet about to accelerate through space. The resemblance to a tomato flower is more than accidental; this nightshade is a member of the *Solanaceae,* or the Potato Family, which includes, of course, potatoes and tomatoes. While not a native, the flowers of Bittersweet Nightshade are attractive, but the football-shaped red berries which follow are best left alone. This shrub is a close relative of Deadly Nightshade (*Atropa belladonna*), the source for atropine, large doses of which are indeed deadly. Unless you plan on brewing up a concoction to coat your poison arrows, it is best to leave this plant be.

> Most people think of European cuisine as being based on natural plants: the tomato sauce of Italy, the "French"-fried potatoes of Belgium and France, the squash and corn purees so dear to the hearts of fancy Seattle and Portland chefs, the fruitful strawberry and pineapple desserts found in patiisieries and in every decent French village, and the chocolate mousse we all crave. In fact, however, before the sixteenth century, no Italian or Frenchman had ever heard of, let alone tasted, these treats. Potatoes, tomatoes, corn, squash, and even chocolate are all indigenous to the New World and were first imported to Europe only five hundred years ago. Pineapple, for example, is indigenous to the Amazonian rainforest while the early Maya drank chocolate, but it was bitter and often spiced with peppers. Truly ancient European cuisine lacked these essential ingredients, but now, happy as we are to indulge, it is of more than passing interest to reflect on the humble origins of many of our native foodstocks. Weeds may inherit some of our earth, but transplants can also provide not only beauty to our gardens but sustenance for our stomachs as well as our souls.

Up the woods below the power line, **Point 2**, the lovely Twin Flower is now out in full bloom; bend down to inhale its sweet fragrance – the effort will be worthwhile. Pathfinder and the yellow Wall Lettuce are plentiful here, along with the reddish-purple Leafy Pea (see below), clinging to other vegetation with its forked tendrils. Also look for the small drooping pearl-white flowers of Salal, the strangely beautiful Inside-out Flower, the large papery white flowers of the Columbia Wind Flower, and the pink-to-white twinkling blooms of Star Flower. The white and green is happily interrupted by a few yellow Broad-leaf Arnicas now sporting their daisy-like blossoms above their pair of egg-shaped toothed leaves mid-way up 1-2½' tall stems. Also here in the first openings is the shrub, Dogbane, with small hanging pinkish flowers and drooping leaves looking like some poor tired pooch's ears listless in the summer heat.

At the power line (550'), in contrast to the woods where white predominates, a Dutch still-life of colors shines in the June sun. The blues of Broadleaf Lupine mix with the reds of Harsh Paintbrush, the whites of Woods Strawberry and Oxeye Daisy, the orange and maroon-spotted Tiger Lily, and the reddish purple of Leafy Pea (*Lathyrus polyphyllus*). This last, a close sibling of the Nevada Pea (*L. nevadensis*), is a perennial plant with clambering vine-like branches capped with tendrils which help it cling to surrounding vegetation. However, our Peas, or Peavines as they are sometimes called, don't always need to rely on near-by plants for support. They can stand on their own as well, and here at the power line, they often do just that. Both Nevada and Leafy Peas have pink-to-purple pea-like flowers and stems with multiple opposite leaflets; they look a lot like the common, but pretty weed, Broad-leaf Pea (*L. latifolius*). Our two west-side native peas can be distinguished by remembering their names – the Leafy species here at the power line has 10 to 16 leaflets while the more common Nevada species (which rarely grows in that desert state) has just four to 10 leaflets. However, just to confuse the amateur botanist, there is some overlap. Since chugging up these steep trails the hiker will only rarely count leaflets, it's fair to simply identify these peavines generically rather than specifically.

Leafy Pea

Nevada Pea

The woods above the power line, between **Points 2** and **3**, harbor a treasure trove this month of twinkling Star Flowers and Columbia Wind Flowers. The pink and white of these blooms is intermingled with the yellows of Broad-leaf Arnica and the Columbia Gorge Groundsel (Senecio). At **Point 3**, on the talus field at 800' elevation, the yellow theme continues, with two Desert Parsleys – Martindale's hugs the stony ground and has ferny blue-green leaves; the Nine-leaf species stands 1½ – 2' tall and sports thrice-divided grass-like spikes of leaves. It's interesting to note how similar the flowers are in these two related plants. It's doubtful that even a trained specialist could tell the two apart from inspecting the flower heads alone. But the leaves are as different as any plant's could be; perhaps they evolved in separate locations, and indeed, Martindale's is most often found on rocky substrates while Nine-leaf grows amongst grasses and must stand tall to gather light. Still, here they are almost next to one another, perhaps reunited on this talus field after millennia spent drifting apart.

It's often hot in June, so climbing the steep and frustratingly loose slope above the talus field is a chore but one brightened by the discovery of several small blooms growing in what seems like the desolate baked soil of the ages. Here, right in the center of the path, Chickweed Monkey Flower smiles its grinning face at the tired hiker as much to mock as to enchant, depending on your state of mind at the time. This tiny but playful plant deserves another look. Its small yellow blossoms have two upper and three lower lobes, just like the snapdragons to which it is related. Inside each tiny bloom, in its throat, lies a single distinct maroon splotch, not the spotted interior of its larger sibling, Common Monkey Flower. Also in contrast to its larger relative, this small monkey flower doesn't need to get its feet as wet; although it often appears on dripping mossy cliffs (as on the Hamilton Mountain Trail on the path up to the Pool of the Winds), it also can colonize drier patches of ground, such as here and on the Starvation Ridge Trail in the openings below the power line. Also here, just above **Point 3**, look for two other tiny plants growing in the hard rocky ground: Varied-leaf Collomia, with its five deep pink petals, and Miner's-lettuce, with its wonderfully strange collar of a leaf through which the tiny white flowers protrude.

As you finally reach the lookout at around 950', take a break to appreciate the succulent jade-green leaves and starry yellow flowers of Broad-leaf Stonecrop, staring you in the face on the rock wall just ahead. This Chicks-and Hens relative is also distantly related to the giant Jade Plants growing in the windows of city shops and homes. The resemblance lies not only in the leaves, which are thick and spongy in all members of the Stonecrop Family (*Crassulaceae*), but often in the star-shaped five-petalled flowers as well, although many indoor Jade Plants rarely flower. Some botanists believe that these succulents gave rise millions of years ago to the more famous family, *Cactaceae*, the Cactuses, which also have spongy leaves that store water and are ideally adapted to the drier parts of our world. Stonecrops (there are five species common in our mountains) also store water in their leaves, although one would think there is little need for that in our wet climate. However, these well-named inhabitants of cliff faces and rocky walls need every molecule of H_2O during our dry summers because the places in which they live are easily drained of water. It's difficult to realize, especially in the incessant mist of winter, but it is totally true that the greatest stress our flowering plants face is drought during the summer.

Other spots of color brighten this lovely spot, including the spiky red of Harsh Paintbrush and the dainty low yellow blooms of Martindale's Desert Parsley with its blue-green ferny foliage sprawling across the rocky ground. However, progressing into the steepening woods between here and **Point 4**, white again dominates. Frequent examples of Vanilla Leaf (see an army of their white spikes standing at attention at 1,200') and the Insideout Flower occur up to about 2,000'. Sweet Cicely can be found in deep woods between 1,350' and 1,800', while the ridge tops for 500' below **Point 4** (the wooded knoll at 1,850') are a favorite haunt for Slender (also called Racemose) Pussytoes, with its 1' tall stems topped by furry dull-white blooms indeed resembling an upturned feline's paw. Small-flowered Alumroot here seems to prefer the margins of wooded areas, while an abundance of Little-leaf Candy Flower sprouts from the rocks and moss at about 1,200' and again at 1,900'. Not all blooms are plain vanilla white here, however. Pink Fairy Slipper is still in bloom and can be admired at about 1,250', 1,675' and 2,200', while the yellow daisy flowers of Broad-leaf Arnica grow in these woods here and there between **Points 3** and **5**.

Interspersed amongst these darker woods are several openings and here can be found a full panoply of colorful blooms, a veritable library of meadow plants for which the Gorge is justly famous. The opening at 1,150' contains an impressionist's palette worthy of Monet: the pink of Taper-tip Onion; the yellows of Martindale's Desert Parsley, Wooly Sunflower and Western Groundsel; the red of Harsh Paintbrush; the blue of Broad-leaf Lupine; and the deep purple of Fine-tooth Penstemon. White is not forgotten here; its representatives include Little-leaf Candy Flower (with occasional streaks of pink) and the raggedy blooms of Field Chickweed. Openings at 1,350' and 1,800' boast similar arrays minus the onion, and add the low blue of Small-flowered Blue-eyed Mary and the wispy yellow of Nine-leaf Desert Parsley.

One new flower, supposedly common but not often noticed near trails in the Gorge, appears in the clearing at 1,150': the unfortunately named Bastard Toadflax. It occurs as a thin-leaved plant growing off the dry sandy soil on either side of the trail. Its head is a cluster of white five-petalled flowers in a ball or spike atop the 3-12" stem. It is our only member of the Sandalwood Family but that is probably not why it gained its unflattering name. Instead, this plant may have engendered animosity because it parasitizes other plants by attaching its roots onto those of its neighbors. While this may not seem very neighborly, it is the only way this little toadflax can survive and it does not appear to harm the host plants to any great extent, although there is no evidence it helps them either. As plants lack nervous systems, we cannot ask them what they think of all this, but it appears likely the host plants are less than sympathetic.

> There are innumerable examples in the plant and animal worlds of parasitism but perhaps even more examples of symbiosis. The former implies that the invading organism obtains all the benefits of the arrangement while the latter means that both organisms derive mutual benefit. True parasitism is rare in our plants however. One example cited, that of One-flowered Broomrape "living off of" stonecrops and saxifrages may be misleading as the flowers of these host species do not appear to suffer even though they apparently supply nutrients to their invader; indeed they may even benefit in some way from the association. There are certainly many more examples of the give and take of symbiosis in the plant world than true parasitism (notwithstanding the strangler figs of the tropics). Here in the Northwest, lupines prepare the soil for paintbrushes by capturing the nitrogen in the air and transferring it to the soil, while fungi living in the ground of all our forests attach their roots (called *hyphae*) to many plants to obtain water and minerals. In return, the plants provide a steady supply of carbohydrates which the fungi could not manufacture on their own. In fact, biologists are now coming to the conclusion that over 90% of all living plants thrive only in combination with such soil fungi. In our often fractious world it is somehow satisfying when such cooperation effects a happy ending.

At about 1,850', on a wooded knoll at **Point 4**, a mix of Sweet Cicely and Slender Pussytoes greets the tired hiker craving a rest. From this juncture to **Point 5**, the start of a scramble up to a clifftop at about 2,220' where there is a brief opening, woods predominate but flowers sprinkle the forest to add interest. Vanilla Leaf, Large-leaf Sandwort, the Inside-out Flower, the Columbia Wind Flower, and both kinds of Solomon's Seals – Plume- and Starflowered, populate these woods throughout June. Not all is white and green however, for the yellow Broad-leaf Arnica is at its best here at this time of the year as well. In the openings, between **Points 5** and **6**, brief though they are, Martindale's Desert Parsley and Field Chickweed thrive along the ridge-tops. There are some steep and rocky sections leading up to **Point 6**, the infamous "catwalk", at 2,660'. Here, at the very top of the Class 3 scrambling route (if you can focus on the flowers), are several excellent examples of Scouler's Valerian, a plant you'll miss if you opt for an escape down to the easier west-side bypass. (I have never been able to convince acrophobic hikers that the sight of the Valerians is worth the exposure.)

The woods beyond the catwalk are filled with Bear-grass (*Xerophyllum tenax*) especially up to the overlook at **Point 7** at about 3,200'. Don't expect to see many blooms most years, though.

Bear-grass

This fickle, but beautifully powerful plant, teases with plentiful clumps of grass-like leaves but infrequent spikes of fragrant white flowers. Beargrass, a member of the Lily Family, can grow to 8' tall and blooms more frequently on open pumice slopes such as those around Mt. Hood; but even there, this relative of the Century Plant may go five to 10 years before all the plants in one area flower at the same time, as if by prior agreement. When they do so, they can transform June meadows into snowy fields and delight photographers and backcountry explorers alike. Here on Ruckel Ridge, a few 4' to 5' plants usually bloom each year, enough to at least appreciate the complex beauty of each star-shaped six-petalled individual flower in a plume that may contain hundreds. Sniff closely to appreciate the delicate lily-like fragrance of this unusual plant. Authors delight in relating all the implements Native Americans could make from the tough leaves, from baskets to blankets to footwear. An urban legend has it that destitute hikers have used the leaves as substitute boots once their threadbare originals were worn through!

Be certain to take the brief side trail to the rocky lookout at **Point 7**. Here you can check out the view northwest to wooded hills and the Columbia. But also be sure to peek down the northwest side of these rocks to be enchanted with one of the most colorful plants anywhere in the botanical kingdom: Rock Penstemon. The shocking magenta-pink of its tubular flowers, growing in mats clinging to vertical cliff faces, has been mistaken at long distance for man-made markers because of its flashy hue.

Just past the catwalk on the right, several Fairy Slippers are still in bloom as well. Also living in these woods are a few deep-blue Oregon Anemones, particularly at 2,900' to 3,000'. It's always nice to stumble upon these species at this time of year as, lower down, they've often disappeared by early May. Yellow Broad-leaf Arnica and the yellow ball-shaped flowers of Cascade Oregon Grape pepper the forest up to about 3,100'. From here to **Point 8** on the plateau at 3,640', the combination of a steep slope and a tangle of firs and hemlocks prevent most flowering plants from thriving.

The top of the plateau is also largely devoid of flowering plants. No matter – you've made it up one of the toughest trails in the Northwest. Take the faint trail across the plateau and cross Ruckel Creek. If you're super adventurous you can head back the way you came, down Ruckel Ridge; most folks, however, will opt for a saner descent down the Ruckel Creek Trail to complete one of the prettiest and certainly most famous of the many loops in the Gorge.

July and August

A surprising number of flowers are in bloom throughout the summer months up the Ruckel Ridge Trail and many of these summer bloomers, in contrast to the preponderance of white flowers in the spring, are quite colorful. The color begins at the parking area, although there it is a weedy yellow.

All three of the weeds growing here throughout the summer are cosmopolitan species and all are members of the Aster Family. They are obliging and malleable, thriving equally well in city and rural environments and under conditions in which weaker-willed flora wilt. Dandelion is well-known to everyone, or so they think, but another weed, even more common along roadsides, also grows here at the parking area, Hairy Cat's Ear. Its tall wiry stems and multiple heads distinguish it from the common lawn Dandelion. It is more fully described in the Munra Point Chapter, as is yet a third yellow weed, Common Groundsel (or Senecio) growing by the side of the highway. However this last barely attracts notice, nor, in most botanists' opinions, does it deserve it. Its tiny yellow heads, ray-less and drooping, strike most folks as just plain ugly, an opinion bolstered by its scraggly-toothed leaves and its tendency to thrive in trashy urban areas. A contrarian, however, appreciates the hardiness and pluck of this unfairly ostracized plant, which has made its way across most continents, can flower throughout even harsh winters, and grows in ground the consistency of concrete.

The little trail off the paved bikeway, turning right at the sign and ascending to the campground (**Point 1**) harbors a number of plants which flower throughout the summer, including the ubiquitous Robert Geranium and the yellow Large-leaf Avens. Some Inside-out Flowers can also be found here, especially through mid-July. In the woods above the campground, Robert Geranium continues to thrive, but a new species now appears by mid-July and will bloom throughout the summer and fall months – Foam Flower. Its name implies the sea, but this plant is found inland at low to mid-height elevations, often carpeting large swaths of the woodlands with its foot-high spikes of delicate tiny white bell-like flowers indeed resembling misplaced foam on the forest floor. Here too is more color, with the yellow of Broad-leaf Arnica and the almost out-of-place bright orange of Honeysuckle, a vine here trailing along the ground just before the junction at **Point 2**, the power line at 550'. This man-made breach holds the two usual suspects of summer low-elevation openings: Pearly Everlasting and Yarrow.

It's hot on the talus at **Point 3** in the summer, and here, at 850', two Carrot Family species thrive in these openings exposed to the full blast of the sun; both are desert parsleys: Martindale's and Nine-leaf. Although both have heads of compact yellow flowers, the former is a low-growing ground-dweller with ferny blue-green leaves, while the latter is taller, inhabits the more grassy areas amidst the stones, and has thread-like leaves thrice divided into three, hence its name. Scramble up the steep sandy "trail" above the talus and you will be rewarded at the lookout at about 950' with a nice view to the north and east, and, looking down, several colorful specimens of Harsh Paintbrush, here as much orange as red, and the yellow Broad-leaf Stonecrop plastered against the rock wall straight ahead. As you enter the woods above this point and progress along the very steep root-infested path, occasional yellow blooms of Wall Lettuce and white-to-pink Little-leaf Candy Flower appear, the latter especially off moss-covered rocks at around 1,050'.

There are several openings between 1,000' and 1,800'. All of these contain pointillist spots of color throughout much of the summer, with the pink-to-purple flowers of Phlox mixing happily with the bright crimson of Harsh Paintbrush and the yellow of Wooly Sunflower. Here too, the peripatetic Yarrow appears; indeed, it seems this found-everywhere plant is never intimidated by a lack of suitable habitat. In the woods between these openings ample specimens of Rattlesnake Plantain are just beginning to stick their weird spikes of tiny greenish white flowers above their mottled leaves, while Foam Flower, Broad-leaf Arnica, the Columbia Gorge Groundsel and Wall Lettuce provide colorful accents. On the ridges and also at **Point 4**, the wooded knoll at 1,850', fine examples of Slender (Racemose) Pussytoes continue to bloom through at least late July. On rocks leading up to this point, and also for some 500' above, watch for Little-leaf Candy Flower, thriving in stony situations where other plants fear to tread.

Above this point, many of the same forest plants continue, but the meadows are less flowery than in the spring. While we could mourn the loss of the Phlox, paintbrushes, blue-eyed mary's, lupines, and penstemons, it is probably nobler to appreciate the views from these openings and more practical to use them as resting points after the steep climbs to regain these ridges. In the forested areas between **Points 4** and **6**, Foam Flower and Broad-leaf Arnica can be found, but a new plant also emerges here, one rather shy and not commonly appreciated, Little Pipsissewa (*Chimaphilia menziesii*). Also known as Little Prince's Pine, this Heath Family plant only grows about 4" to 6" high and has bell-shaped creamy-white flowers which nod above thick and dark green leaves. It can be found hiding amongst the forest litter at 2,350' and 2,650', on the right side of the trail. Little Pipsissewa is a smaller version of the regular Pipsissewa or Prince's Pine (*C. umbellata*) and can be distinguished from it by its fewer flowers per stem (one to three instead of three to 15), its fully-toothed leaves (as opposed to toothed mostly above the middle for its larger relative), and its shy and lonely nature – it only appears here and there, while regular Pipsissewa most often grows in clumps. As is typical of their Heath Family lineage, they prefer the acid soil so common in Northwest forests and do not appear to need an abundance of light. Botanists have found that both species depend on the hyphal roots of fungi to help them thrive in this dark environment.

Little Pipsissewa

At the catwalk (**Point 6**), Scouler's Valerian is still blooming through much of the summer, while at the dip in the trail just beyond two specimens of the relatively rare White-vein Pyrola can be found, one first on the left, then another to the right of the trail. Also a member of the Heath Family (some botanists separate these plants into their own family, the Wintergreens), this species sports pretty greenish-white bell-like flowers atop a 6-12" stem. However its main attraction, especially when not in bloom, is its irresistible leaves marked by white stripes outlining its major veins. White-vein Pyrola never forms colonies and is always found in deep woods; it again is thought to depend on the mold of the forest floor and on the affiliation with fungi living and working their nourishing magic deep within the acid-rich soil.

In the woods above the catwalk, Bear-grass may be found but count yourself lucky to find one in bloom. Much Pipsissewa can be found as well, along with an occasional Mountain Arnica. Although you may find many leaves of these species, few flowers are in bloom at any one time, even though this is their season.

> It takes a lot of energy to produce a flower. But if plants don't flower, how are they going to reproduce and spread? A balance must be achieved and many species, especially those not situated to receive great quantities of light, jealously guard their energy production by blooming sporadically. This is particularly true of perennial plants, which can bloom one season and not the next. Also, some plants spread by underground runners and what may look like a single specimen which is not flowering may in fact be only part of a plant whose next-door neighbor on the same root system is in bloom. Thus it is common to encounter masses of Arnica or Pipsissewa plants not in bloom and imagine that a bountiful display is around the corner. Yet such a display rarely occurs as these plants will commonly either flower sequentially or not at all, leaving the amateur botanist disappointedly gazing over a field of leaves with but one or two blossoms to appreciate. All the more reason to learn plants from their leaves alone.

Between **Points 7**, the rock-strewn viewpoint at 3,200' and the "Thank-God" plateau (**Point 8**), mercifully reached after a taxing climb, not too many flowers can be found here in summer. Early in July, however, Oregon, or Mountain, Boxwood can be found still in "flower", although its blooms are so tiny as to require magnifying-glass scrutiny. This shrub is good to know as it appears at mid- to high elevations throughout the Cascades. Its small stiff leaves and tiny maroon flowers appear especially above the rocky lookout at 3,200' and continue almost to the plateau. If you've made it this far, congratulations; hopefully the steep grind has been compensated by the still-plentiful flowers and the all-encompassing views.

September and October

Queen Anne's Lace

The fall months bring cool air and a greater appreciation of changing color and mood to the Gorge, but there are still flowers to be thankful for. At the parking area, weeds hang out like criminals hiding amongst the tall grasses. Hairy Cat's Ear and White Clover invade this neighborhood, along with the ubiquitous roadside weed of summer and fall, Queen Anne's Lace (*Daucus carota*). Believe it or not, this 2-3' high Carrot Family species is the progenitor of the various cultivated carrots enjoyed by so many people around the world today, although I would not recommend eating *these* roots – they have a quite uncarrot-like bitter taste. Many texts claim that this wild carrot was named after England's Queen Anne, who proclaimed its head of lace-like flowers more delicate than the finest lace woven in that age. The truth of this attribution is difficult to judge, but the claim that the Queen also wore this flower in place of lace doesn't make too much sense.

Introduced from Asia and encountering no native pathogens in other areas, Queen Anne's Lace has spread around the world and now represents one of the commonest of weeds. Nonetheless, it's not an unattractive plant, with ferny parsley-like leaves and an umbrella-shaped head (technically called an *umbel*) of clusters (*umbellets*) – small umbrellas of tiny lacy-white flowers, so typical of the Carrot Family – the *Umbellifereae* – also called the *Apiaceae*. There are too many food and spice plants in this family to list, but some among them include celery, parsnip, parsley, caraway, dill, and fennel. Do not be tempted to pick these types of plants to eat, however, for among them are also some of the most poisonous species known, including Poison Hemlock. As the fall season progresses, note how the dry stalks supporting the seeds of Queen Anne's Lace begin to curl inward and eventually form a "bird's nest". Although I've never seen a bird actually inhabiting such a structure in this plant, the similarity to a wren's or robin's nest is striking.

Traveling up the paved bikeway to the signpost, one is accompanied by the equally ubiquitous pink-to-purple Robert Geranium, another attractive invader. In fact, since both this geranium and Queen Anne's Lace bloom prolifically in late summer and fall, it is possible that their pollination strategies involve insects most active at these times of year. Take a right at the sign and ascend to the campground (**Point 1**). After several switchbacks you reach the openings at the power line above (**Point 2**), where Pearly Everlasting remains in bloom throughout the autumn months, although its "lasting" flowers begin to look a bit weary and droopy late in the season. The woods above **Point 1**, the campground, are mostly devoid of blooms but those above the power line, at 550', are still replete with another weed, Wall Lettuce, with its five-petalled small yellow flowers lining the trail on both sides, particularly above the talus (**Point 3**), at 1,450' and again a bit beyond **Point 4**, the wooded hilltop at 1850', and again at 1,900', like an invasion of foreign soldiers in formation guarding their newly-claimed territory.

Sadly, the many openings between 1,100' and 1,850' are devoid of blooms during the autumn months, except for the well-named Pearly Everlasting. Even this is absent from the open areas above **Point 4**, and again just below and above the cliff above **Point 5** at 2,220'. These colder months may even find early snow. Bare branches at the higher elevations in the Gorge, like prickly hands reaching out from the frozen earth, gasp for a last breath of icy air. Despite the stillness, these woods and meadows still possess a severe beauty in the fall: cooler weather and gorgeous leaf colors, not only on the many deciduous alders, oaks and maples, but among those fallen leaves that litter the ground as well. These mottled patterns of green, gold, orange, brown, and red compensate for the lack of flowers during these months and render even a bloomless trip up Ruckel Ridge an adventure to savor in memory during the long and wet winter to come.

Down Ruckel Creek

Since most hikers ascend Ruckel Ridge and descend Ruckel Creek, the following description has you proceeding *down* the Ruckel Creek Trail; thus the references to left and right should be reversed if you plan on only going *up* Ruckel Creek. In addition, the **Points** on the map proceed downhill. Take my advice, and, if you can, make the round trip. Loops are usually more fun than up-and-down-the-same-way trips. But in this case, most folks will want to do the loop in the direction I've described; descending Ruckel Ridge is a hazardous chore, even for experienced hikers. One outdoor magazine rated this circuit "one of the toughest trail loops in North America".

Caution: The Ruckel Creek Trail, though a smooth and easy-to-follow route when compared to Ruckel Ridge, is still quite steep in places, especially above 2,400'. It is not recommended as a first hike for anybody.

February and March

Crossing the plateau at 3,640', often snow-covered through these months, head down to Ruckel Creek and cross on nicely placed rocks or a convenient log, then head at about a 30 degree angle and to your left up the slope and to the left to intersect the Ruckel Creek Trail at a signed junction with Trail # 405, **Point 9**. Heading down, this trail may seem steep but not as steep as what you just came up. The first flowers at this time of year will probably not appear until the muddy area and small streams between 2,800' (**Point 11**) and 2,680' where, by late February in some years, and certainly by early March, yellow Stream Violets brighten the mucky ground on either side of the trail. You know it's early spring when you greet the first Trilliums, between 2,600' and 2,500' on the right; these continue sporadically in open woods, especially on the right, between **Points 11** and **12** (the first large meadow); their three crisp white petals are particularly attractive this early in the year, when they are often compressed and just beginning to unfold. Up this high, don't expect to see these beauties until mid-March, although further down, Trilliums may be out earlier that month. Many juvenile plants unfold their leaves and flowers slowly so as to protect their sensitive inner parts, particularly their sexual organs, from the cold. This process is a favorite subject for videographers, who shoot a blossom unfolding in real time, then speed up the playback so we can see an animated flower opening to the world in just a few seconds.

There are six major openings, or meadows, and several smaller ones, as you proceed down the Ruckel Creek Trail. It's probably confusing to try to label these as the smaller meadows could be taken for what I call "major" ones. In addition, some of the smaller openings are being compressed by the growth of trees. Nonetheless, I will try to identify the major openings and, if it appears puzzling as you hike down the trail, you can take comfort in knowing that some of the same flowers appear in many of the different openings. Meadows, numbered 1 and 2 lie between 2,450' and 2,375' and are separated by only a few hundred yards and by first a left switchback, then a right one, down a steep slope. In these openings, sometimes as early as late February, Glacier Lilies nod their bell-shaped yellow heads toward the cold ground and may even push their way up through what appear to be holes in the lingering snow cover. In fact, the heat of these small lilies actually melts the snow and creates a space for these blooms to receive air and light. In the second small meadow, these lilies are accompanied by the small daisy flowers of Gold Stars, here in swaths of yellow carpeting this field.

The third major opening, at 2,250' (**Point 12**), is called the "upper hanging meadow" in the Lowe book. It can be identified by a stream two-thirds of the way across. In this large opening Gold Stars are coming into bloom as early as mid-March. Another flower here is a tiny plant that even most amateur botanists might overlook, Spring Whitlow-grass (*Draba verna*). Like a foreign spy, this introduced species, no more than several inches high, hides among the grasses on either side of the trail in many Gorge meadows. Its tiny, fleshy spoon-shaped leaves cluster in a basal rosette while its short stem is often leafless. As is so often true in the native plant world, a close look rewards the careful, patient hiker. Each of the four tiny white petals is cleft in two, deceiving the brief observer into thinking there are actually eight petals. This amusing Mickey-Mouse-ears effect is hidden until you stoop to inspect these inconspicuous plants, whose genus you may think was named *Draba* because of their drab appearance; untrue – the name is from the Greek word for "bitter". Their four petals in a cross-like shape and their six stamens, yellow in this species, betray their membership in the large Mustard Family, and it is unlikely that even if they were sent here as spies, they have committed any treasonous acts.

The next large opening, called the "lower hanging meadow" in the Lowe book (our **Point 13**) cannot be mistaken. As you enter the meadow and curve to the right, note that by early March the tiny Small-flowered Blue-eyed Mary has bloomed and forms a nicely colored counterpoint to the Gold Stars liberally sprinkled about the opening. These

Gold Stars may be out in some years as early as mid- to late February. The same duo can be found in meadows numbered 4 and 5, at about 2,150' and just ¼ mile further along the trail. These two openings below the lower hanging meadow, the first, number 5 at 2,150', the second, number 6 at about 2,100', each contains masses of Gold Stars by late February through March and by early March, some Small-flowered Blue-eyed Mary's as well. From this point, the trail enters the woods and another mid-February to March Mustard Family plant begins to appear, Oaks Toothwort. Its pale lavender-to-pink four-petalled flowers can be seen growing especially on the right (uphill) off well-drained slopes sporadically between 2,100' and 1,800', then in greater profusion below **Point 13**.

At **Point 14**, 1,850', take the short side trail to the right out to the lookout for a nice view (on a clear day) to Mt. Adams and the Columbia River below. Here, as early as late January in some years, and certainly by mid-February, the spectacular deep-purple clusters of Columbia Kittentails are a delightful discovery, growing at the overlook on its north-facing slope. Several more can often be found as you make the left switchback on the main trail just after this side trail and even several more at 1,340' on the right. This plant defies convention, and common sense, often blooming as early as January at lower elevations. I found a single Kittentail in flower on the Mt. Defiance trail at about 400' on December 27th, 2004! Not even experts can explain what winter-loving insects pollinate this Snapdragon Family early bloomer.

Between **Points 14** and **15** (the "Indian Pits"), the forest holds many examples of the yellow Evergreen Violet between 1,600' and 1,500' while Oaks Toothwort re-enters the lightly-forested slopes especially on the right (uphill) between 1,400' and 1,200'. A brief opening at 1,200' affords a view down to the river and a last chance to glimpse Gold Stars. At the pits themselves, in an opening at about 875', Little Western Bitter Cress may be the sole flower in bloom until late March. The pits themselves, however, offer some interest. On the left of the trail, these round stone-lined pits are said to be of mysterious and primeval origin; some suppose they were used by Native Americans to steam camas roots. More cynical observers believe they were dug by the inhabitants of Cascade Locks to attract tourists.

Below the pits, forest travel resumes, and here, many Oaks Toothwort and yellow Evergreen Violets flourish, particularly from 850' to 700'. Here too, by early March, are some nicely-formed Trilliums, which continue in open woods all the way to below the power line, **Point 16** at 490'. Both above, then below the power line, this same trio continues, but is soon joined, in mid- to late February by Robert Geranium, growing on either side of the rocky trail at and below **Point 16**, where the switchbacks down toward the stream begin. It is often here that knee- and foot-sore hikers begin to wonder how it is that they are descending more feet than it seems they ascended earlier that day. Does this trail actually go below the river? Soon, however, Ruckel Creek can be heard, then seen, at this time of year in full wash. As you make the final right switchback by the stream, watch for more Trillium and Robert Geranium. Don't fail to peer over the north wall as you come to the paved bicycle road to see the plume of Ruckel Creek rush through a narrow gorge beneath the bridge.

Walk back west along the road and see the invading Robert Geranium spilling over the retaining rock wall to your left. Despite the earliness of the season, another plant is in bloom here, Colt's Foot; it grows both on top of the wall and below it by the left side of the road. This early bloomer is a member of the Aster Family and sports a mop of dull white-to-pink brush-like flowers at the summit of 2' stems. This misbehaving species is topsy-turvy: Its huge deeply-cleft leaves, looking like something out of an Amazonian forest, actually appear *after* the flowers have wilted. This unusual plant can also be found just before hitting the bikeway as it thrives along the streambed as well. I have seen Colt's Foot flowers emerging from the icy earth as early as mid-February in warm and sunny years.

Osoberry

Further along the road, another early bloomer is in flower just before you make a right turn and return to your starting point at the trail sign. Indian-plum (*Oemleria cerasiformis*), also called Osoberry, can begin to bloom in early February. Indian-plum can also be found in city fields and uncultivated parts of parks. Here, a nice specimen can be found on the left of the road. A 5-10' tall shrub in the Rose Family, this plant is a true late winter jewel, its upright bright green leaves accompanied by a hanging tassel of pearly-rose "earrings" composed of cheery drooping bell-shaped flowers worthy of Vermeer. Unfortunately, later in spring and throughout the summer, without its droplets of flowers, this shrub is lost in the green anonymity of the understory. Smell a leaf or a flower; the odor resembles cat urine, but no matter – a true harbinger of spring, it is all the more welcome after a dreary winter of grays and browns. Native Americans ate the plum-like fruit later in the spring but only in emergencies or as a medicine for colds and influenza. While the species name promises "cherry-like in form", the taste is far less rewarding.

It should come as no surprise that plants often provide compounds which have profound effects on humans. Both the Plant and Animal Kingdoms derived from a distant shared ancestor and many of the genes and chemicals common in the plant world are present in animals, including *Homo sapiens*. Think of caffeine, nicotine and a host of medicines now in widespread use, including digitalis for heart conditions and aspirin (from willow trees). What is even more surprising, however, is the potential plants may hold for future use in treating illnesses now stubbornly resistant to current therapies. Plants are stationary and lack the option of fleeing from predators, such as insects and mammals who might eat them to the point of extinction. Therefore, many plant species have evolved static defenses to ward off potential threats by producing toxic chemicals (such as natural insecticides) or other substances that must taste putrid to animals who might pillage and destroy them. However, when we humans eat such plants, as for example, blueberries, broccoli or other fruits and vegetables we all know are "good for us", we also ingest small quantities of these chemicals. Scientists thought that these fruits and vegetables were helpful to us because of their anti-oxidant effect, as oxidation aids in the destruction of cells and hastens the aging process. However, recent research has shown that, in fact, the benefits of these healthy plant foods can also derive from a small stress such marginally harmful substances pose to our immune systems and other cells in general, a response termed "hormesis". Such small stressors, as with other procedures known to delay aging and its dire consequences – think cancer and Alzheimer's Disease – such as fasting every other day (ugh!), promote the production of a protective chemical in our bodies called "brain-derived neurotrophic factor", which protects cells from oxidative damage and prevents them from forming the destructive proteins known, for example, to be associated with many types of dementia. Thus your mom's admonition to eat your veggies may have been based on more than just assumptions and old wives' tales. Plants may be good for us in ways we might never have imagined.

After easily surmounting the concrete barrier you arrive back at the parking area, perhaps exhausted, but proud of completing this rigorous loop. Even completing this circuit this early in the season, you've not only gotten a ton of exercise, but hopefully, you've also been able to identify some of the late-winter/early-spring flowers which cheer us on despite the frequent showers at this time of year.

April

April showers bring May's flowers must have been an East Coast jingle. Here in the Gorge, and especially on the Ruckel Creek Trail, April's flowers most often precede May's occasional sprinkles, although this month also brings its fair share of rain. Even if it's gloomy out, and especially in showery weather, Ruckel Creek is a good choice in April as the floral display can cheer up even a dull day. Remember, we're starting from the plateau in describing these blooms, assuming you're hiking the loop.

By this month, although most of the snow should be gone from the plateau, there are still no plants in bloom up this high. As you descend from the trail sign at **Point 9**, notice some clumps of Bear-grass, not yet in bloom, especially on your right. These clumps are especially prominent at 3,150', as you descend to a talus field on your left (**Point 10**). Be patient and try this trail again in May and June to see if any Bear-grass blooms appear – it's worth the wait. Here too are the tiny red flowers of Oregon Boxwood (Mountain Box), almost invisible amongst its attractive leathery leaves. Now, on the right between 2,900' and 2,750', then again at 2,575', a few Trilliums are in bloom, many suffused with the wine-colored hue that comes with age and causes us to wonder why we don't age as gracefully. At 2,900' a single high-living Fairy Slipper can be found on the right, and, beginning about 100' downhill from the talus field, and continuing down to 2,525', a number of Oregon Anemone's lend their deep blue to the green forest, again especially on your right.

It's at the muddy stream areas at 2,800' (**Point 11**) to 2,680' that a different population of plants appears – those which flourish in the wetter areas of our woods. Here, Stream Violet (also appropriately called Wood Violet) makes a strong showing, along with its significant other, Candy Flower, the latter particularly prominent after mid-month. Unfortunately, between this point and the first meadows, flowers are mostly absent. Botanical compensation occurs, however, beginning at the first of the six meadows mentioned above in February and March. This first opening, a small grassy enclosure at 2,450', twinkles with the yellow of Gold Stars, Martindale's Desert Parsley, Glacier Lilies and Chickweed Monkey Flower along with the tiny blue of Small-flowered Blue-eyed Mary and the crimson of Harsh Indian Paintbrush. (Several specimens of this last species here in this meadow are actually closer to a blazing orange than red.) We appreciate these smaller meadows as much or more than the larger ones just ahead, in part because they frequently offer a more compressed floral display, not unlike islands of color embedded within a sea of brown and green.

Common Monkey Flower

The second small meadow, at 2,375', is simply stunning as it happily repeats the plumage of a similar color-packed display and adds a new bloom to the April flower show. This, a larger sibling of Chickweed Monkey Flower, is variously called by several unimaginative common names, including Common and Yellow Monkey Flower (*Mimulus guttatus*). This, our most frequent Monkey Flower, is unmistakable, with its double-lipped trumpet-shaped snapdragon flowers (it *is* a member of the Snapdragon Family), the lower lip prominently marked with a series of reddish-brown spots. It is a much more robust plant than its smaller relative, Chickweed Monkey Flower, growing in favorable habitats up to 2½' tall and with flowers reaching a length to make a cultivated garden plant envious – up to 2". In these upper meadows however, these Monkey Flowers are usually only 1-2' tall, with a floral tube of about 1" in length. These plants commonly hang off the wet slopes to the right of the trail, although in the two larger meadows described below, they are also found growing in flatter areas as well, although always associated with dampness. The roots of this (and many other) Monkey Flowers are not susceptible to a root fungus which can thwart the growth of many other species in such damp soil. As the steams and boggy areas dry up later in the spring, Common Monkey Flower withers as well. *Mimulus* means "clown" or "little actor", and *guttatus* means "spotted". While I can see the spots, it's still difficult to imagine clowns or monkeys when viewing this gorgeous flower; a more powerful imagination is possibly in order.

The third opening, the "upper" hanging meadow (**Point 12**) is actually at about the same elevation as the lower, 2,250'. Beginning with Glacier Lilies as you enter the opening, especially on the right, the hiker can find nice examples of Prairie Star on the left and Chocolate Lilies at the far end of the meadow on the right. Indeed, you may feel you're at a tennis match with all the head-turning! Several new flowers are now out in this opening that deserve mention. The first is actually a vine, although one that most often trails on the open ground rather than climbing trees. Big Root (*Marah oreganus*), also called Manroot or Oregon Cucumber, is indeed a member of the Cucumber Family, though I wouldn't advise eating the roots or fruits: The Latin name of the genus means "bitter". Here, this plant's stems trail along the ground for 3-5' down the left side of the trail in the middle of the meadow, displaying white tubes of flowers along the way which flare out at the ends into a four- to six-petalled star. Big Root's stems can also be found in moist farm fields at low elevations, often climbing up other vegetation in thickets or even up fence posts.

A more colorful bloom can be found in glorious profusion at the stream after the middle of this meadow, Few-flowered Shooting Star (*Dodecatheon pulchellum*), also called, more appropriately, Pretty Shooting Star. Its masses of magenta flowers are ornamented with five swept-back magenta petals united at the base in a yellow collar bisected by a purple wavy line, all ending in a tapered pointed column containing five fused stamens (the male parts) surrounding the pistil (the female part). The entire effect is of a pink comet roaring through the heavens or of a colorful bird about to take flight. No outdoor traveler can forget the sight of massed Shooting Stars here, (or in the lower meadows of the Starvation Ridge Trail). Gardeners and indoor plant enthusiasts will note the similarity of all species of *Dodecatheon* (which means "twelve gods") to the closely related cyclamens of supermarkets everywhere.

While the Latin names of plant species and genera appear difficult to English speakers, there is an important reason for their use. Common names are too frequently colloquial and numerous to satisfy the demands of a scientific nomenclature. However, there is more to a plant's name than meets the casual eye. By grouping species, genera, and families together, we can learn more about their relationships and hence, their evolutionary descent. Although Linnaeus is generally credited with creating the binomial system used for the past three hundred years, as is usual in scientific work, others had preceded him and he drew on their work. In the late 1600s, the Englishman John Ray published the first scientific study of botany. Prior to that time, plants were mainly characterized by their morphologic similarities or their value as sources of food or medicine. That plants reproduce sexually was only generally accepted by the early 1700s. Linnaeus' contribution, that species could best be classified by their sexual parts rather than by the shapes of their leaves or flowers, formed the basis of the manner in which we think about plants these days. This binomial system, although constantly being updated by DNA analyses, has survived relatively intact over three centuries. Nonetheless, present-day research has led to some reclassifications in botany and to more definitive, and sometimes surprising, revisions to our understanding of the mysteries of plant relationships.

Big Root

Proceeding along the mostly-level part of the trail between meadows, there are few flowers in bloom. However, you will next come to the lower large hanging meadow (**Point 13** – actually at the same elevation as the "upper" one), where familiar friends ornament its grassy reaches. Here, Gold Stars carpet areas of the opening while Glacier Lilies, Common Monkey Flower, and two batches of Chickweed Monkey Flower on the left all add their yellow charm. The pink three-lobed petals of Small-flowered Prairie Star atop 6-12" tall spindly stems can be found scattered on the left of the trail (downhill) here as well. One new flower appears this month in the wet areas especially on the right, Western Saxifrage. Neither showy nor occurring *en masse*, as some of the other common species in this meadow do, it is nonetheless a good-to-know plant frequently found growing off moist banks and cliffs in the Gorge at low to mid-elevations. From a base of egg shaped plump toothed leaves, its 4-10" tall thick stem is topped by a raceme (a cluster of blooms sticking out separately at the top of an unbranched stem) of small white five-petalled flowers.

Saxifrage means "rock-breaker" and, indeed, many species in this genus, and in the entire *Saxifragaceae* Family do grow off stony substrates. At times in the Gorge, such as on the rocks below Multnomah Falls and at the mossy cliffs next to the Eagle Creek Trail, Western Saxifrage can be found hanging bravely from the wet rock faces. Here, however, in these grassy openings on the Ruckel Creek Trail, it is mainly found in moist drainages and not on rocky substrates. One other member of this family, Northwestern, or Grassland, Saxifrage, can be found in this meadow as well, also frequenting the wetter areas. This species shuns rocks altogether, preferring damp grassy slopes. It also has a cluster of tiny white flowers atop 1' tall stems but it can be distinguished from Western Saxifrage by its basal rosette of leaves which have no, or very minute, teeth ("entire") and a more compact, conical inflorescence. Like good siblings, these two saxifrages often cluster together.

Descending slightly, the trail next enters a fifth meadow at 2,150', this one smaller but no less attractive. It displays some of the same blooms as the first several meadows, including Glacier Lilies and Chickweed Monkey Flowers. Also, on the left of the trail in its usual rocky habitat, Martindale's Desert Parsley can be found. This low Carrot Family species often spreads its lacy blue-green leaves and tiny sprays of yellow flowers along the stony soil lining trails at low to mid-elevations in the Gorge. The next, and last, fairly large opening, only a few minutes' walk down the trail, at 2,100', offers a last opportunity to see Common Monkey Flower, at least on the Ruckel Creek Trail; take some time to admire its intricate architecture, shaped by evolution to enable pollination by long-tongued bees and moths. Small-flowered Blue-eyed Mary, Harsh Paintbrush, and a different Desert Parsley, the Nine-leaf species, inhabit this lovely meadow. This last is taller than Martindale's, with thin grass-like leaves very different from the ferny foliage draping most other Carrot Family plants. You won't be wasting your time lingering here as a slower pace and some downhill rest stops not only help the knees but enable you to savor these meadows, backlit by the afternoon sun like some luminous oil on canvas but seen in all its three-dimensional beauty. These are among the best of the best in the Columbia Gorge.

There are some smaller openings between 2,050' and 1,900', the coda's to the symphonies above, where the last of the Gold Stars this season can be found, along with some Martindale's Desert Parsleys and Small-flowered Blue-eyed Mary's, and, in some years, Harsh Paintbrush. In the woods down to the lookout at **Point 14**, a few Oaks Toothworts are still in bloom, especially on the slopes to the right of the trail. At 1,900' just as you leave a small opening, watch for the spectacular deep reddish-pink blossoms adorning the shrub of Red-flowering Currant. At the lookout, at 1,850', head out on the short right-hand branch trail to the overlook to catch a glimpse of deep purple Kittentails. These beauties must crave views over the river as they are most frequently found on north-facing slopes in semi-open woods and clearings.

Joining them are excellent examples of Western Saxifrage, here in a drier environment than its usual quarters, and a new and, for the Gorge, rather unusual species with only a Latin name, Smooth Douglasia. A few plants are either inconspicuous enough, or found in such difficult-to-reach habitats, that they have not attracted common names. It's a pity for this low plant with beautiful rose-colored flowers can be found often enough on high-elevation ridges and clinging to steep rock faces in the North Cascades and Olympics, but with a disjunctive population here in the Gorge (also see it on Table Mountain). In fact, a number of species are found at low to mid-elevations in the Gorge which are more common in the higher mountains of the Northwest. It may be that the more severe weather, greater amounts of moisture, and mountain-like terrain here mimics in some manner the alpine conditions further north.

As you descend the switchbacks between **Points 14** and **15** (the Indian pits), woods prevail. Here, throughout the month, Fairy Slipper, or Calypso Orchid, can often be found hiding shyly in deep forest on either side of the trail. Specimens of great beauty are especially prominent between 1,800' and 1,450', springing up and thriving off the plentiful mosses hereabouts. (Actually, most of these "mosses" are actually *Lycopodiums* or Club Mosses.) There is a healthy patch of these jewels of the forest at the left switchback at 1,550'. Trilliums litter the more open sections of the forest, many turning burgundy by later in the month. Large-leaf Sandworts send up tiny white five- pointed blooms in the deep shade between 1,700' and the pits, perhaps seeking relief from the sun. Oaks Toothwort continues to display its small four-petalled lavender-to-pink blossoms on hillsides along the trail, particularly from 1,550' down to the pits. The colorful flowers of Red-flowering Current are hard to miss in the all-too-brief openings at 1,550' and 1,450'. Entering the large rocky opening at 1,400' (with tricky footing on the foot-sore talus-strewn trail), fine examples of Harsh Indian Paintbrush can be found on your right and left. Further along, a single Chickweed Monkey Flower is growing on the right alongside the trail at 1,450', accompanied by Martindale's Desert Parsley and Small-flowered Blue-eyed Mary.

Descending the frequent switchbacks between here and the pits, mostly in the woods, a number of Oaks Toothwort continue to bloom, along with a goodly number of Fairy Slippers, particularly notable on a small plateau at 1,200', and a few purple-turning Trillium, especially in a clump at 1,100'. Also at 1,225', note a last purple Kittentail at a switchback on the north slope. In these woods, Cascade Oregon Grape is also just beginning to bloom. At the Indian pits (**Point 15**), 875' in elevation, be sure to concentrate on the tricky rock-strewn trail as well as appreciate the flowers. Here the familiar yellows of Martindale's and Nine-leaf Desert Parsleys mix harmoniously with the blues of Small-flowered Blue-eyed Mary and the red of Harsh Paintbrush. In contrast to the upper meadows, however, several new species are found here, especially late in April. The white blooms of the shrub Serviceberry sparkle in the distance off to the left beyond the moss-covered rocks while Chocolate Lilies hang their yellow-mottled brownish heads to the left and right of the trail just past the mid-point of the opening, not too much beyond the pits themselves. A single specimen of the yellow American Wintercress spreads its raceme of yellow four-petalled flowers on your right near the end of this rocky opening.

Less conspicuous species also should not be overlooked here. Two of these belong to the Phlox Family: The pretty pink Varied-leaf Collomia inhabits both sides of the trail as you enter the opening; and the well-named Midget Phlox (*Microsteris gracilis*), a tiny plant up to only about 8" tall, can be found growing right on the rocky soil of the trail at the beginning of the opening. Its small hairy leaves seem to feather out in all directions, like some old man's unruly hairdo, but its nanosized flowers beg a closer look. Unlike most of its bigger cousins, this phlox has beautiful pink flowers with five cleft petals often suffused with a deeper pink at their bases – a colorful micro-cosmos on the head of a tiny flower. This and

Showy Phlox are our only phloxes with cleft petals. The real show however in this opening comes from the large and mysteriously-fragrant Mustard Family plant, Rough Wallflower. Its bright yellow-to-orange four-petalled blossoms are often a garden-like ½ -1" in diameter and stand 1½ -2½' tall. Masses of these blooms can be found by the third to fourth week in April on the right side of the trail beginning just past the pits. I promise you will not soon forget their spicy-sweet odor.

Between **Points 15** and **16** (the power line, here at 490') you will mostly be in open forest. Much Oaks Toothwort is encountered, along with the occasional burgundy Trillium, yellow Evergreen Violet (usually growing off banks of mossy substrates), and the ubiquitous but pretty weed, Robert Geranium, especially common shouldering the rocky stretches of the trail between 650' and 600'. Meadowrue and Fairy Slipper are abundant in the woods just above the power line, from 700' to 500'. Little-leaf Candy Flower thrives off the mossy rocks in small openings between 800' and 500'. The power line itself sports many of the same flowers as in March, including Cascade and Shining Oregon Grapes, Woods Strawberry on the right, several Chocolate Lilies, the Nevada Pea on both sides, and the omnipresent Dandelion.

As you travel down the seemingly endless switchbacks below the power line you are accompanied by a host of plants to ease your fatigue. Oaks Toothwort, Trillium (turning blood-purple), Evergreen Violet, Meadowrue, Fairy Slipper, and Hooker's Fairy Bells are all plentiful enough for recognition by most amateur botanists. Heading down the switchbacks within earshot of the creek, some Trillium and Evergreen Violets are still out, along with a few Fairy Slippers still hanging on. Oaks Toothwort is fading here, while Plume-flowered Solomon's Seal is just beginning to bloom late in the month and Robert Geranium seems to be everywhere, especially down the trail after the right turn beside the stream. Just as you hit the paved roadway, look to your right to see several specimens of Scouler's Valerian. By this time, you may want to test its putative sedative value, but you really should not pick the flowers. At any rate, it's the European Valerian that has been used for centuries as a calming agent. We're unsure whether our several American species carry any medicinal properties.

Along the paved bike path as you head back to the trail sign, you can easily spot the huge leaves and fuzzy flowers of Colt's Foot growing on top of and beneath the rock retaining wall on your left, along with the pink flowers of Robert Geranium spilling over this same wall, and even making use of crevices in the wall itself – quite the opportunist! Early in the month Indian Plum (also called Osoberry) can still be found in bloom on your left as you make the right turn to head back to your cars. Robert Geranium accompanies you to the parking area off I-84, while Dandelion and Purple Dead Nettle see you into your vehicle, safe and sound after enjoying this premier loop in the Columbia Gorge.

May

Flowers bloom in continuous and glorious profusion during this bountiful month and a trip up Ruckel Ridge and down Ruckel Creek continues to be a May tradition among Gorge lovers everywhere. Heading down from the plateau at **Point 9**, a few wine-colored Trilliums bidding goodbye to early spring can be seen on your left at about 3,350' to 3,200' in open woods. As you approach the talus field at 3,150' (**Point 10**), look right and left to see clumps of Bear-grass with perhaps some stalks just beginning to flower late in the month. If so, breathe the perfumed air to lighten your steps down this steep trail. Burgundy- tinted Trilliums inhabit the woods below this point early in May down to about 2,300', and are particularly abundant at approximately 2,875'. Even a few Fairy Slipper Orchids are still out on your left at 2,950', while on your right between 2,900' and 2,500', the deep blue of Oregon Anemones graces the deeper woods. Salal also makes a high-elevation appearance both above and below **Point 10**, especially on your right.

The small stream areas between 2,800' (**Point 11**) and 2,680' are particularly mucky at this time of year but this is partly compensated by an array of four plants now in bloom. The moisture-loving yellow Stream Violet is joined here by its frequent companions, Meadowrue and Candy Flower. Here, persisting into the early part of the month, the familiar Bleeding Heart also enjoys this moist environment, particularly on the left side of the trail. However, a new, and fairly unusual Carrot Family species joins them in this location, one not that commonly found in abundance this far west, the Fern-leaf Desert Parsley (*Lomatium dissectum*).

Fern-leaf Desert Parsley

Before you say "Not another desert parsley to learn", let me assure you that this *Lomatium* is fairly easy to distinguish from the more common Martindale's and Nine-leaf species found in open areas throughout the Gorge. For one thing, it's a much larger bushier plant, growing here up to 4' tall. Found on the left side of the trail by the second wet area at 2,700', it gets its name from the finely-dissected fern-like leaves that grace its thick hollow stem. Its flowers can be yellow or purplish-brown, the latter giving rise to another common name, "Chocolate Tips". Our plants here are mostly purple and consist of bunches of tiny flowers in compact balls on stem-lets (technically, "pedicels"). "Umbels", umbrella-like inflorescences from which branch many smaller "umbellets" are characteristic of the Carrot Family, called both the *Umbelliferae* and the *Apiaceae*, as are lacy leaves. Although this desert parsley is often found further east, it definitely does *not* live in the desert. As so often happens in taxonomy, the first-named species of *Lomatium* was a dry-land plant, but subsequently-discovered members of this genus have been found thriving in a variety of other habitats as well.

> Several of our Northwest species appear in crowds because they extend underground stems in all directions, with flowers popping up from these extensions *en masse*. This is true of our common Vanilla Leaf and the Inside-out Flower. But some other massed plants, such as Trillium and Bleeding Heart, lack this underground root system. How do they flourish in such localized profusion? One answer is the manner in which they are dispersed. Trillium and Bleeding Heart seeds have been shaped by evolution to attract ants, whose sweet tooth draws them inexorably to an appendage, called an eliasome, on these seeds. The eliasome is composed of a delectable oily and sugary substance. The insects, driven by an admirable parental instinct, carry off the entire seed to their hidden nests, where they detach the eliasome and feed it to their young larvae (probably snitching a bite for themselves as well). To the ants, the rest of the seeds themselves are useless, so they deposit them in a scattering at the nest's entrance. Hence, come spring, many of these seeds germinate and bring forth the profusion of Trillums and Bleeding Hearts we enjoy almost as much as the ants appreciate their sweet gift. In this manner, the plants can colonize new areas, while the ants can nourish their young, another example of cooperation in nature our own species might wish to emulate.

The trail below these muddy areas becomes less steep, all the better to enjoy the Star-flowered Solomon's Seals just emerging after mid-May from their long winter's sleep; these sparkle in the dark woods between 2,600' and 2,300'. Fringe Cup also flowers after the middle of the month, particularly at several additional moist areas between 2,500' and 2,300'. The intriguing starburst blooms of the Inside-out Flower appear at this time as well between 2,600' and 2,100' on either side of the trail, along with the ever-present Vanilla Leaf, at times forming a continuous and lush understory beneath the tall and dense Douglas Firs and Western Hemlocks.

The trail from 2,500' to the meadows levels out some before opening up to the first of six openings that glorify the Ruckel Creek route. The first meadow, a small but colorful one at 2,450', is still ablaze with many of April's blooms, including Small-flowered Blue-eyed Mary, Chickweed Monkey Flower, Harsh Paintbrush, and Few-flowered Shooting Star. A few Gold Stars may persist early in the month as well, but are usually gone by the 15th. A new bloom, however, colors this meadow pink at this time of year, Rosy Plectritis, with a ball of tiny rose-colored flowers atop 1' tall stems. Be sure to glance up to the right as you pass from this first meadow to the second one in a tiny opening mid-way. Here, just above the trail, a magical duet of violet Shooting Stars and deep blue Menzies' Delphiniums play out their colorful song in a hanging meadow, a prelude to the flowery melodies to come.

Delphiniums, perhaps better known as larkspurs, are frequent in the Gorge, which offers a home to five species, many quite difficult to distinguish. All share the same characteristics of garden and cut-flower larkspurs: five deep-blue sepals which are commonly mistaken for petals, the upper one with a long backwards-pointing spur, and pale blue to white less-conspicuous inner petals in the center. The spur contains nectar, a treasure reserved for insects and hummingbirds with the longest of tongues and beaks. The leaves of all larkspurs are often deeply cleft in a palmate pattern — that is, as the diverging fingers of a palm, as opposed to pinnately cleft leaves, which branch off their stem like feathers off a common axis. The larkspur so beautifully displayed here and in the large hanging meadows to come is Cliff (Menzies') Larkspur (*Delphinium menziesii*), the most frequent species west of the Crest. A more eastern species, Upland Larkspur (*D. nuttallianum*), and the less- frequently encountered June-flowering Nuttall's Larkspur (*D. nuttallii*) are usually found east of the Crest in drier grassy habitats. We have already met the giant of the family, Tall Larkspur (*D. trollifoiliium*), on the Angel's Rest Trail. I know this gets confusing, but remember that Nature and evolution did not create plants for people to easily remember; fortunately, most of the larkspurs you will meet between Angel's Rest and Starvation Ridge will be the friendly Cliff variety. These beautifully-shaped and deeply-colored species are all said to be poisonous to livestock and at least toxic to humans so don't eat the flowers.

Cliff Larkspur

Upland Larkspur

Tall Larkspur

Resuming your down- and up-hill trek, you will soon encounter the first hanging meadow, at 2,250', **Point 12**. Many of the same flowers enjoyed in the upper meadows are happily repeated here, but the

Few-flowered Shooting Stars at the stream near the middle of the opening are astoundingly plentiful. Also here, especially early in the month, are some Small-flowered Blue-eyed Mary's (they also appear earlier in this meadow as well) and several Chocolate Lilies after the mid-point of the opening on the left. Several members of this colorful menagerie of plants appear in May as well; these include the yellow Nine-leaf Desert Parsley and, late in the month, the yellow daisy-like flowers of the 2' tall Western Groundsel (Senecio) at the far end of the meadow uphill on your right.

Point 13 is the lower hanging meadow, also at 2,250'. Here, both Western and Northwestern (Grassland) Saxifrage continue to bloom from late March through mid-May, especially on the banks and in moist areas on your right. On the left, the intricate three-cleft pink petals of Prairie Star mix nicely with the yellow of the remaining Glacier Lilies, still blooming until mid-month. Both Monkey Flowers, Chickweed and Common, are abundant on the right side of the trail, the latter especially hanging off seeping banks past the midpoint of the meadow. Larkspur and Large-flowered Blue-eyed Mary generously contribute varying shades of blue, the latter down-slope on the left, while further along, Rosy Plectritis and Shooting Stars add pink and violet to this magical joining of hues and textures. **Point 13** is a perfect point to rest and admire the botanical diversity of mid-elevation meadows in the Gorge.

Not far beyond this large opening, you will encounter two smaller grassy meadows at 2,150', then 2,100'. These continue the May floral display, with Nine-leaf and Martindale's Desert Parsleys, Harsh Paintbrush, Small-flowered Blue-eyed Mary, Chickweed Monkey Flower, and the small white, yellow-eyed Common Cryptantha, all found in both openings. The lower meadow adds large specimens of Common Monkey Flower as well. Tiny openings at 2,000' then 1,950' display Harsh Paintbrush and Field Chickweed.

You now enter open woods where the steep descent down the switchbacks from 2,100' down to 1,400' is accompanied by Fringe Cup; Large-leaf Sandwort, especially in denser forest clustered around the bases of large Douglas Firs; masses of Cascade Oregon Grape, many in bloom between 2,000' and 1,500'; Evergreen Violets, particularly off mossy banks on the right at 1,900'; Sweet Cicely, at least early in the month; and, also early, Fairy Slippers, especially at 1,750', 1,600' and 1,350', more on the right than the left of the trail. A few Oaks Toothwort, with their pink-violet four-petalled blooms, hang on through early May, especially just below the last meadow, at around 2,000' on your right; they can also be found lower down as well — look for several on the right just before you enter the rocky opening at 1,500'. The broad, papery-white 1" to 1½" blooms of the Columbia Wind Flower glisten in the dark forest here and there, particularly on the ridges at 1,970' and 1,850'. Also preferring ridges is the 1' to 1½' tall Slender (Racemose) Pussytoes; you can find its fuzzy dull-gray flower heads on these same ridges, as well as at the overlook at 1,850' (**Point 14**), and again on a ridge at 1,600'. These really do look like the upturned paw of some feline. The overlook, at 1,850', with its fine view of snowy Mt. Adams to the northeast, also displays a nice collection of Broad-leaf Stonecrops, with yellow starry flowers and jade-green leaves plastered against the stony walls to your right.

The woods below this point, also replete with steep switchbacks, contain the usual Vanilla Leaf and Cascade Oregon Grape, but at 1,650' on the right, look for a nice specimen of Spotted Coral Root, a saprophyte thriving sufficiently off the humus and fungi of the forest soil as to have no need for direct sunlight. Its 1' tall reddish-purple stems and leaf remnants, along with its orchid-shaped flowers with a white maroon-spotted lip, bear not the slightest trace of green chlorophyll. At 1,450', another new "plant", really a tree but well-deserving of mention here, can be seen through the open forest and downhill on you right as you enter the rocky opening at 1,450'. (Another can be found just below this opening as well.) This is the familiar and well-loved Pacific Dogwood, a 15-50' tree found throughout the Gorge in springtime. As with its low-lying cousin, Bunchberry, its actual flowers are rather inconspicuous but its showy white bracts twinkle brightly in the low to mid-elevation forests here. Petals are used to attract pollinating insects, but in this case, modified leaves just under the flowers, called bracts, usually green but here white, do the job. A similar example of such botanical subterfuge occurs in the paintbrushes, whose red bracts often hide the actual flowers.

Native Americans used the wood of the Pacific Dogwood for bows and arrows, and may have contributed to its common name, claiming that the berries were unfit even for dogs. I have not tested that assertion. These Dogwoods are not as colorful as the many pink garden varieties (stemming from the Eastern Dogwood) so common from March through May in our cities, but still they lend a bright air to our lowland forests and possess a trait not found in their cultivated cousins: They often re-bloom in the fall.

The talus-strewn openings at 1,550' and 1,450' boast a number of species throughout the month of May, including the yellow Martindale's and Nine-leaf Desert Parsleys; Harsh Paintbrush, generally on your right; several Chocolate Lilies up until midmonth; Chickweed Monkey Flower; Small-flowered Blue-eyed Mary; Robert Geranium; and right at the start of the first opening, Miner's-lettuce, with its tiny white-to-pink flowers atop a stem which pierces the single stem leaf.

Once back into the forest later in the month, Broad-leaf Arnica begins to show its yellow daisy blooms down from about 1,400' to the Indian pits (**Point 15**) at 875', providing welcome bits of color to the darker woods. On the rocks just before the pits, at about 1,050', Small-flowered Nemophila offers a bit of light blue above its nicely formed club-like leaves. At the pits, the two familiar Desert Parsleys again put in an appearance, accompanied by bountiful specimens of the Serviceberry shrub across the mossy talus field to your left. The yellow form of Western Groundsel is on your right, while Miner's-lettuce appears on both sides of the trail and the four-petalled Wallflower continues its odiferous presence on your right as you chug slightly uphill and leave the meadow. Before you do, note the two specimens of the first penstemon to bloom in the Gorge – Fine-tooth Penstemon, its blue-to-purple tubes in a vertical cluster at the summit of its 2' tall stems.

The forest between 900' and 500' is replete with Plume- and Star-flowered Solomon's Seals, Hooker's Fairy Bells, the Inside-out Flower, Arnica's, Meadowrues, Small-flowered Nemophilias, Sweet Cicely's and a multitude of Star Flowers. At the power line (**Point 16**, 490'), white Woods Strawberry flowers trail along the hardened ground, along with several Dandelions and profligate masses of the imported yellow-flowered shrub, Scotch Broom. The woods beneath the power line are also filled with Star Flowers, Solomon's Seals (mostly the Plume-flowered species), Vanilla Leaf, Fairy Bells, Alumroot, Arnica's, Inside-out Flowers, and, on the level areas (there are at least a few), some Salal. As you finally reach the right turn by Ruckel Creek and proceed down the trail besides the stream, you can wonder why this trail is named after a creek it never reaches until this point but appreciate the pink Robert Geraniums and white fluffy-topped Scouler's Valerians as you reach the road, particularly on your right.

On the paved bikeway, more Geraniums are growing beneath and directly out of the constructed rock wall to your left, industriously making use of tiny patches of soil in the chinks of this barricade, built to prevent even more land from cascading down over the pavement. Also making good use of this wall are some Broad-leaf Stonecrops, these rock-lovers seeming to plaster their leaves and yellow flowers right off the stones. Further on, Oxeye Daisy is just out on your right, along with much Scotch Broom. A lonely Large-leaf Avens displays its yellow five-petalled flowers about 50 yards before reaching the wall before the Freeway, as if to bid a May goodbye to the by-now exhausted hiker, while more weeds greet you at your vehicle, including Dandelion and the similar-looking

Hairy Cat's Ear. Remember not to disparage these weeds too severely, especially after experiencing the gorgeous native floral displays above. After all, they probably did not choose to invade our country, even though they apparently thrive in this plant democracy.

In reflecting back on the bounty of blooms experienced on this Ruckel Ridge-Ruckel Creek circuit, most outdoor travelers will be happy that we have a Columbia Gorge, with its unique combination of microclimate and geography so beneficial to our flowering spring plants yet so close to the urban areas where most hikers live. No finer loop will be found hereabouts.

June

The month of June does not disappoint the flower seeker on either the Ruckel Ridge or Ruckel Creek Trails as botanical magic continues throughout this month. Heading down from the plateau, almost immediately you encounter one of the prototypical plants of all western American mountains, Bear-grass (*Xerophyllum tenax*). It grows here and on down to below the talus field (**Point 10**) at 3,150', where it is particularly abundant. Springing up from a clump of wiry grass-like leaves, 3-4' tall stems are topped by an unmistakable 6 to 10" plume of tiny white flowers. For some reason, most authors have failed to note its lovely fragrance, a sweet odor that perfumes the forest when Bear-grass is in bloom. Do not be too disappointed, however, if most of the wiry clumps of basal leaves fail to produce flowering stalks. Bear-grass blooms irregularly, especially in these forest habitats. Have patience – every five to seven years or so members of this species decide, somehow, to all bloom at once. Do they all take a vote early that spring? More likely, these inhabitants of pumice, talus and hardened soil, like their close relative, the Century Plant, store up water and nutrients for a number of years before being able to afford the energy to bloom. When they do, be there – the sight and smell are unforgettable!

Bear-grass

A favorite of northwest photographers, particularly in association with the native Rhododendrons around Mt. Hood, Bear-grass belongs to the Lily Family. A close inspection reveals that each small flower in the plume is composed of six tepals and this, in common with the grass-like leaves, are indicative of monocots such as the Lily, Iris and Orchid Families, thought to be distantly related to the grasses and to be of more ancient origin than the dicots, which comprise the majority of flowering plants. Bear-grass leaves are so strong (*tenax* means "tough") that Native Americans made all sorts of baskets from them, many of these watertight.

The next bloom to greet the Ruckel Creek explorer is the yellow Evergreen Violet, several of which live in a clump on the right at 3,400'. One can also still find the delightful blue Oregon Anemone from 3,050' to 2,750' on the right. Proceeding down the steep trail to **Point 10**, Bear-grass is frequent. Although the streams at **Point 11**, 2,800' to 2,680', are by now drying up, there is still sufficient moisture to support a veritable water garden of moisture-loving plants. One first encounters Stream (Wood) Violets, especially early in June on the right, along with Candy Flowers on either side of the trail. False Lily-of-the-Valley, Meadowrue (mostly female), Bleeding Heart, and Youth-on-age are common, along with the lacy-leaved Columbine, with its crimson and yellow swept-back tubular flowers. Between the first and second wet area, find two specimens of the relatively uncommon American Wintercress on your left, with a raceme of small yellow four-petalled flowers atop 1½' stems. Several new flowers appear here in June, including two at either end of the size spectrum: the 3-5' tall Cow Parsnip, with its monstrous cleft leaves big enough to hold a small backpack and its huge compound umbel of white flowers, at times a foot wide; and the foot-tall Little (Small-flowered) Buttercup, its shiny yellow nano-petals barely noticeable to any but the most careful observer. Here, several specimens can be found on your left just past the second wet area.

As you progress down the now less-steep trail, the woods are punctuated with blooms, plentiful, though less colorful than in more open or moister areas. Fringe Cup is present in abundance, while Star-flowered Solomon's Seal carpets the forest in fragments of bright green and twinkling white. Large-leaf Sandwort inhabits the darker wooded regions and a few Columbia Wind Flowers wave their large white petals (really sepals) in the forest breezes. The lovely Twin Flower (is Twin Flower ever preceded by any adjective other than "lovely"?) trails over the ground in drier spots on both sides of the path between 2,750' and 2,500', while the Inside-out Flower makes an occasional appearance. Star Flower is also a frequent companion in these woods. Bits of color are not wanting entirely: Fringe Cups are turning pink with age here and there and, even more delightfully, Little Wild (Baldhip) Rose brightens the forest deeps, particularly on your right at 2,640' and 2,475'.

Are you one of those hikers bored by unbreached forest who can't wait to come to the meadows? Well, even though you had your chance earlier in the spring, these openings on the Ruckel Creek Trail are still stupendous throughout June and into the early summer months. These famous openings hold many discoveries for the June traveler anxious to hone botanical skills. The first small meadow, at about 2,500', now filled with taller grass more straw-colored than green, carries many holdovers from May: the fluffy pink balls of Rosy Plectritis and many yellow and maroon-spotted Common Monkey Flowers (at least early in the month), these last covering swaths of seeping grassy walls to your right. The second opening, also a small meadow, at 2,375', displays the same two blooms but adds Big Root, better known as the Oregon Cucumber, trailing along the ground on your left, with four to six small white petals here and there along its ground-hugging stems.

These upper meadows, at around 2,400 – 2,500' display a foreshadowing of what's to come: All these openings described below display an astonishing flower show of blazing blue, purple, pink, yellow and white. Blue-eyed Grass, 1-2' tall, covers swaths of the meadows left and right, intermixed with deep purple Ookow (Ball-head Cluster Lily – *Brodiaea congesta*). Its 2' tall stems are topped by a cluster of six to ten purple flowers, each with six tepals, six yellowish stamens, and several grass-like leaves, thus marking it as a member of the Lily Family. Ball-head Cluster Lily is more commonly referred to as Ookow, from the Native American language, although no one seems to know what that word really means. Herald-of-summer (Farewell-to-Spring), appears frequently in all these openings, with its pink petals showing off splotches of red at each petal's center.

Ball-head Cluster Lily

Happily, Cliff (Menzies') Delphiniums continue to brighten the twin larger meadows with their blue-purple spurred flowers. Yet a new plant, Heart-leaf Buckwheat (*Eriogonum compositum*), lives up to its name here, with blue-green heart-shaped leaves and buckwheat-shaped masses of dirty-white clusters of flowers. This plant is also found on high ridges on Table Mountain and, lower down, on rocks above Highway 14 on the Washington side of the Gorge just east of Stevenson. It can be seen here on your left downhill in mid-meadow and is part of the large Buckwheat Family and thus a relative of the cultivated buckwheats and rhubarb. Unless you are starving (unlikely in the Gorge), don't eat the stems.

Heart-leaf Buckwheat

The first large "hanging" meadow (**Point 12**), slightly higher than the small one preceding it at 2,250', continues to display May's blooms, although in lesser profusion. These include the yellows of Common Monkey Flower, Nine-leaf Desert Parsley and Western Groundsel (Senecio); the blues of Small- and Large-flowered Blue-eyed Mary's, Fine-tooth Penstemon, and of Lupines smiling from the edges of the meadow; the pinks of taper-tip Onion and Rosy Plectritis; the red of Harsh Paintbrush; and the cheerful whites of Big Root, Common Cryptantha and both types of Strawberry found in the Gorge – Woods and Broad-petal or Virginia, Strawberry. Although the five-petalled flowers are essentially the same, Woods Strawberry has yellowish-green leaves with bulges between the veins while the Broad-petal Strawberry has blue-green flat leaves.

Western Saxifrage continues to bloom early in the month near the drying stream in this meadow. This large opening also contains a beautiful clover unusual in its occurrence this far west. This, the Sand Clover (*Trifolium tridentatum*), is small at just 6" high, but its intricate head of 10 to 20 individual pea-flowers bears close inspection: Each of the tiny flowers is tri-colored, with a pink "banner" as background and two white "wings" flanking a dark purple "keel". All clovers are members of the Pea Family and contain this striking miniature architecture. This particular clover, though related to those invading your lawn, is a native so do not regard it harshly.

The second large meadow (**Point 13**) also at an elevation of 2,250', harbors many of the same plants as the upper opening. One difference is the persistence of Chocolate Lilies just past the midpoint and near the end of the opening. New here this month are the sky-blue ball-shaped heads of Blue Field Gilia, also known more descriptively as Blue-head or Ball-head Gilia. These beautiful 1' tall plants inhabit the middle reaches of this meadow and thrive amidst the long summer grasses here, on Table Mountain's higher reaches, and on the lower meadows of Starvation Ridge.

Another new plant is in bloom in this meadow, about three quarters of the way through the opening – Sticky Cinquefoil (*Potentilla glandulosa*). It has glands along its 1-2' stem which secrete a sticky substance, although it is uncertain if it can digest any insects unlucky enough to get trapped there, as many catchflies can do. This cinquefoil, a member of the Rose Family, has flowers with five pale-yellow petals and sticky toothed leaves. It can be found in grassy openings along the lower reaches of the Starvation Ridge Trail as well.

Sticky Cinquefoil

The two smaller meadows that follow, at 2,150' and 2,100', contain miniature pastures of Rosy Plectritis; Lupine; Western Groundsel; and, in the last one, Harsh Paintbrush. Blue-eyed Grass and Hyacinth Cluster Lily plus Ookow also continue to brighten these lower meadows and will also persist throughout summer. There are openings as well further down the trail, especially between 2,090' and 1,900' and, although these are quite small, they are also quite welcome. They each proudly display gardens of Lupine, Common Monkey Flower, Harsh Paintbrush, Western Groundsel, Taper-tip Onion, Nine-leaf Desert Parsley, Fine-tooth Penstemon, Field Chickweed, Wooly Sunflower, Spreading Phlox, and masses of Common Cryptantha. The lowest of these openings also contains the balls of bright purple flowers which dazzle many low-elevation meadows in the Gorge, the Ball-head Cluster Lily, more commonly known as Ookow. We have four brodiaeas, also known (erroneously) as hyacinths and (correctly) as cluster lilies, in the Gorge, including the purple Harvest Brodiaea east of Dog Mountain, the white and green-striped Hyacinth Cluster Lily both here and on Dog Mountain, and the beautiful blue-striped Bicolored Cluster Lily found on the Wygant Trail.

> Since plants are classified by their sexual parts, such as the number of stamens and the shape and position of their ovaries, it is good to know how to identify these organs and how they actually function. All these parts are commonly found in the center of a flower. The male parts are collectively named the stamen, and consist of two parts, a thin filament at the summit of which is the seed-bearing anther. The pollen seeds are usually generated lower down, then ascend the filament and depart out of the anther. The female structures are called the pistil, generally a three-parted system consisting of an egg-bearing ovary at the bottom surmounted by a thin style topped by a cleft stigma through which pollen enters, descends the style and fertilizes the eggs. While this over-simplified summary holds for many flowers, many others have played out evolutionary variations on this theme, not all of which can be enumerated here. Two outstanding ones, however, deserve mention here in the Northwest: Aster family flowers adhere to this architecture, but must be observed closely, as they commonly have hundreds of flowers in their single-appearing head, often of two types – disk flowers and ray flowers. A close look, especially with a magnifying glass, reveals that each tiny flower has its own set of stamens and pistils. Penstemons also have typical sexual parts but their stamens protrude far beyond the reach of their pistils in order to avoid self-fertilization, which would prove harmful in restricting the genetic variability needed to promote healthy populations. Many plants have evolved strategies, some mechanical, others seasonal, to avoid the same flower fertilizing itself. For those lusting after more information about botanical sex, check the more detailed X-rated texts in the bibliography.

Although further openings will appear lower down, take a few moments to appreciate the colorful and dazzling floral displays in these meadows, some vast and others confined, because, below they will mostly be missed. Nonetheless, many flowers are in bloom below **Point 13**, especially in the open rocky woods and in the plentiful, though smaller, meadows on the lower reaches of the Ruckel Creek Trail. In the interstices between forest and more open areas between **Points 13** and **15**, Small-flowered Alumroot flourishes, seemingly preferring rocky surroundings. The wooded areas between 2,000' and 1,400' continue to harbor typical shade-loving plants such as Large-leaf Sandwort, Twin Flower (a patch on the right at 1,900' is particularly lovely), yellow Broad-leaf (Mountain) Arnica, Scouler's Valerian, and Plume-flowered Solomon's Seal, at least early in the month for this last. It is not difficult to find Star Flower at these elevations as well, along with the Inside-out Flower and blooming specimens of Little Wild Rose. A nice specimen of Spotted Coral Root is on your left at about 1,925'. On mossy rocks at 1,900', 1,875', 1,700' and 1,450', Little-leaf Candy Flower thrives, seemingly growing out of the rock and moss without the benefit of soil.

The lookout at 1,850' (**Point 14**) is bejeweled by masses of yellow star-shaped Broad-leaf Stonecrop on its basaltic north-facing walls, accompanied by the furry Slender (Racemose) Pussytoes. Here, one can look steeply down the vertiginous cliffs to the river below, across to Table Mountain to the north and, on a clear day, to the snows of Mt. Adams to the northeast. Below, the open forest displays a strangely disjunctive population of Robert Geranium between 1,800' and 1,750'. This European import usually appears much lower but its appearance here serves as testimony to this plant's pretty but invasive nature. A left switchback at 1,600' reveals a prototypical example of the Little Wild Rose. At 1,500', notice a clump of white Columbia Wind Flowers brightening the woods. Adding some color are occasional examples of the narrow endemic, Columbia Gorge Groundsel, an Aster Family species about 1' tall with yellow daisy-like flowers and lobed and toothed roundish leaves that appear almost frilly.

Between **Points 14** and **15** there are several small rock-strewn openings which provide inviting floral islands within this forested sea. At 1,600', then again at 1,200', the orange-yellow Wallflower blends nicely with the blues of Blue Field Gilia and Lupines, the satisfying yellow-on-yellow of Wooly Sunflower, the crimson of Harsh Paintbrush, the purples of Fine-tooth Penstemon and Ookow, the deep orange and maroon of the Tiger Lily and the deep pink of Taper-tip Onion. White is not overlooked: The raggedy petals of Field Chickweed, the yellow-centered blooms of Common Cryptantha and the tiny white daisy flowers of White-flowered Hawkweed are present in abundance as well. At the 1,400' rocky opening find Harsh Paintbrush, diminutive pink Varied-leaf Collomia, Robert Geranium, yellow Nine-leaf Desert Parsley, Lupine, Miner's-lettuce, Small-flowered Blue-eyed Mary, Fine-tooth Penstemon, and the fuzzy Blue Field Gilia. Another small opening, this at 1,100', sports the wooly dull heads of White (Birch-leaf) Spirea and the impressive mottled-orange flowers of the Tiger Lily, both on the left. The woods in between continue to hold healthy specimens of yellow Broad-leaf Arnica, a plant in denial that only white blooms thrive in deep shade.

The mysterious Indian Pits, at **Point 15**, 875', harbor few botanical enigmas. Here, Lupines cheerfully sprawl across the mossy, rocky landscape, along with Varied-leaf Collomia, Heart-leaf Buckwheat, Harsh Paintbrush, Nine-leaf Desert Parsley, Cleavers, and Miner's-lettuce. Aside from clumsy hikers falling into the pits (unlikely), the greatest danger lurking in this stony field comes from the Poison Oak lying in ambush along both sides of the trail. Although there are few secrets at this supposedly holy site, one species, both novel and curious, does appear here this month: the Long-beard Hawkweed (*Hieracium longiberbe*). This Aster Family plant, three to four specimens of which will be found on the right at the end of the opening, rises 1-2' tall and at its summit appear a cluster of yellow heads consisting solely of rays; there are no disk flowers, much as in the common Dandelion. What makes this plant unusual is its extreme hairiness. It looks for all the world as if a more common hawkweed, such as Scouler's, tumbled into a vat of wool or at least grew a shaggy beard. Most botanists believe such a fleecy covering protects a plant from harsh ultraviolet rays or from losing too much water to evaporation. Whatever its purpose, this unusual endemic is singular within its genus, and being different, probably deserves our utmost appreciation.

Long-beard Hawkweed

The woods below this point in June are replete with Star Flowers, the Inside-out Flower, Columbia Windflowers, Little Wild Roses, the yellow Columbia Gorge Groundsel, Broad-leaf Arnica, Robert Geranium, the purple Leafy Pea, and Twin Flower. Several smaller openings between 800' and 600' display the familiar Yarrow and White-flowered Hawkweed. Another choice specimen of Spotted Coral Root is located at 590' of elevation on your left, while the jaunty color of the Orange Honeysuckle vine blazes both on the ground and above in the trees between 600' and 400'. At the power line (**Point 16,** 490'), Oxeye Daisy, Woods Strawberry, Tiger Lily, Lupine, and Miner's-lettuce enliven the blight of electrical poles. Beneath, forest predominates, and here the tired amateur botanist should still be able to recognize the Inside-out Flower, Twin Flower, Himalayan Blackberry (still too early for the fruits), and, on the switchbacks down to the creek and beyond, the omnipresent Robert Geranium. Along the road, the obligatory weeds, Oxeye Daisy and Hairy Cat's-ear, decorate the paved pathway, but several natives appear as well: a few Little Buttercups, with their undersized waxy yellow petals on the right; the ubiquitous shrub, Snowberry; and the yellow Large-leaf Avens in a field, also on your right. At the parking area, a field-course in weeds is being held this month, starring Oxeye Daisy, Common Groundsel, Self-heal, and Hairy Cat's-ear. Spot quizzes, however, will likely not occur.

Overall, you can't go wrong completing this loop in June, although it's a jewel any time of year. Ruckel Ridge/Creek distils the best of the Gorge's floral wonders in these late spring months; no Portland-area outdoor explorer should miss these botanical opportunities.

July and August

Mountain climbers have now left this famous loop to summer hikers while they explore the jagged spires of ranges further north poking holes in the Washington sky. All the better, for the heat of these months does little to dissuade a new crop of blooms from appearing up and down the Ruckel Creek Trail. From the plateau at **Point 9** down to the first rocky gap at **Point 10**, 3,150', Bear-grass rears its tall white plumes, fragrant in the gentle drafts at this elevation. Foam Flower, the archetypal woodland breed, tosses its white spray of dainty flowers throughout these shady thickets but is especially plentiful at 3,400', 3,100', and 2,900', as is Bunchberry at 2,900' on your right, at least in July. Two Heath Family plants, Pipsissewa (Prince's Pine) and One-sided Pyrola (*Pyrola secunda*) dominate the understory down to approximately 2,750'. One-sided Pyrola (pyrolas are also known as wintergreens) grows to just 6" tall with wavy oblong leaves and miniscule bell-shaped greenish flowers, all on one side of the stem. Indeed, *"secund"* means one-sided. Lost amidst the green floor of the forest, not many hikers ever notice this tiny plant, yet it comprises a predominant part of the flora in certain coniferous Northwest woodlands and makes a superb identification triumph for the amateur naturalist.

One-sided Pyrola

In the formerly waterlogged areas at **Point 11**, 2,800' to 2,680', our old friend, Candy Flower, with its pink-striped white petals, cavorts with its moisture-loving companion, the yellow Stream Violet. But by early summer, a new species has emerged that will dwarf the other vegetation in this damp area – Cow Parsnip, found here on the left of the trail. Its Latin name is derived from Hercules and indeed, this plant is as mighty as an ancient Greek god, often exceeding 10' in height, although here it usually reaches around 4-5'. Its umbrella-like clusters of tiny white flowers may span a foot in diameter while its deeply-cleft leaves can reach a length of 1½'. This monster, the largest of our flowering native herbs, may seem like a relic of some Paleolithic tropical forest, but in fact it's quite a prosaic and recurrent resident of stream-sides and other damp domains in the wet west-side thickets of the Gorge. Although some flowering trees and shrubs may exceed Cow Parsnip in height, it remains the size champion of the plants we consider "wildflowers" in the Northwest.

> Aren't herbs the things we put into food to spice things up? Yes, in common usage. But to a botanist, the term *herb* means any flowering plant without woody parts above the ground. In most such plants, the stems are supposed to die back to the ground each year, although this is not always the case as in certain species, such as Wall Lettuce and Pearly Everlasting, remnant stems may persist over the winter. Herbs are to be distinguished from *trees*, which are defined as single-stemmed woody plants greater than 33' in height (who gets up there to measure?), and from *shrubs*, which are woody plants, usually multi-stemmed, and under 33' in height. These definitions often lead to designations that are counterintuitive: While every outdoor traveler would agree that dogwoods and maples are trees and rhododendrons and roses are shrubs, not even skilled amateur naturalists can fathom why the tiny Twin Flower or the equally diminutive Pipsissewa are classified as shrubs while shrub-like "herbs" such as Fairy Bells and certain penstemons, are not. A close adherence to the definitions, however, leads to these necessary distinctions. In their apparent disregard of common sense, perhaps botany and quantum physics share at least a small common bond.

The woods between **Points 11** and **12** contain scattered examples of Foam Flower, here in its trifoliate form (see the Nick Eaton Chapter). Several examples of Rattlesnake Plantain are present, especially at 2,650' and 2,475', both on the left. This rattlesnake doesn't bite; in fact, its plain greenish flowers on a single upright stem must be closely observed to realize that it is a legitimate member of the Orchid Family. Pipsissewa is in abundance between 2,700' and 2,400', while a few high-living examples of the yellow five-starred Wall Lettuce are beginning to bloom by late July. An occasional Broad-leaf Arnica bequeaths more yellow to these woodlands and a smattering of Inside-out Flowers struggle to survive into mid-July.

Between **Points 12** and **13**, all the meadows, grand and small, are adorned by the combinations of deep pink Taper-tip Onion (note the aroma), blue Lupine, red Harsh Paintbrush, yellow daisy-like Wooly Sunflower (principally the lower large hanging meadow), the small-flowered White Hawkweed and the familiar white-topped Yarrow. The woodlands below, from 2,200' to 1,800', display examples of Rattlesnake Plantain, Foam Flower, Wall Lettuce, and some Twin Flowers at 1,900', at least until mid-summer. A fine specimen of Little Pipsissewa, with its waxy pink flowers drooping towards the forest floor, can be seen at 1,950' on your right. The lookout at **Point 14** continues to display the starry yellow flowers and jade-green fleshy leaves of Broad-leaf Stonecrop plastered against the rocky terraces rising in tiers towards the sky. Down the switchbacks in the woods below, Pathfinder (Trail Plant), Foam Flower, and Wall Lettuce predominate. Watch carefully so you can appreciate several nice specimens of Rattlesnake Plantain at 1,550' and 950' on your left.

Nature has provided several small openings for your amusement to break the monotony of the forest; these, at 1,600' and 1,200', are teeming early in the summer with everyone's (well, my) favorite daisy, the Wooly Sunflower, and its less spectacular relative, White-flowered Hawkweed, which often prefers the margins where forest and meadow meet. The open talus field at 1,450' displays pink Varied-leaf Collomia, crimson Harsh Paintbrush, blue Lupine, and white-to-pink Miner's-lettuce. At **Point 15**, 875', the Indian Pits appear, hopefully not filled with trash. Yellow plants dominate this stony landscape, including the umbels of Nine-leaf Desert Parsley, the tall daisy-shaped blooms of Western Groundsel, and the curious wooly stems and Dandelion-like flowers of Long-beard Hawkweed. The small white-to-pink flowers of Miner's-lettuce also thrive here in this rocky opening.

The forests below the pits harbor an occasional Pathfinder (Trail Plant), its tallish stem supporting a brace of tiny dull-white flowers, and a score of Rattlesnake Plantains on your left about 20 yards after the pits. However, the European invader, Wall Lettuce, is now a ruling member of the understory, with its small but bright yellow five-petalled flowers defying the native plants of its new home, especially just above and below the power line, our **Point 16**. A bevy of other introduced plants inhabit the sunny gash created by this artificial breach in the forest, including Oxeye Daisy, Dandelion, the similar Hairy Cat's-ear, and Queen Anne's Lace. Several natives seem to exist in harmony with these foreigners, including Self-heal, Pearly Everlasting, and the elegant orange and maroon-spotted Tiger Lily, examples of which can be found on your left at the towers and just below.

Along the paved road, similar plants appear, especially the ubiquitous Queen Anne's Lace and Robert Geranium. However, a new species can now be recognized in these manufactured lowlands, Common St. John's Wort (*Hypericum perforatum*). Growing 2' tall, this pervasive and widespread weed is not unattractive, with lustrous yellow flowers, fuzzy stamens protruding from their centers, and purple-to-black dots often spotting the margins of its leaves and petals. Unfortunately, it has become a serious pest in fields and pastures. All city dwellers should be familiar with its form as seen from a passing car as, similar to today's adolescents, it seems to thrive in the concrete milieu of shopping mall lots and roadside grassy areas. And, as with our present-day youth, this opportunist may even pop up in the most unexpected and unwelcome places, such as our gardens and the chinks in highway abutments. This type of plant may also be familiar to gardeners because a close *Hypericum* relative carpets front lots in Portland and Seattle with similar bright yellow flowers capped by scores of fluffy stamens. We do have several native *Hypericums* in the mountains of the Northwest, generally inhabitants of bogs and marshes, but these are rarely encountered in the Columbia Gorge.

Picking your way up Ruckel Ridge, then crashing down Ruckel Creek, each jarring step may make you wonder why? But for most mountain travelers, the answers are as typical as they are numerous. The serene forests, the views over the mighty Columbia, the diversity of the floral life, and the sense of accomplishment are reasons enough, even in the swelter of summer. Yet there is more – perhaps the universal need many hikers feel to suffer slightly in order to achieve a goal. Nonetheless, the suffering on this celebrated loop is minor compared to its ample rewards, not the least of which is the opportunity to sample and identify some of the most diverse floral displays in the western Gorge.

September and October

Autumn is a peaceful season in the Gorge, illuminated by the changing colors of the leaves and the slow regression of much of its plant life. Yet, blooms do remain to be appreciated by the late-season explorer on this challenging loop. Stately Bear-grass blossoms perfume the cooling breezes down from the plateau between **Points 9** and **10**; it is especially prominent at the 3,150' talus field (**Point 10**), an agreeable spot to rest or take a lunch break in contrast to the dark and chilly woods on the plateau above. Foam Flower still spots these upper woods with its sprays of tiny white flowers, although by October one sees mostly its multiple seed heads. At **Point 11**, 2,800', and down to 2,680', the streams have dried up by now, though sloppy mud persists throughout the year. Hanging on is the stubborn Candy Flower, examples of which may still be appreciated into November at these moist higher elevations. This species carries great genetic diversity in its blooming times, auguring well for its persistence in a future whose climate is uncertain.

The woods between **Points 11** and **12** continue to hold examples of Foam Flower, Pipsissewa and the yellow Wall Lettuce. For an imported "weed", this last has slowly crept upward in elevation and now can be found in city yards, roadsides, woodlands, and forests from northern California into southern British Columbia and from sea level up to 5,000'. Indeed, this Northern European import has established itself in more diverse habitats in the Western Hemisphere than on its original continent! While not un-pretty, it is, from an ecological point of view, unwelcome as it has displaced a number of native species.

The meadows at **Points 12** and **13** now lack the colors of spring and summer but still have some flowers on display and offer a chance for the interested hiker to identify plants less well-known than the flashier blooms of earlier months. The chief one of interest here, conspicuous in both large hanging meadows and in several of the smaller ones below, is Hall's Goldenweed, an Aster (Daisy) Family plant with scraggly yellow rays and a yellow disk. Typical of the species, the specimens here are massed, with upright to arching stems 1-2½' tall. The flowers are attractive in a way, perhaps mainly because there is scant competition from other plants at this time of year, but the most fascinating characteristic of Hall's Goldenweed (also called Hall's Haplopappus in order to confuse us amateurs) is its longevity. You can find this plant in flower at the end of August, but return in early November and you're just as likely to see it again still blooming happily. Its pollinators are evidently Eskimo bugs.

Common St. John's Wort also continues to display its yellow star-shaped flowers, particularly in the lower hanging meadow (**Point 13**), but, by mid-September, most other species have retired for the winter. One often unnoticed plant, however, should be mentioned: the diminutive Autumn Knotweed (with the unpronounceable Latin name of *Polygonum spergulariaeforme*). This Buckwheat Family relative of urban weeds, such as Lady's Thumb and Dock, grows just 3-6" tall and often sprawls across the dry stony ground it appears to favor. The pink cluster of flowers grows in a spike-like raceme (stalked flowers blooming from the top of a stem) but you'll need to be super-observant to notice these tiny blooms at the middle to the end of the lower hanging meadow on either side of the trail. Never common, this knotweed also populates upper dry meadows and stony ground on Table Mountain's West Ridge Trail.

Autumn Knotweed

The lookout at **Point 14** is now devoid of flowers, although the switchbacks in the woods continue to harbor plentiful examples of Wall Lettuce. In fact, as should be apparent by now, many of the latest blooming species in the Gorge, as elsewhere in the mountain world, are members of the Aster Family, one of the largest families in the Plant Kingdom. (The largest is now thought to be the Orchids, with over 23,000 species.) Perhaps one reason for its relative dominance, besides its habit of mimicking a single large flower by combining disk and ray flowers, is its ability to bloom late in the flowering season, thus avoiding competition with the multitude of spring and early summer bloomers.

The openings beneath the lower hanging meadow, at 1,600' and 1,200', display Pearly Everlasting, living up to its name as it endures the increasingly chilly days, and White-flowered Hawkweed, a 2' tall branching Aster Family plant with small daisy flowers. At the rock-strewn meadows at 850', our **Point 15**, Miners-lettuce and yellow Long-beard Hawkweed may be in bloom through mid-September but little else brightens these openings.

At the power line (**Point 16**, 490'), not much unusual is happening in the fall. Pearly Everlasting puts on its usual show, but be on the lookout for a vibrant midget, Grass Pink; a specimen lies close to the rugged ground on your left just before you pass under the electrical poles. This diminutive species displays hot-pink flowers with white spots dotting its five petals. A relative of carnations in the Pink

Family, it bears little floral resemblance to its more flamboyant cousins, but its spark of color brings a measure of joy to the bleak power line gash. Below, Wall Lettuce continues its invasive ways, while along the paved road, Robert Geranium tumbles down the rock wall to the left while Hairy Cat's-ear and Queen Anne's Lace accompany the weary hiker along the asphalt path.

Common Chicory

At the parking area, one is greeted by the unfortunate drooping yellow-and-black bud-like flowers of Common Groundsel (Senecio), along with much Queen Anne's Lace. One last gleam of color, however, also inhabits this grassy and weedy domain, Common Chicory (*Chicorium intybus*). Standing 3-4' tall, with much-branched stems, this Eurasian import may be considered a weed, but its sky-blue 1" wide flowers are of such a cheerful hue that it attracts pleasant adjectives despite its foreign origins and invasive nature. As many people know, the roots of Chicory have been ground, roasted and employed as a coffee substitute for centuries. Less well-known is this species' close relationship to the common salad plant Belgian Endive. Indeed, its leaves have been used in France for winter salads, and, if it's good enough for the French to eat…well, you know the rest.

Completing the Ruckel Ridge/Ruckel Creek loop in autumn may not prove as arduous a task as earlier in the season because many hikers will be in better shape during these post-summer outings. Nonetheless, this celebrated round-trip is still a formidable undertaking, so be proud of accomplishing this premier loop in the Columbia Gorge. Although the flowers may not be as abundant in the fall, the season still blazes with color and graces the outdoor enthusiast with cooler temperatures and many cloudless days – a winning combination at any time of year.

The Flower Finder: Ruckel Ridge and Ruckel Creek Loop

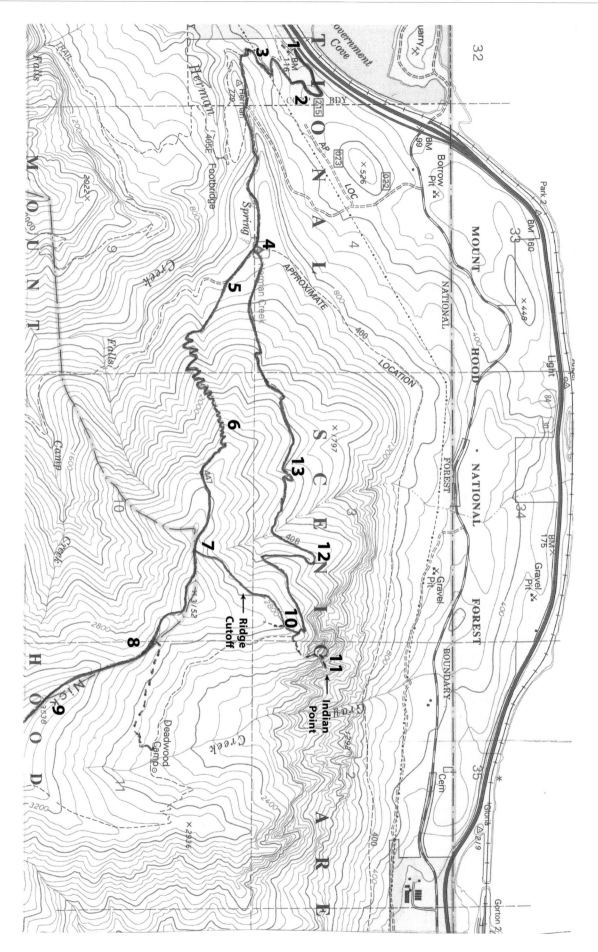

Nick Eaton Ridge/Gorton Creek Loop

The combination of an ascent up Nick Eaton Ridge (Trail #447) and a descent of the Ridge Cutoff (#437) to connect with the Gorton Creek Trail (#408), then a descent of this latter trail, comprises another choice loop in the Columbia River Gorge. Although this loop contains the combination of woodland and meadow environments so unique to the Gorge, the Nick Eaton neighborhood is replete with so many trail systems that a multitude of round-trip loops are possible. You could ascend Nick Eaton, then choose to descend the Deadwood Cutoff (#422), scary sounding but really quite easy, to join the Gorton Creek Trail farther upstream. Another loop takes you up Nick Eaton, then down the west side of the ridge on the steep Casey Creek Trail (#476) to join the Herman Creek Trail (#406) upstream from where you began. Alternately, resolute and hardy hikers may elect to complete the 13-mile loop, continuing along Nick Eaton Ridge to the point where it joins the Gorton Creek Trail and swings back to the north. Here, in the Nick Eaton area, there are no bad choices.

The present description takes you up the Herman Creek Trail (#406), onto the old Herman Creek Road, then past the unused Herman Camp and up Nick Eaton Ridge. We stop arbitrarily at a rocky viewpoint at around 4½ miles in, unless you chose to complete the entire loop to the junction of Trails #447 to #408. Our route then proceeds *back down* the ridge past the Deadwood Cutoff (#422) to the first cutoff you encountered on the way up – the Ridge Cutoff (#437), then down the Gorton Creek Trail to rejoin the Herman Creek Trail at Herman Camp. This loop combines the advantages of views (including some of Mt. Hood), flower gardens, and broad well-graded trails, a blessing for those of you who have already experienced Ruckel Ridge. In addition, another benefit of exploring this area is its position practically on the Pacific crest. Here, the ecosystem benefits from its western exposure to our infamous rain yet is spared exposure to the full fury of these storms. As a result, a mix of plants can be found – not truly dry east-side species, but rather a different proportion of flowers. The deep rich blue of the Oregon Anemone is more frequent here, along with the sky-blue of Blue Field Gilia, as are the vivid pinks of the Taper-tip Onion and the cheery Rosy Plectritis. Thus, whichever loop you choose, and in almost any season, there's much to enjoy on Nick Eaton Ridge.

NOTE: The numbered meadows beginning at about 2,300', are almost contiguous due to the 2017 Eagle Creek fire. Trees remain to separate them but these are much reduced in size and scope, making some of the meadows, particularly numbers 2 through 4, almost contiguous. As the shrubbery will eventually re-populate these in-between woodlands over the next several seasons, they will once again become more distinct. In the meantime, you can correlate the numbered openings with the switchbacks beginning at about 2,200'; almost just after each of six switchbacks is the corresponding meadow.

CAUTION: The short side trip described to Indian Point involves a rough steep path, though one most hikers will find easy to follow. However, the ascent of the loose pile of rocks at its end that marks Indian Point itself should be reserved only for those comfortable with Class 3 rock climbing, and even then, care must be exercised as loose rock, the hallmark of the Cascades, is much in evidence here. Even should you not attempt the climb, be careful with children here as the trail, though level near its end point, deteriorates into a mass of talus. Hand-holding of younger kids here should not be taken as a sign of over-protection.

Up Nick Eaton Ridge

February and March

Starting the Nick Eaton Trail this early in the season, you may have to park at the gate beside the old highway just past the Forest Service Work Center if the weather has been icy, although in many years, the gate seems open most of the year. If it's closed, bypass the gate on its left side, then head up the paved road for a few minutes to reach the picnic area and the start of the Herman Creek Trail. Note, on your trip up the road, a concrete set of stairs on your right which used to lead to a Forest Service structure whose current remains are in a pleasant state of crumbling neglect.

Despite how early it is in the season, there are a considerable number of flowers out in these youthful spring months beginning right at the parking area by the road, our **Point 1**. However, the first one you will encounter here is really not a wildflower at all, but a garden violet (*Viola odorata*), a common plant, often called the English Violet, now escaped from many city gardens. This comely and graceful violet *is* a violet color, in distinction to the mainly yellow violets which are Northwest natives. We do have a blue-purple violet in the Gorge, the Long-Spurred, or Early Blue Violet, but these violets at the gate and up the paved road probably spread from someone's long-ago garden or perhaps from the Work Center adjacent to the gate.

These cheery violets chaperone you as you plod up the road, accompanied in early March by Robert Geranium on the right during the first 100 yards, along with the tiny white flowers of Hairy Western Bitter Cress as early as mid-February, and two fine specimens of Indian Plum on your right, also 100 yards up the paved road. The first shrub to flower in the Northwest, Indian Plum displays its bright green leaves in upright flight while its greenish-white flowers hang in pendant sprays, decorating the woods with February blooms while all other eye-level plants are still shivering in bare-stemmed anonymity. Fortunately, another example of this early bloomer can be found at the dormant picnic area, **Point 2**, 180', on the left between the outhouse, apparently oblivious to its ignominious setting.

The Herman Creek Trail begins at the sign at the west end of the picnic area and proceeds in a series of switchbacks to the power line, at about 400' elevation. By March, many yellow Evergreen Violets are beginning to bloom and a shoot of a Trillium appears on the right early in March; by mid-month there will be many white flowers of this species on these switchbacks. The power line, **Point 3**, shows only some droopy Pearly Everlastings, hold-outs from last season. However, you next proceed past the "cool caves", a series of rocks with deep openings on your left just after the power line (which emit blasts of winter-bred cool air in the hot summer months). About a quarter-mile up the trail, just past the right turnoff marking a spur to the Pacific Crest Trail, you encounter left, then right, two semi-open switchbacks (now mostly overgrown) between 500' and 600'. Here, that harbinger of spring, Oaks Toothwort, can be found enjoying its role as one of the few late-February bloomers in the Northwest. Its pale pink-to-violet four petals often nod downward in false modesty on dry hillsides such as these far above the roaring Herman Creek.

Up the road that comes after the trail, many Evergreen Violets are blooming by late February to mid-March; they are especially prominent just before **Point 4**, the derelict Herman Camp at 960', now simply a widening in the road and the point at which the Gorton Creek Trail rejoins your route should you complete either of the two east-side loops described above. Note a few white Trilliums in the open woods as you tramp up this road, both before and after

Point 4. A few hundred yards after Herman Camp on the road, and about 1½ miles from the start, find the sign on your left indicating the Nick Eaton Trail, **Point 5.** As you begin what at first is a gradual climb, selected examples of Trillium appear, accompanied by Evergreen Violets. These latter, preferring a mossy substrate, are particularly apparent on your left. The trail steepens after a short distance but the slope is eased a bit by the presence of more violets, in flower through March up to about 1,800', and by companies of Trillium between 1,200' and 1,400', most prominent on the right. Oaks Toothwort is not to be outdone here, and climbs higher than all other early spring blooms, appearing on the steep portions of the trail between 1,300' and 1,900', then again by late March just before and after the openings between 2,300' and 2,700'.

Following the deep forests, three small openings at 1,700', 1,850' and 2,000' are welcome even in the absence of buds or blooms this early. However, the six meadows which follow, between 2,300', **Point 6,** and 2,750', partly compensate with several flowering species. Regrettably, the first and last of these are botanically barren, although they do offer choice views down to the Columbia and out to Mt. Hood on the rare clear day. The second through fifth meadows, however, display the most precocious of all the Aster Family plants, Gold Stars. Cute to a fault, these diminutive yellow daisies sprout from rocky meadows as early as mid-February following sunnier winters, and to add to their allure, they often persist throughout the spring.

> Plant associations are common in nature and often stem from a natural synergy between species. Indian paintbrushes are frequently found in association with lupines. The combination is neither accidental nor merely for the benefit of our human appreciation of the red and blue color combination. Their alliance rather involves the ability of lupines to gather nitrogen from the atmosphere and transplant it into the soil. Most other plants, including paintbrushes, can't accomplish this trick and hence rely on lupines to perform it for them. Perhaps the lupines get something in return, but this is less clear; at least it doesn't seem to hurt them. This association may be crucial for the paintbrushes, which prefer rocky open areas in the mountains, regions where nitrogen is, but for the lupines, scarce in the soil. In a similar fashion, Gold Stars and Grass Widows most often can be found living together, as can Glacier Lilies and Spring Beauties. Perhaps they are simply good neighbors who enjoy each other's company, but the alliances are so frequent as to raise the chance that some symbiosis may be occurring here as well.

Despite their good looks, however, and their starry name, these plants are not the featured actors in this early spring drama, for they are outperformed in late March by a magnificent display of Grass Widows (*Sisyrinchium douglasii*) in the fourth meadow, where they can be found just up the hill on your left after a right switchback. This ravishing member of the Iris Family has foot-high grass-like leaves slightly taller than the stunning magenta ¾" flowers, each bloom sporting six satiny petals (actually three petals and three sepals, collectively called "tepals"). As soon as the days lengthen, a signal stimulates the growth of these beauties, which thrive in wet soil which will dry out later in the year. One author, a staid botanist, waxed eloquent about these plants: "The reigning queen of the genus…a slope covered with these sprightly bells, sensitive to every whisper of the wind, is one of the floral delights of early spring" (Mathews). The tepals' satiny sheen, sparkling in the spring sunshine, also gives rise to another common name for this species, Satin Flower. For more Grass Widows, also check out the Starvation Ridge and Wygant Trails. Perhaps they're called widows because they bloom so early; if I were a flower, they wouldn't be widows for long!

Grass Widows

The woods above these openings are sprinkled with the occasional high-living Oaks Toothwort, occurring here around 2,850' on your right. Just after **Point 7,** the trail junction with the Ridge Cutoff Trail, a large rocky opening occurs. In late March, look to your left as you enter it to see the first blooms of Hairy Manzanita, a 3-6' tall shrub related to the many chaparral species more common in California. Its alternate egg-shaped blue-green leaves are evergreen, but the drooping pinkish-white flowers come out in March and can last until May. A bit of Spring Whitlow Grass can be found here also a short distance up the trail through this rough opening. Manzanita's seeds, like those of Fireweed, can lay dormant in the soil for years until awakened by the heat of a forest or ground fire. Such long-lived seeds are common in the plant world, as evidenced by the sprouting of foreign weeds after long sea journeys or the recent discoveries of Paleolithic spores and seeds 30,000 to 40,000 years of age which miraculously bloom once planted by modern hands.

As you start up the steep trail to the ridge above, between **Points 8** and **9**, several Oaks Toothwort are beginning to appear on your left at about 3,100' by late March, although they look a bit haggard, perhaps ill from a higher altitude than that in which they usually flourish. Beyond this point, little else is blooming during these early spring months and indeed, this part of the trail is most often blanketed in snow through mid-April. Not to worry, however, because as you retrace your steps back down the ridge to the cutoff and the round trip, more blossoms are guaranteed to cheer you through February and March, turning-points in the Pacific Northwest's makeover from the fierce gray of winter to the light-filled blue of later spring and summer.

April

Red Dead-nettle

The Nick Eaton/Gorton Creek circuit offers a breathtaking array of flowers in April, with the addition of several species not as commonly found west of the Crest. The paved road is now un-gated but note that the parking area at **Point 1** presents several new flowers to identify this month that you are missing further up the trail. To your left as you drive up the road (slowly – it's largely one-way), a large specimen of Red Elderberry throws white spikes of tiny flowers and opposite pairs of toothed sharply-pointed leaflets above its 10' tall stems. Several plants familiar from urban locations grow here as well, both immigrants from distant lands. The beloved and adorable Woods (or Common) Forget-me-not (*Myosotis sylvatica*) carpets the grassy ground around the gate, no doubt escaped from a local garden. Diminutive orange-centered deep blue flowers sprout in a panicle (branches from the main stem with secondary branches supporting the flowers) atop foot-high stems. These hard-to-forget flowers may be related to a plant which left a foul taste in the mouth, hence their name. A less disagreeable etymology holds that this European garden import's name derives from the habit of giving your lover this flower as a remembrance.

Woods Forget-me-not

A second foreign weed, less lovely but more widespread, is Red (or Purple) Dead-nettle (*Lamium purpureum*), a 6" high Mint Family species which can be found at the parking area and up the road to the picnic area, **Point 2**. Preferring trampled waste ground, this plant, and its close sibling Common Dead-Nettle (*L. amplexicaule*), originated in Asia but both are now cosmopolitan weeds, inhabiting far-flung regions of the globe as diverse as Florida and Siberia. Red Dead-nettle's reddish-purple heart-shaped leaves are roughly corrugated and its rose-purple flowers, helmet-shaped on top with a forked lip below, crowd about its summit. Common Deadnettle differs from the Red species in forgoing any stems to its upper leaves and having green, rather than reddish-purple, leaves. These plants are not deceased, despite their name; nettles are supposedly plants with stingers. These two are related to the true nettles but lack prickles, hence are "dead", at least in that sense.

Driving up the paved road between **Points 1** and **2** (the parking lot at 180'), watch for a flowering

Dogwood tree on your left just at the start, and examples of Robert Geranium, Candy Flower, and Hairy Western Bitter Cress as you gently ascend. Starting up the actual Herman Creek Trail from the parking lot in forested switchbacks, Trillium (now turning burgundy by mid-to-late March), Evergreen Violet, Fairy Slipper, Big-leaf Sandwort, and Meadowrue are now all in bloom. By late April, the two Solomon's Seals, Plume- and Star-flowered, are beginning to turn their tiny buds into spikes or bursts of white flowers. At the power line, **Point 3**, Dandelion (a description can be found later in this chapter, as if anyone needed one), Woods Strawberry, and Shining Oregon Grape decorate the otherwise barren ground. Just above, at the "cool caves" on the left as you re-enter the woods, Big-leaf Sandwort predominates, while further on, on the trail and then beside the Herman Creek Road, Fairy Slipper and Evergreen Violet grace the forest.

There are several semi-openings by twin switchbacks (becoming progressively overgrown as the deciduous trees gain in stature). These can be found between **Points 3** and **4** at 500' to 600'. Here in April, the strangely-colored Chocolate Lily nods its splotched blossoms from 1-2' tall stems. White Woods Strawberry and Small-flowered Blue-eyed Mary litter the dry hillsides above the trail. Look again for the combination of Fairy Slipper Orchids and the yellow Evergreen Violets on the ancient road as you approach Herman Camp, at **Point 4**, and beyond up the road to the turnoff onto the Nick Eaton Ridge Trail proper at **Point 5** (1,000'). Up this trail, clumps of Evergreen Violets dominate the floral landscape, especially on your right, to be replaced later in early April by Stream (Wood) Violet, also yellow. Fairy Slippers are particularly abundant at 1,050', 1,320', 1,590', and 1,800', the last on the left. Cascade Oregon Grape covers the Douglas Fir forest floor from 1,000' to 1,600' elevation while at 1,750', three fine specimens of Red-flowering Currant appear down and to your left.

One new bloom makes its first and auspicious appearance this month; outdoor enthusiasts welcome its presence from this point in the Gorge to its farther reaches east of the Crest. The Oregon Anemone (*Anemone oregana*) shares, along with yellow Broad-leaf Arnica, the distinction of providing the most color in deep woodlands in the Gorge. Its deep Prussian blue startles the dark forest visitor; how does it obtain sufficient light to manufacture this remarkable hue? This is as yet unanswerable. Sitting atop 1' tall stems with tri-cleft leaflets are the 1-2" flowers with what appear to be five petals, although technically these are sepals (see the sidebar). Stellar examples of this unforgettable flower can be found by mid-month about 300' past **Point 5** on your left, then several more in another 100' on your right. Don't despair if you find no more on the Nick Eaton Trail; they will reappear at about 2,300' in occasional spots, then in full array at 3,000' and beyond, all the way to the top. They can also be found along the loop on the Ridge Cutoff and the Gorton Creek Trail.

Oregon Anemone

The parts of a plant we usually notice are the petals, the showiest and most colorful elements of the flower. Petals are defined as the inside ring of modified leaves which, through evolution, have taken on bright hues in many plants to attract the attention of pollinators such as birds and insects. Indeed, plants which are wind pollinated, such as the grasses and certain groundsels, contain flower parts which are less than striking. However, some plants have foregone petals entirely, yet still depend upon animal pollination. Bunchberry and the dogwoods have bright white bracts, which are actually modified leaves that support the flowers, and which form the most conspicuous part of these plants. The anemones have what appear to be colored petals, but these meet the definition of sepals – the ring of modified leaves outside the petals, which also support the flower structure. By definition, if petal-like structures occur in just one row, they are considered to be sepals, not petals, hence their designation in the anemones, which lack a supporting row of sepal-like structures. While sepals are usually green (think of the

> green leaf-like structures behind the flowers of your garden roses), they can also evolve petal-like pigments and masquerade as petals. From the daises through dogwoods to the anemones, deception appears to be as common in the Plant Kingdom as in the Animal one.

There are six anemones in the Cascades, all members of the Buttercup Family, including the mop-top Western Anemone, or Pasque Flower (*A. occidentalis*), common in higher-elevation meadows just as the snow melts. Several others, including the Cut-leaf (*A. multifida*) and Drummond's (*A. drummondii*) species only occur on high mountain ridges and are largely absent from the Gorge. However, Lyall's Anemone (*A. lyallii*), a smaller light-blue to yellow flower, can be found occasionally in moist woods here, and we have already been introduced to the Columbia Windflower (*A. deltoidea*), a white version also in the anemone genus. So attractive is the Oregon Anemone that some misguided hikers attempt to transplant it to their city gardens. Rooted out from its habitat, however and planted in an artificial setting, it soon withers and dies, apparently homesick for the deep shade and comforting environs of its original abode.

In the forest between **Points 5** and **6**, Cascade (Mountain) Oregon Grape is now coming into bloom and Oaks Toothwort continues its spring show, commonly found on slopes above the trail poking its diminutive four-petalled pink-to-purple flowers above its 6-8" stems; several typical specimens can be found on your left at 1,850' and 1,900'. Small-flowered Prairie Star puts in an appearance wherever some light reaches the forest floor. Trilliums are scattered in open woods here, especially bordering forest edges at 1,900' and 2,200'. A real botanic shock, however, occurs at these same woodland edges, where the rare and incredibly-shaped Dutchman's Breeches graces hikers with an appearance far west of its preferred range. Hikers will appreciate the dainty white-to-pink flowers, related to Bleeding Hearts, and looking for all the world like a pair of upturned 18th Century pantaloons.

Another new plant appearing this month is Sweet Cicely (*Osmorhiza chilensis*). Its tri-partite leaves are its main distinguishing feature as its flowers, despite its membership in the usually decorative Carrot Family, are rather inconspicuous.

Sweet Cicely

It appears to grow taller as spring and summer progress, its stems branching out and offering at their ends, single large seeds greatly exceeding in length their tiny flowers. At first, it appeared to botanists to be an identical plant to one in Patagonian woodlands, and, in my travels, the *Osmorhiza* I saw there looked identical. But leave it to scientists and DNA to complicate things and prove that, though similar, these are two different species. In retrospect, it would be strange for identical species to show up at different ends of both the north and south parts of two continents, though disjunctive populations do exist, but not *that* far apart. The "sweet" in the name supposedly derives from a somewhat licorice taste to the roots. I've never tested that attribution and have no plans to do so in the future.

Two small openings occur at 1,700' and 2,000'. By mid to late April, these contain a wonderful mix of colorful species: yellow Glacier Lily, dirty white Northwestern (or Grassland) Saxifrage, and pink Rosy Plectritis (see May below), pinkish-white Prairie Star, and the sky-blue puffs of Ball-head Waterleaf. In between, a semi-opening at 1,850' is flower-less at this time of year. Next however come six major blossom-filled openings, beginning at **Point 6**, 2,300'. I would note that, since the 2017 conflagration, the wooded areas delineating the distinct openings are minimal, if they exist at all. It is best, for the present, to simply count the switchbacks as these correspond to each numbered meadow:

a. The first, at 2,300', displays Goldstars, Little Western Bitter Cress, Prairie Star, and Small-flowered Blue-eyed Mary.

b. Gold Stars and blue-purple Cliff (Menzies') Larkspur, particularly on your left, dominate the second meadow at 2,350', along with the Long-spurred Violet.

c. The third opening, at 2,400', contains a pleasing mixture of reds, blues and purple, with Harsh Paintbrush, Larkspur, Long-spurred Violet, and both Small- and Large-flowered Blue-eyed Mary's.

d. At 2,550', the fourth meadow rewards not only with terrific views down to the Columbia and out to Mt. Hood to the south, but with a gorgeous display of Grass Widows (see February/March above), at least until mid-month. Prairie Stars, low-lying yellow Martindale's Desert Parsley, and Larkspur complete the garden-like display.

e. The fifth opening, at 2,625', boasts Prairie Stars and two brilliant crimson Harsh Paintbrushes on your left as you enter the meadow.

f. Martindale's Desert Parsley and Small-flowered Blue-eyed Mary bid you goodbye to the last opening in this series at 2,750'. No need to fret – there will be more meadows further on.

As you progress above these openings and plunge back into deep forest, note, at the vey top of the Ridge, immense boulders, randomly strewn about, as if some giant was playing a losing game of marbles. These silent sentries of stone pose a mystery: Why would the largest of rocks be at the very top of a ridge? Ask a geologist; they probably have a better answer than I could propose.

Just before **Point 7**, the Ridge Cutoff trail junction, a dip in the trail at 2,850' presents several examples of the yellow Evergreen Violet partnered with the anomalously-named tiny white Big-leaf Sandwort. Just past this junction, a large sandy and steep opening occurs at 2,960' to 3,000', where Hairy Manzanita can still be found displaying its pendant pink-white drops of blossoms on the left as you enter this rocky meadow. Later in April, look for yellow Nine-leaf Desert Parsley on your left and Woods Strawberry on both your left and right; both of these plants are best found around the mid-way point of this opening.

The woods beyond, where the trail levels off and even takes several dips, contain snowy-white specimens of Trillium, especially in the open forest before and after a brushy area at 3,030', midway between **Points 7** and **8**, the junction with the Deadwood Cutoff. Here also are many examples of Candy Flower and both the Evergreen and Stream Violets. While both Violets are yellow, recall that the Stream species has thinner, more heart-shaped leaves which end in a taper-tip. This Violet seems to replace the Evergreen variety as you progress along the trail towards **Point 8** at 3,040'.

Beyond the Deadwood Cutoff, as you labor up the steepening hill towards the ridge-top, shimmering white Trilliums partially ease the struggle, and at rocky east-facing viewpoints along the ridge, Evergreen Violets again welcome you to your highest destination, **Point 9** at about 4½ miles and 3,550'. You could also choose to continue another ¾ mile to a second rock pile but you would only gain a further 100' of elevation and few views. Either way, it may be all downhill from here but even more blooms await as you contemplate completing this loop in April, one of the finest flower-viewing months in the Gorge.

May

May brings out the best of Nick Eaton's blossoms, and this, combined with the improving weather, makes this loop one of the Gorge's premier treks in this beautiful month. By May the paved road to the picnic area will be open, but you may choose to walk the short road, beginning at **Point 1**, as it's a brief stretch and there are flowers (well weeds really, though pretty ones) to greet you at the parking area and up the road to the trail parking area. Amidst the grasses by the fence, the aggressive Himalayan Blackberry continues its American juggernaut, overtaking native plants faster than branches of Starbucks can open. Its pretty white papery flowers belie its fierce thorny nature, but all is forgiven with August's crop of juicy berries. Forget-me-not's and those purple English Violets still grace this civilized district.

Up this road, pink Robert Geranium and tiny white Cleavers make an appearance, the latter clinging to other vegetation to support its weak stems. Not parasitic, the leaning habit of this species doesn't appear to harm the obliging plants which lend their support. The bristles on the stems of this species help the plants attach to, and gain support from, neighboring foliage; this habit is mimicked in the twin seeds, which are festooned with tiny burs which stick to passing animals and hikers' socks, thus gaining a free ride in spreading to new habitats.

At the picnic area, **Point 2**, where at this time you will be able to park, Miner's-lettuce and Robert Geranium are in bloom, the latter particularly prominent just past the first bathroom on your left, apparently not minding these surroundings. Up the trail from the parking area in switchbacks below the power line, Small-flowered Nemophila is sprouting tiny light-blue flowers above neatly-cut club-shaped leaves immediately on your right, while Robert Geranium accompanies you for at least the first 100' of trail. Two specimens of Baneberry are on your right within the first 100' of trail within 50' of each other. This Buttercup Family plant shows spikes of thimble-shaped white plumes above 2-3' tall stems. Queen's Cup has just emerged from its long winter slumber by late this month at these lower elevations and can be found here, halfway up before the power lines, at about 250' elevation. Sweet Cicely is on your right at 290' and, throughout this section, Vanilla Leaf, Meadowrue (particularly on the right), Stream Violet, Fairy Slipper, Plume-flowered Solomon's Seal, and Columbia Windflower all participate in this low woodland garden. Hooker's Fairy Bells toll frequently on your left between 200' and 350', while new sprouts of Fringe Cup rim the trail on your right between 300' and 350' while Nevada Pea lends a deep pinkish note just before the power line.

The manufactured gash of the power line, at **Point 3**, 400', displays Dandelion and Woods Strawberry at first; as you move up the dirt track under the poles, you can find Broad-petal (Virginia) Strawberry and Small-flowered Forget-Me-Not, with its delicate but tiny orange-centered sky-blue flowers on your right. Also here are two common weeds, Red Clover and the pretty Oxeye Daisy, also on your right. As you enter the woods here, be sure to feel the blasts of cold air emanating from the mini-caves to your left. Cooled in winter, the air inside these blunt-ended gaps in the talus remain frigid through much of the summer, providing refreshing gusts of air for a welcome chill after a scorching day hiking in the Gorge.

Here, a few Fairy Slippers persist into early May, growing right off the mossy boulders of the caves, along with multitudes of Little-leaf Candy Flowers. Small-flowered Alumroot grows around these stones, preferring rocky, partially shaded habitats, while in the woods following, Vanilla Leaf, the Columbia Wind Flower, Plume-flowered Solomon's Seal, white-to-pink Star Flower, and the occasional reddish-purple Leafy Pea complete the forest flora.

A brace of switchbacks between 500' and 600' in a partial opening hosts fields of Small-flowered Blue-eyed Mary's, nicely contrasting with the white Woods Strawberry, White Hawkweed and the strangely-mottled Chocolate Lily, this last hanging on until mid-month. Most of these blooms are to be found on your uphill side as you ascend these semi-open switchbacks. The flashiest splash of color, however, is provided by the vivid crimson-orange of Harsh Paintbrush. This is the most common paintbrush in the Gorge, although the Common Paintbrush is more plentiful throughout the Cascades. Harsh Paintbrush thrives in low to mid-elevation openings and can be distinguished from its usually taller cousin, Common, or Tall, Paintbrush, by its forked leaves and cleft colorful bracts; Common Paintbrush has linear leaves and bracts.

However, there is nothing really "harsh" about this plant, although it got its common name from the modestly stiff bristles which clothe the stem. You, of course, recall that the "brush" is really composed of colored, modified leaves called bracts; the true flowers of all paintbrushes are hidden inside these bracts and are not only tiny, but also nondescript, generally just dull green tubes. To compensate, the plant tricks pollinators with colored bracts that are far easier to maintain than showy sepals and petals. Paintbrushes are also partly parasitic on the roots of neighboring plants, which not only makes them nearly impossible to transplant, but may explain how they get away with having the most dazzling hues among all our wildflowers. Plants may seem rooted, stationary and largely defenseless but not all in the perpetual arms race of evolution is as it seems.

Compared to such blazes of color, it may be hard to appreciate the miniature orange-centered white flowers tumbling down the dry hillsides on these switchbacks, but they occur in such abundance that it's just as hard to ignore them. These are the Large-flowered, or Common Cryptantha (*Cryptantha intermedia*) and they were not named in sarcasm. Compared to their brethren in the genus *Cryptantha*, their flowers *are* actually enormous. Why they have not attracted a more facile common name is a mystery as they are quite widespread, but amateur botanists persist in calling them "Popcorn Flowers", and indeed they almost resemble a floral version of that edible treat.

However, the true Popcorn Flowers, in a different genus (*Plagiobothrys*) but in the same family (the Borage Family, home to the somewhat similar Forget-me-not's), are generally found further east. Common Cryptantha's can be found throughout the Gorge on low to mid-elevation sunny banks and hillsides; look for them especially on the switchbacks near the power lines going up Starvation Ridge.

Above these switchbacks, the trail continues mostly on the level, with Vanilla Leaf to accompany you, along with, the Inside-out Flower, and a few Fairy Slippers, at least early in the month. The path then breaks out into a broad opening where an old road enters from your left. The English Daisy is here, the small white flower looking for all the world like a dwarf of the more common Oxeye Daisy. English Daisies populate urban lawns and grassy roadsides throughout the spring and summer months, at times coloring entire schoolyards and park grounds entirely white.

This man-made breach in the trail is also home this month to another familiar Aster Family invader: The Dandelion, perhaps the most detested flower in the world. We probably should not despise any living thing, but it's difficult to admire a plant that mars your otherwise perfect lawn, and greedily consumes nutrients meant for your prized azaleas. Perhaps the best we can offer in its defense is that the Dandelion rarely interferes with crops the way Russian Thistle might in Eastern Oregon, and it hasn't been shown to cause allergies like its Aster family cousin, Tansy Ragwort. Yet examine the ray flowers closely of a single flower and you can appreciate the symmetry its DNA has generated; better yet, after the flower has turned to seed, puff out a few of its fluffy seed heads and stare intently at the beautiful complexity of their downy architecture. Even in the scourge of a weed, splendor can reside.

> Until recently, the mechanisms creating the body plans of plants and animals remained a mystery. In the mid-1990's however, several scientists began to unravel the workings of a group of genes, called *Hox* genes, which appeared to orchestrate the arrangement of legs and antennae of the fruit fly. It came as a surprise, then, to our rather exalted sense of self, when comparative biologists discovered that these genes are operative in *all* complex species of plants and animals, from fungi, spiders and worms to fish, people and dandelions.

Large-flowered Cryptantha

Indeed, although we think of ourselves, with typical hubris, as the most complex of creatures, even if we are related to all animals, in fact many plant and animal species boast a larger number of genes and chromosomes than *Homo sapiens*. We have perhaps 23,000 active genes according to the most recent count, yet some plant species, like lettuce, contain up to 40,000. Scientists now believe that the enormous number of genes in many plants holds the key to why there are so many botanical species. An abundance of genes allows for greater flexibility in the face of rapidly changing environments. Hox genes tell the developing plant or animal embryo where on its body the head, arms and legs or its leaves, flowers, or seeds should go. Mutations in this set of genes can lead to gruesome abnormalities in insects but are fortunately rare in mammals. When we see insects, plants or animals, it's important to recognize the shared heritage central to all living things. When some New Age prophets tell us that we are all one with the earth, they may be more correct than they realize.

Progressing up the road, many woodland species appear. Plume- and Star-flowered Solomon's Seals peek out at you here and there, along with some Fairy Slippers early in the month. Just past the opening at about 750', note three specimens of the Cascade Penstemon, with blue-purple tubes and sharply chiseled leaflets. North of Mt. Rainier, this penstemon is more often pinkish-red. Also prominent here are Vanilla Leaf, Fringe Cup, the yellow Stream Violet, the Inside-out Flower, Columbia Wind Flower, and Sweet Cicely. Lending some color are vines of the purple Nevada Pea and the prickly shrub, Little Wild (Baldhip) Rose.

Since the 2017 fire, a native but weed-like plant has taken up residence alongside the road, beginning just past the opening noted above. Thyme-leaved Speedwell (*Veronica syrpyllifolia*) is a tiny plant but it grows in mats shouldering the roadside, especially on your right, up to, and even past, Herman Camp. Although just 3-5" tall, with light blue flowers marked by darker blue streaks, it is prominent because of its habit of forming mats which creep over the ground and it has displaced the usual panoply of roadside plants since that conflagration. I believe it will give way to more usual flora as the woodlands revert back to their more normal state. This species may look familiar to gardeners as a weed plaguing lawns and urban gardens but it is actually a native closely related to our other Veronicas in the Figwort Family.

Just before and after Herman Camp, at **Point 4**, a plethora of Varied-leaf Phacelia (*Phacelia heterophylla*) lines the road, then proceeds up the Nick Eaton Ridge Trail itself in much greater abundance than before the 2017 conflagration. Textbooks do not attribute its surprisingly teeming plentitude from 650' to well above 3,000' from the fire itself, as would be true for Fireweed, for example. Perhaps alterations in soil conditions and light penetration have encouraged its resurgence. Indeed, at many points on this trail, even above 3,000', this phacelia grows majestically, exceeding 5' in height in areas between 2,550' to 3,000'. Its name betrays its leaf structure as its three-parted upper leaves give way below to a variety of oval and simpler linear shapes, and even to an occasional whorl of leaves beneath the dull whitish-gray flower heads, often curling as if a hand was bending its fingers.

Hooker's Fairy Bells can be found in abundance at these same elevations. As you turn up the Nick Eaton Trail at **Point 5** (1,000') and proceed at first on the gentle gradient, Red-flowering Current points the way, along with the Evergreen Violet, chiefly on the right, and the similar, but more upright Stream Violet, mostly on the left. By mid-month, more yellow color can be found with the daisy-like Broad-leaf (Mountain) Arnica, beginning to bloom here mostly from 1,100' to 1,600'.

Blue Oregon Anemone appears at 1,100' on your left, again at a right switchback in these woods at 1,200', and in groves tinted with its delightful hue at 1,400', 1,650', and 2,100'. The yellow Cascade Oregon Grape forms an almost continuous understory in parts of the forest between 1,100' and 1,900'. Less colorful flowers are also present, including white-to-pink Star Flower, very common at 1,100' to 1,250'; the tiny white flowers of Bigleaf Sandwort at 1,250' to 1,350'; Miner's-lettuce, with its diminutive white blossoms topping a stem which pierces its single round leaf; and dispersed throughout the open woods, Trilliums, burgundy below 1,650' and white above.

Up the steepening slope, a gaggle of typical forest species appear, including the Inside-out Flower, Meadowrue, Hooker's Fairy Bells, and Sweet Cicely, all common between 1,300' and 1,900'. However, several unusual and curious plants inhabit these woods as well. Two Orchid Family saprophytes, shorn of their green chlorophyll, dwell in these dark woods. Striped Coralroot makes an appearance at 1,050' on your right. Soon thereafter, its close sibling, Spotted Coralroot appears on your left at 1,120'; another example can be found again on your left at 1,300'. A different and somewhat unusual species, though plentiful enough in the Gorge, is Scouler's Valerian, a 1-1½' tall plant with a fuzzy mop of white flowers at its summit. It is particularly common between 1,500' and 1,700' in open woodlands on both sides of the trail.

An additional, and spectacular saprophyte, Ghost Orchid, also called Phantom Orchid, is plentiful up the hillside to your left. It is a spooky white, with not a spot of color to be seen. It lives off the humus and fungi of the forest floor beneath, and needs neither sun nor other above-ground plants to sustain it. Simply oxygen and the bacteria and fungi within the forest soil give it life.

There are two small openings at 1,700' and 2,000'. Actually, they are the same opening, simply encountered at two different points on the switch-backed trail. Both contain yellow Glacier Lilies early in the month, scarlet Harsh Paintbrush, the pink tri-cleft petals of Prairie Star, the white Mickey Mouse-eared Field Chickweed, Woods Strawberry, and Grassland Saxifrage, all familiar inhabitants of these open Gorge meadows. A regular denizen of these low-to-mid-level fields and grasslands is another example of a species lacking a decent common name, Rosy Plectritis. This omission seems quite unfair in this case as this plant is not only colorful, but common as well. Pink waves of this Valerian Family flower can be appreciated by all Gorge walkers, not only here but in low-elevation openings on Starvation Ridge and Dog Mountain. Each foot-tall single stem has egg-shaped opposite stalk-less leaves and is topped by a balled head composed of perhaps 20 to 30 tiny rose-colored five-petalled flowers. Entering these lower fields on the Nick Eaton Trail and coming upon this sea of pink after struggling up the steep shaded woods behind fills the heart with joy for the light and color, or at the least for the brief stretches of level ground these openings offer.

The six meadows beginning at 2,300', **Point 6**, will again be described in an outlined fashion, which hopefully will clarify in which opening the various plants are now in bloom:

a. The first meadow one encounters is replete with Gold Stars early in the month, though these may be withering to senescence by mid-month. Harsh Paintbrush is happily present here along with the small but fascinating Chickweed Monkey Flower, with its single maroon splotch neatly tucked inside its bright lower yellow petals. In mid-meadow, and early in the month, striking purple Grass Widows bend gently in the breeze, their orange-yellow stamens protruding beyond the five rounded petals. Here too, a close sibling of the Grass Widows appears early this month, as if invited by its relatives to partake in a festival of May blossoms. Called Blue-eyed Grass (*Sisyrinchium idahoense*), this 1' tall plant displays a subtle intensity to mark its Iris Family origins. Grass-like leaves sheath the stem, with one of its leaf-like bracts often extending above the flower, thus appearing to be a continuation of the stem itself. The common name well reflects the appearance of these plants, with their blue-purple petals and

Blue-eyed Grass

golden "eyes" peeking from the side of grassy stems. You can distinguish Blue-eyed Grass from Grass Widows by several features: The former are blue-purple, the latter, a deep satiny magenta; moreover, Blue-eyed Grass generally has more pointy tepals (the three petals and three sepals combined, as they are really indistinguishable) than do Grass Widows. Both, however, favor grassy habitats at low to middle elevations east of the Crest which are wet in winter and early spring, then dry throughout late spring and summer (called "vernal" by botanists). This meadow, and all the following are awash in blueish-purple hues, almost a complete coloring, as if a deranged Persian wove a rug using a single, but beautiful stain. Large-flowered Blue-eyed Mary's combined with Cliff (Menzies') Larkspurs and Long-spurred Violets create as dazzling a show as can be found in all our meadows.

b. At 2,350', the second meadow amplifies the patchwork of varied purplish hues begun below. Harsh Paintbrush, Cliff Larkspur, Gold Stars, Chocolate Lilies, Long-spurred Violets and Woods Strawberry lend their tones to color this stony opening. Also here are the dainty three-pronged pinkish-white petals of Prairie Star, poking up its 8-12" tall stems on both sides of the trail. In all the meadows now described, a show of Large-flowered Blue-eyed Mary's, the finest aside from a plateau below Munra Point's summit, now inhabits and, indeed, dominates, these upper openings with masses of Large-Flowered Blue-eyed Mary's as if, from a distance it appears all the green has been obliterated and the sole hue to be marveled at is the deepest of blue imaginable. This show will continue until mid-June.

c. The third large open field, at 2,400', displays many of these same blooms, but adds the yellow, low-lying Carrot Family plant, Martindale's Desert Parsley and the tiny Midget Phlox, as tidy yet delicate a species as grows in the Gorge. Found growing in compacted soil, often as here right in the middle of the trail, its tiny star-shaped pink petals lie close to the rocky ground and are often cleft like individual bunny ears. Their pink hue deepens towards the center of the flower. In addition, a nice specimen of the shrub Serviceberry can be seen on the left as you pass the mid-point of this opening.

d. Just before entering this fourth meadow, at 2,550', look to your right to see another example of the white-flowering shrub, Seviceberry. In these semi-open woods, before you come upon the broad open area, an unusual plant lurks beside the trail, especially on your left – Ballhead Waterleaf. Deeply cleft blue-green leaves seem to support a fuzzy ball of purple blossoms, a spiky inflorescence seemingly out of place in these woodsy openings. You might think this species arrived from another planet, so different is its appearance from the usual five-petalled flower atop a stalk. However, this apparent alien is actually a member of the Waterleaf Family, well-represented in our Northwest mountains and forests. Harsh Paintbrush, Prairie Stars, Small-flowered Blue-eyed Mary, Larkspur, and masses of Large-flowered Blue-eyed Mary's complete this magnificent array. By late May into mid-June, this meadow brims full with the mats of blue, yellow and pink as if some adventurous artist spilled paint across a green backdrop, so dense are these blooms; an unforgettable scene.

e. At 2,625', the fifth meadow is filled with deep blue-purple Cliff Larkspur, crimson Harsh Paintbrush, Prairie Stars, Small-and Large-flowered Blue-eyed Mary's, Long-spurred Violets, Chocolate Lilies, and yellow Nine-leaf Desert Parsley. The slight changes in elevation are probably not as important as the germination histories and slope exposures in determining the modest differences in plant communities in these meadows.

f. The last opening, at 2,750', is one of the larger meadows and continues the general scheme of Larkspur; Violets; Harsh Paintbrush; Martindale's and Nine-leaf Desert Parsleys; Large- and Small-flowered Blue-eyed Mary's; but adds the 3'-tall Western Groundsel (Senecio), yellow here but white just further east; and the bright orange-yellow of Rough Wallflower, a 2' tall species with one of the most appetizing fragrances in all our Northwest plants. It's worth bending down to sniff the sweet-spicy fragrance. Several specimens can be found along the trail toward the end of this field on your right.

The woods between these openings contain abundant examples of Candy Flower, joined by Stream Violet at areas damp in spring, and much Star-flowered Solomon's Seal. Trillium is still present, especially in the open woods around 2,450' and 2,600', the latter turning a satisfying wine-red hue by the middle of this month. Note the immense boulders strewn about on the crest of the ridge. A dip in the trail at 2,850' reveals White-vein Pyrola on your right, with striking white venation decorating its ground-hugging leaves, and Small-flowered Nemophila, a diminutive light blue flower hiding shyly in the deep forest and barely topping its club-shaped leaves.

Another large meadow appears at 2,960', just after the sign pointing down to the Ridge Cutoff Trail at **Point 7**. Here, on your left as you enter this opening, Hairy Manzanita comes into view. This Heath Family member can grow up to 10' tall, but here, it is a 5' tall shrub with drupe-like flowers that mimic the blooms of many other Heath Family plants, such as the blueberries and heathers common to high Northwest elevations. Unfortunately, the berries of Manzanita are mealy and barely edible. The gray-green leaves and flaking reddish bark distinguish this Manzanita from other shrubs in the Gorge, while its stature differentiates it from others in its genus, such as the low trailing Kinnikinnick and Pinemat Manzanita. There are over 30 species of Manzanita, most found further south in California, and several conspire to constitute the brush around Mt. Shasta and on the eastern slopes of the Sierra, perhaps California's revenge for our own combination of Devil's Club, White Rhododendron, and Salmonberry. Manzanita's seeds, like those of Fireweed, sprout particularly aggressively after a forest fire, and may lie dormant in the soil for years, awaiting a wake-up call from the heat of these conflagrations.

Also in this opening, which is disappointingly steep, you can find the low-lying yellow Martindale's Desert Parsley on rocks about mid-way through, and its sibling, the taller Nine-leaf Desert Parsley here as well. Within this uppermost of meadows, there exists a triplet of yellow daisy-like flowers. Contrast the tall multi-headed Western Senecio, from the beginnings of the buds and flowers of Wooly Sunflower, shorter than the Senecio, with blue-green multi-branched leaflets all covered in a wool-like coating, thus rendering them almost whitish; these should be compared with the third daisy-like plant, Mountain Arnica, just coming into bloom in late May. This taller flower is in-between in height compared to the Senecio and the lower-lying Wooly Sunflower. The Arnica has several pairs of opposite leaflets, with sharp regularly-spaced teeth in its leaves. All three will persist well into June at these heights.

Above this meadow, in deeper woods, much Candy Flower flourishes between 2,975' and 3,000', while Woods Strawberry and Western Groundsel can be found on your left up a grassy slope at 3,000', and Serviceberry as well. The woodland trail here is mostly level, with slight ups and dips. Fairy Slippers, Oaks Toothworts (early in the month – at the brushy area just before a dip in the trail), yellow Stream Violets, and Trilliums, these last also at the dip, dot the forest. The real show comes, however, after the dip, at about 3,000', where left and, mainly right, acres of common and Small-leaved Candy Flowers and Miner's Lettuce combine with Star-flowered

Solomon's Seals, with their twinkling racemes of white blossoms, all carpet the woodlands. Beyond this point, and up to **Point 8,** the junction with the Deadwood Cutoff, Stream Violets alternate with Sweet Cicely and, early in May, the glorious deep blue of Oregon Anemones, surely the most beautiful species named after our state.

Starting up the last grunt of the steep slope to the top of Nick Eaton Ridge, these gorgeous blue beauties continue for the next 250', accompanied by forests of Trilliums, with Stream Violets shouldering the path. Cascade Oregon Grape is plentiful, as is Wild Ginger on your left at 3,150' and again at 3,300' (look closely beneath the leaves to glimpse the weird brownish-red flowers). Meadowrue, Scarlet Columbine and Hooker's Fairy Bells lead the rough way to the top of the ridge and the viewpoint at **Point 9**, a site at some rocks where a view east can be had and the trail levels off at 3,550'. Here watch for a nice example of Serviceberry, the bush with large white raggedy blossoms. A well-deserved resting spot, this point marks our suggested turnaround point, although you could continue along the ridge with minor ups and downs another ¾ mile to a second rock pile, though with few views.

Whether completing the loop down either the Ridge or the Deadwood Cutoffs, or simply going back down the entire way you came up, Nick Eaton will live up to its reputation as one of the premier trails in the Gorge, especially in the month of May, when, it seems, all the wildflowers possibly growing are at their best in this middle-of-the-Gorge spot. It may leave a hiker drained physically and leave some travelers sore the next few days, but all that is easily compensated by the views, the sunshine, and the wildflowers that bloom *en masse* during this radiant month.

June

Although you may believe that the garden show winds down in the month of June, and indeed, there is a slight reduction in the bounty of blooms as the month progresses, the floral diversity continues, with several new and fascinating floral additions this month. The expansive views and variety of flowers continue to render Nick Eaton a rewarding loop any time of the year, but during late spring and early summer, this is especially so as the blossoms just keep blooming and the weather keeps improving, a felicitous combination for any outdoor adventurer in the Northwest.

The start of the trail in June now moves up from the main road, **Point** 1, to the picnic area as you can drive the paved road, although you can still catch a glance of the cultivated violets on your right along the road early in June as you hurry by. At the picnic area, **Point 2**, where you can now park, there are still some Red Dead-nettles to be found amongst the outhouses, picnic tables, and detritus left behind from multiple construction projects. The paved road saves five minutes of walking and, more important, allows you to start on a trail instead of a road, a more aesthetic beginning to this trip.

> While trying to identify flowers when you're driving is generally not recommended, sometimes it is difficult not to notice colorful blooms on the side of the road. And, if you're merely a passenger, there's no reason not to try. The carpets of pink which color the median on I-84 in April (Dove-foot Geranium), the bright magenta blooms lighting up the shrubs heading back west on your left around milepost 39 later in that month (Red-flowering Current), the waves of purple on top of the butte on the south side of the Freeway just past Exit 28 (Oregon Flag or Tough-leaved Iris) in early May, the giant plants with white topped clusters on the south side of I-84 between mileposts 19 and 22 in May (Cow Parsnip), all cry out for identification. However, species identification can be tricky even at the slower pace of a walk. At 65 miles an hour, it's probably best not to attempt complicated taxonomic analyses, especially if you're the one in command of 4,000 pounds of hurtling metal.

Starting up the trail by the new sign, Robert Geranium makes its presence known immediately, along with the Inside-out Flower, while Vanilla Leaf and Fringe Cup are bowing out by early this month, fascinating to seed lovers but appearing dull to the majority of hikers captivated more by color than process. Much Self-heal is apparent almost immediately and will accompany you all the way to the junction with the Gorton Creek Trail. However, a new species now appears, Queen's Cup, about halfway up this lower part of the trail on the way to the power line. Found more frequently on your left, this Lily Family plant is commonly considered a summer species and often appears in July at higher elevations. Here, however, it sprouts its parallel-veined elliptical leaves by mid-June and sends up 6-10" stems topped by a pure-white six-petalled flower, the largest of the individual white blossoms to be found in our darkly-shaded west-side forests, matched only by Trillium.

At the power line, **Point 3**, white flowers of Woods Strawberry portend the delicacies to come later in summer, while blue Lupines bend gently in the breeze, perhaps snubbing their cousins in the Pea Family next door, the weeds White and Red Clover. Oxeye daisies proliferate here, adoring the abundant sunshine. Several Tiger Lilies break the monotony of weedy clovers in this artificial opening. Just after passing under the power lines, note a fine example of Mock Orange, a 3-4' tall bush on your left. Its white flowers have a vague citrusy odor

If it's hot, head into the woods after passing beneath the power pole and poke your head and hands into the semi-caves on your left to feel the cool drafts left over from winter's chilling blasts. While cooling off, notice that Small-flowered Alumroot, with its dainty panicles of tiny white bell-shaped blossoms, seems to grow right off the rocks, sharing its stony habitat with Little-leaf Candy Flower. The winsome Queen's Cup graces these woodlands just after this rocky stretch as well. As you progress up the gentle incline in the forest, Vanilla Leaf appears, although on its last floral legs.

Quite soon after entering the forest & even opposite the caves, Tiger Lily appears and will happily accompany you in wooded openings and deep forest up the road here and there past Herman Camp, and even after you take the left turn to ascend the actual Nick Eaton Ridge Trail. Above about 2,000' note that it may be in bud, though by mid-month, even certain of those as high as 3,200' show a mixture of bud and bloom.

These 2"-wide gorgeous orange blossoms atop stems that can grow to 6' tall and whorls of linear leaves well-spaced up these stems, attract almost too much attention. Each equally rivals or surpasses the lilies so common in grocery stores around Easter, although, to be honest, they evoke only the faintest of perfumes. In fact, these and the Calypso Orchid are the two most-picked (illegally) Gorge wildflowers so it is not uncommon to see their high stems cut just a few inches below what was the flower itself. It is true that the common stock of many garden lilies originated from the five or so species of orange "Tiger Lilies" which thrive in Oregon, Washington and Southern California. However, they cannot be transplanted in your front yard. They only flourish within a tightly-limited ecosystem, an exacting complex of soil acidity, fungi, bacteria and even viruses, which cannot be duplicated in urban settings. And please don't pick them to show to Aunt Sophie when you arrive home; by that time, their glory will have melted away as drastically as a soft ice-cream sundae from the East Wind Café in Cascade Locks left in the car for hours on a 100-degree day.

Two overgrown switchbacks between 500' and 600' contain patches of woods and several semi-openings. In these, Woods Strawberry holds court, with the first inklings of a true summer plant, White-flowered Hawkweed, mostly on your uphill side. Although it is 2' tall, the flowers of this Aster Family plant are microscopic compared with its overall stature. Look closely to see its multiple tiny white daisies with a yellow center. Here also, you can find Chocolate Lilies early in the month, although they are by now mostly ageing, yet in a peculiar way: The petals seem to fall off sequentially, leaving many plants with two to four normal-looking petals and strange gaps in between, sort of like your kids when they were losing their baby teeth. However, unlike your kids' teeth, these petals won't be replaced, at least not until next spring. Another hawkweed can be found in these sunny openings, Scouler's Hawkweed, although this one bears much bigger dandelion-like blooms topping 2' to 3' stems. A few Harsh Paintbrush specimens brighten these openings as well.

Just as you take a first left turn on these switchbacks, note on your left the striking tubular flowers of the Cascade Penstemon. Found from low elevations through the alpine, this penstemon brings a deep purple hue to break the uniform dominion of brown and green hereabouts. Note its deeply serrated leaf edges, one of its main points of differentiation from similar penstemons trailside in the Gorge. Thus, it can be distinguished from the similar Cardwell's Penstemon, which usually grows at higher elevations, leans more towards a pinkish color, and has leaves which are either smooth on their edges or, more commonly, are just a bit saw-toothed. Although the authoritative *Northwest Penstemons*, by Dee Strictler, mentions a "lightly-bearded staminode" (the fifth but sterile stamen) extruding well beyond its more fecund brethren on the bottom of the floral tube, you must squint and peer closely to observe it both in this species and the related Fine-tooth Penstemon to be encountered further up the trail.

As you round the second switchback, this to your right, more examples of this colorful Penstemon grace these partial openings. Note that it appears to favor hillsides; thus always find it, right or left, by gazing uphill. Its roots may not tolerate too much standing H_2O. Look west here as well to catch a nice view of the gully-streaked hills bordering the west side of Herman Creek, whose roar can still be heard in the distance.

Up the road, Vanilla Leaf is largely bowing out while the Inside-out Flower and Sweet Cicely remain in flower through the middle of this month. Varied-leaf Phacelia grows in abundance alongside the trail from the road on up to above 3,000'. Its prominence here began after the 2017 fire, which may have altered soil and light conditions to allow this 4-5' plant to thrive hereabouts. It bears a curl of dull white-to gray fuzzy flowers subtended by three-parted leaves which change below to simpler oval shapes. Occasionally, a whorl of leaves subtends the flower cluster as well, thus explaining the Latin name of *Phacelia heterophylla*. The hetero- part implies great variability in this plant's leaf shapes. However, as reminders of the conflagration grow more dim, its presence appears as of the present (2024) to be waning.

This plant, prior to 2017, was not uncommon but the Eagle Creek fire appears to have brought out this species, in the Waterleaf Family, to great abundance. There are three possible reasons (or a combination of them) which could explain its present prominence: The fire eliminated much of the understory of shrubs, allowing this species to dominate; the conflagration also opened up more light as some of the leaves and needles of the trees (even those which survived) were burned; and, like a number of species, the heat of the fire helped its seeds to spring to life, just as those of Fireweed do. In fact, many such heat-activated seeds can lie dormant for over 40 years until a blaze heats the earth sufficiently to allow them to spring to life.

Indian Pipe

Another invader, this an imported plant, is Thyme-leaved Speedwell; it also populates the roadside, though it is thinning out by mid-June. It forms a mat of creeping leaves topped by pale blue flowers reaching above the basal leaves just 3-5" in height. Its pretty pale blue flowers, though tiny, bear a close look as they are steaked by darker blue guide lines. It was prominent earlier in the month but by mid-June, it persists in the main, only beyond Herman Camp.

In an opening where an old road enters, you may spot the English Daisy, a diminutive version of the common Oxeye Daisy, on both your right and left in clumps of grass. Here too are scads of Scouler's Hawkweed accompanied by abundant Self-heal. At **Point 4**, Herman Camp, more Vanilla Leaf can be found, as well as on up the road to the turn-off sign for Nick Eaton Ridge. Here, and up the initially low-gradient trail at **Point 5**, admirable examples of Star Flower are much in evidence, their low pinkish-white flowers of 6 to 9 petals illuminating the grounds of these dim woods between 1,100' and 1,600' of elevation. Within a few steps up this ever-steepening trail, late in the month, Tiger Lily is in bloom, especially on your left. While it rarely grows in clusters, from 1,100' to 1,600', a number of 4-5' specimens lighten the forest with their unmistakable orange hanging blossoms with burgundy spots. Gangs of Columbia Gorge Groundsel grow in batches (unusual for this species) just up the actual Nick Eaton Trail after about 5 minutes from its start for about another 400'. A troop of twinflowers closely follows the ancient stone trail marker just up this trail on your left.

At 1,100' on your right, the ghostly white saprophyte, Indian Pipe (*Monotropa uniflora*), appears. This phantom of the forest, a 6-10" tall cluster of waxy stems with bent or nodding bell-shaped flowers at top, is a member of the Heath Family. Roots of Indian Pipe gain their nourishment from fungi which live on the roots of conifer trees. Don't be surprised to find blackened remnants of this plant as it ages later in the summer. As it turns to seed, it also turns a deadly black, adding to its aura of mystery and its ghoulish reputation.

All along the way, where even a tiny sliver of light reaches the ground, the ever-present Varied-leaf Phacelia appears. It has assumed predominance alongside many of the Gorge's trails since the 2017 fire, from moderate elevations up to 4,000'. It is especially prominent from 1,300-3,700' and will be seen in all the openings mentioned below as well. Although as the remnants of the blaze recede, former flora begin to dominate.

As the trail steepens, Sweet Cicely (1,500' to 1,900'), Scouler's Valerian (especially at 1,350' and 1,400'), Broad-leaf Arnica (particularly around 1,400'), Meadowrue (1,100' to 1,450'), and the Inside-out Flower (1,200' to 1,550') all make their expected appearance. Scouler's Bluebell grows in prolific clusters, like miniature starbursts, between 1,100-1,450', here more white than blue. The Little Wild (Baldhip) Rose puts in an occasional burst of pink blossoms throughout the wooded portions of the trail all the way to the very top of the Ridge. Along the way, particularly in early June, the lily-white "petals" (actually sepals) of the Columbia Wind Flower brighten the forest depths.

A new and most unusual species now appears at 1,900' on your right: White-vein Pyrola (*Pyrola picta*), a plant as remarkable for its leaves as its flowers. Growing atop 6-12" high reddish stems, its blooms, in flower from June through much of the summer, are a strange greenish-white. They appear to be upside-down bowls with a greenish curving style (the female organ) often protruding down beyond the flower. Its leaves, in a basal rosette, are egg-shaped, and are striped in a riveting pattern of white exactly tracing its veins. These markings, similar but not identical to Rattlesnake Plantain, may help pollinating insects locate these two species, which grow in an otherwise anonymity of green close to the ground on mossy substrates. (For ways to distinguish these two plants, which possibly cause the most confusion of any two species in our forests, see the Mt. Defiance Chapter.)

White-vein Pyrola

However, the leaves of White-vein Pyrola may not even be its most fascinating feature. This plant is considered by some scientists to be an example of evolution in action. Early botanists discovered a version of this species with no leaves at all. At first thought to be a new species (*P. aphylla*), it was soon recognized to be a variant of White-vein Pyrola which was becoming increasingly dependent on fungi and bacteria in the soil of its dark home. Perhaps this plant is on its evolutionary way to becoming a full saprophyte like its Heath Family cousins, Indian Pipe and Pinesap, whose own needs for chlorophyll have been eliminated as they became less reliant on the sunlight now blocked by firs and hemlocks, and more dependent on nutrients from the welcoming soil.

The small openings on the switchbacks at 1,700' and 2,000' exhibit Rosy Plectritis, at least early in the month, along with Harsh Paintbrush, the three-pronged pinkish flowers of Prairie Star, Menzies' Larkspur (Delphinium), bright purplish-blue Cascade Penstemon, Large-flowered Blue-eyed Mary, and the Mickey Mouse-eared white blooms of Field Chickweed. The next meadows, beginning at **Point 6**, are again listed in order:

a. The first meadow, at 2,300', is announced at its commencement by a lush and dense array of Wooly Sunflower, which will proceed to be a dominant feature of this first opening. Frequent displays splash over these meadows composed of a patriotic trio: red Harsh Paintbrush, blue Lupine, and white Field Chickweed, but adds yellow Wooly Sunflower as well. This last can be distinguished from the many other yellow Aster Family plants by its wooly deeply in-cut grey leaves. It predominates in this opening in particular. At a small side trail down to a convenient resting spot, note first two specimens of Sticky Cinquefoil, pale yellow in flower and with green sepals protruding in between. Feel (gently) the stem to appreciate its common name. Several examples of the Nootka Rose occupy the far north ground beyond a downed and fire-blackened log at this spot, accompanied by a slew of Wooly Sunflowers. The Rose, our largest native of the *Rosa* genus, carries pink petals large enough to garner the envy of city gardeners; unfortunately, it lacks the scent of its more urban-bound cultivars. This meadow, and all the subsequent ones are bejeweled by carpets of Blue-field (Blue-head) Gilia, its sky-blue cluster of flowers rivalling the azure heavens above. Also populating these openings are clusters of Common Cryptantha, with small white flowers with an orange "eye" at their centers. A clutch of Heart-leaf Buckwheat downhill on your right garners attention, with its crinkly grayish leaves and large compacted head of dull white flowers. Varied-leaf Phacelia fills the interstices between more colorful plants; I suspect that as tree cover grows back, this plant may well be diminished in abundance.

b. At 2,350', the second opening adds to the above species the blue-purple Cliff (Menzies') Larkspur; unbroken fields of Large-flowered Blue-eyed Mary; the pinkish-white cleft petals of Prairie Star; the white Woods Strawberry, this last particularly prominent on the uphill, or left, side of the trail; and the ever-present Wooly Daisy shining its sunny face almost to mock your upward travails and uncertain footing amongst the marbly pebbles littering the trail. Several more examples of the Nootka Rose are here on your left as well. Herald-of-summer now appears accompanied by a veritable lawn of Blue-field Gilia; combined with Spreading Phlox, a particularly striking combination of pink, red and blue here and in all the following meadows as well. In many years, the sky-blue of the Gilia predominates, with fields populated with its charming ball-head of sky-blue compound flowers appearing as a mass of cerulean seamlessly merging with the azure skies above. However, it is the parades of Fine-tooth Penstemon which rule here; note the slimmer tubes and less deeply indented leaves which distinguish this Penstemon from the deeply serrated leaves and larger flowers than the Cascade Penstemon you encountered below. Indeed, the plenitude of these Penstemons constitute the densest population of this species in all the trail-sides within the Gorge. The few intervals where the Gilia is absent feature Birchleaf Spirea, a 1-2' tall plant with a furry mop-head of dull white flowers. Larger shrubby pink versions are cultivated in city gardens throughout the summer months. To add even more botanical interest, Nootka Rose bushes rein as you approach the left side-trail to the 2,450' viewpoint and persist as you make the right switchback leading to the next meadows. They rival garden roses, with subtle pink flowers as large as those in any cultivated garden.

c. The third opening, at 2,400', adds to the Larkspur: Blue-field Gilia; Large-flowered Blue-eyed Mary; and Birch-leaf Spirea; the two yellow umbrella-headed Desert Parsleys: Martindale's, with its ferny blue-green leaves embracing the rocky ground, and Nine-leaf, with three-parted filamentous leaves again divided into three's, each plant standing as tall as 2½'. Wooly Sunflower and Nootka Rose bushes persist here, primarily on your left.

d. Perhaps this fourth meadow, at 2,550', is the most prolific, teeming with such colorful blooms as all those mentioned above. It is especially flooded with Gilias, combined with the Lupines and Spirea's noted above; all will persist through July and early August. It is as if an artist spewed cans of blue paint across these meadows, so thick are they with the sky-blue of the Gileas. Also present are Small- and Large-flowered Blue-eyed Mary's, the latter forming a blue carpet so abundant that hardly any green can be spotted. At the very beginning of the open space, note a single example of Sticky Cinquefoil, its pale-yellow blossoms held high above viscous 2' stems. Nootka Roses remain throughout the month.

e. At 2,625', the fifth meadow finds many fine examples of Wooly Sunflower, Larkspur, Gilia, Spirea, Harsh Paintbrush, Prairie Star, sharply-pink Taper-tip Onion and even a few surviving Chocolate Lilies, lasting longer at these elevations than their brethren below. An unexpected treat is the phalanx of Thread-leaf Phacelia's, (*Phacelia linearis*) with pastel hues of violet on its five petals lined with darker blue-purple lines. These beauties grow in congregations of color on your left 2/3rds of the way across this opening. But it is the Fine-tooth Penstemon that is most outstanding here and up through not only meadows e and f but at the topmost opening at around 2,900-3,000', where it is as abundant as a politician's campaign promises.

f. This last opening, at 2,750', continues to display Gilea; Spirea; Taper-tip Onion; Larkspur; Harsh Paintbrush; and the two desert parsleys mentioned above, the Martindale's species growing right off rocks on both sides of the trail, while the Nine-leaf species towers above, but with similar yellow flower clusters and grass-like leaves. Here, and in the topmost opening ahead, the Fine-tooth Penstemon puts on its utmost dazzling display; you walk up a model's runway of this most splendid pack of penstemons trailside in the Gorge. Although gone are the Blue-eyed Mary's and Wallflowers of earlier spring, these meadows nonetheless still provide the color of sun-loving blossoms now, and will continue to do so throughout the summer months.

The woods above these meadows, up to **Point 7**, the junction with the Ridge Cutoff Trail, contain armies of Broad-leaf Arnica, yellow and daisy-like; Candy Flower; and one more specimen of White-vein Pyrola, this at 2,700' on your right. The large, steep, and rocky opening just beyond, at 2,960', is now thick with Taper-tip Onion, at least through the early weeks of June, its appetizing odor permeating the meadow and its deep pink color contrasting well with the battalions of blue-purple Fine-tooth Penstemon and the yellows of Martindale's and Nine-leaf Desert Parsleys. Here too, a common weed has penetrated these heights from its origins alongside railroad right-of-way's, roadsides, and abandoned city lots – Sheep Sorrel (*Rumex acetosella*), a relative, believe it or not, of the common garden rhubarb. This 10-18" tall straggly plant has visibly arrowhead-shaped leaves and reddish-brown "flowers" that, as with the related docks of urban roadsides and fields, resemble seeds more than an inflorescence.

The leaves of Sheep Sorrel have a tart taste not unlike Oregon Wood Sorrel due to the presence of oxalic acid, not unpleasant in a mixture of salad greens. The runty flowers of this plant bear the distinction of being dioecious, a trait this plant has in common with only a few others in the Northwest, including Meadowrue and Goat's Beard. Dioecious species carry their female and male flowers on separate plants whereas most other species combine male parts (such as stamens) and female parts (such as styles and ovaries) on the same plant, usually on the same flower. Thus this pernicious weed of European origin should offer some elements of interest to the amateur botanist, despite its undistinguished appearance.

Sheep Sorrel

> Of what evolutionary advantage could it be to divide female and male parts onto separate plants? For one thing, this disconnection could help to avoid self-fertilization, sort of the same idea as not marrying your sibling because lethal genes in a family lineage can merge and become dominant, leading to dire consequences for the offspring. Recent research, for example, has demonstrated that inbred plant progeny have stunted growth patterns, lower amounts of chlorophyll and few to any flowers. Then why are the majority of species monoecious? Perhaps because nature loves thrift. It may have been easier for evolution to shape plants with combined sexual parts than to disassociate them, at least in the majority of cases. The presence of dioecious species, however, reminds us to be wary of overarching generalizations and the hubris they can engender in the human world. Fortunately, such hermaphrodism is rare in our species, but not unheard of in the reptilian and piscine realms, where it may either confer benefits as yet undiscovered or be a result of random innocuous mutations. The text explaining all these mysteries remains unwritten. Authors are desperately needed.

Beyond this opening at 2,980', Candy Flower; Stream Violet; and Birch-leaf Spirea, a foot tall with a brushy mop-head of dull-white florets predominate. However, at a brushy area at about 3,000', Lupine can be found on your left, continuing into the woods thereafter and accompanied by Small-flowered Alumroot, occasionally right in the middle of the thin trail. As the trail trends subtly downward through a veritable forest of 5'-tall Fireweeds (not yet in flower) it crosses several logs and thins perceptibly, though it is not too difficult to follow, even though dense vegetation often obscures the true path. From this point onward, and following a slight dip in the trail, Scarlet Columbine begins to appear, especially on your right as you do the ups and downs of the thinning trail and will continue to brighten your way even at times up the steep last 500' of on-and-off-again trail to the top of the Ridge. Star-flowered Solomon's Seal and Miner's Lettuce still blanket the woods beyond as the trail dips, then takes a turn to the right leading to **Point 8**, the junction with the Ridge Cutoff Trail. A few feet beyond the trail junction, three specimens of Scarlet Columbine grace the trail on your right. Just beyond, by mid-month, a surprise awaits: Tiger Lilies will also be found close by, many already in glorious bloom by mid-to-late June, again mainly on the right. From this point, and on up the steep trail to the top of the ridge, look for the hidden purplish-brown flowers of Wild Ginger, matted against the forest floor beneath its large heart-shaped dark green leaves.

Vanilla Leaf still hangs on at these heights as well, though only to mid-month. A gathering of Scarlet Columbines brightens the way as well. At the top of the ridge, 3,550', the Inside-out Flower and Vanilla Leaf assume control all the way to **Point 9**, the usual stopping point for most day hikers. Here, Serviceberry and Ocean Spray straddle the boulderly ground. However, you may continue along this ridge to a second grouping of rocks (but precious few views) ¾ mile further. If so, you will also find choice examples of Oregon Boxwood, although its diminutive red flowers may by now have gone to seed. Nonetheless, the leaves of this shrub are distinctive: Thick, leathery, and occasionally rolled over, they bear fine teeth along their edges. Learn this 2-3' tall shrub well as it appears repeatedly at mid-to high elevations throughout the Cascades.

Now finished with the ascent, you can relax at either rocky destination and enjoy your June lunch while contemplating one of the many round trips possible down to your starting point. A terrifically scenic loop, combined with an energetic work-out, make Nick Eaton worth many return trips.

July and August

What more agreeable way to beat the heat of a summer's day than to stroll up the largely-shaded Nick Eaton Trail and reach its cooling heights? Along the way, the flowers of late spring may be fading but summer's blooms, if not more plentiful, are no less enchanting. You need not stop at the parking area, **Point 1**, as you can drive up the paved road to **Point 2**, the picnic area. At both spots, Robert Geranium belies its weedy status with charming pink blossoms atop well-dissected leaves, while the purple garden violet (called "English Violet" by many) continues its long-lasting flowering. One new species, out by early August, Chicory, surmounts these low-lying plants with its 2½-4' tall stems topped by 1-2" sky-blue dandelion-type flowers with squared-off tips. It can be found by the sign marking the trail at **Point 2** as you enter the forest. A common Aster Family weed of summer roadsides and fields, Chicory nonetheless delights with its stature and the intense hue of its blossoms, which happily last all summer and into the fall. Although not recommended, Arabians and early American settlers often dug up the roots and ground them into a coffee substitute, a practice continued today in both Europe and the Middle East. Its leaves have also been used in salads.

Progressing up the trail, you encounter Robert Geranium (cleft leaves rounded at their ends and a protruding style shorter than ½" with brownish-red stems); this should be distinguished from Filaree, another geranium but with sharper endings to the leaf tips, a deeper red shade on all its stems and an outsized style up to one inch long. Both these Geraniums are pretty pink invaders from Eurasia and the Robert species will continue its cheerfully intrusive presence up to the power line. Self-heal is plentiful, its purple flowers atop 6-12" stems; so abundant is this plant near the beginnings of trails and in trampled earth, that many botanists assumed it must be an imported weed, but newer study shows it to be a circumboreal species native to many northern lands.

About halfway up this trail, at about 250' elevation, Queen's Cup makes an appearance, with six white petals (actually "tepals", a combination of petals and sepals common in the Lily Family) and two or three elliptic leaves often reclining on the ground in seeming support of the 6-8" stems. Despite being a lily, I've not been able to detect much of an odor from these otherwise handsome plants.

At the power line, **Point 3** at 400', the common weeds Dandelion and both Red and White Clovers deserve no introduction. However, blue Lupine now hides some of the stony overworked ground, at least through late July, accompanied by the compound umbrella-like flower clusters of queues of Queen Anne's Lace. This royal weed is so dense here and up the trail for the first ¼ mile, that it may be taking over more native flora. Queen Anne's Lace has a habit of curling its clusters of blooms inward as the summer progresses, so much so that at times it can resemble a sphere. More often, an opening is left at top, which most authors compare to a bird's nest. While it does resemble a small bird's dwelling, I have yet to see any passerine move in with her brood and set up housekeeping.

Just before you enter the woods and after you've passed under the power line poles, a clump of Tansy Ragwort can be seen on your left. As you re-enter these woods just above the power line, notice on your left the "cool caves", their gusts of chilly air a welcome haven on a blistering summer's day. Here, you will find the delicate flowers of Small-flowered Alumroot, its 2' tall stems adorned by sprays of dainty white blooms resembling miniature bells, the stamens reaching beyond the tiny petals like diminutive silent clappers. Cultivated varieties, mostly colored red, often decorate urban gardens. On your right, just after the cave-like rocks, Queen's Cup adorns the forest floor, while on your left, purple Self-heal thrives in the dappled shade.

By mid-August, the floral show is mostly devoid of blossoms but a new plant assumes prominence here (and often at other low-lying trampled areas of the Gorge, such as at the approach to the Mt. Defiance/Starvation Trails), Spotted Knapweed. While its frayed shrub-like appearance and thin spiny leaves betray it as a foreign invader, and its affiliation with the Thistle Tribe of the Aster Family, its blooms do sport pretty pinkish-purple flowers supported by a curiously spotted and highly visible bract.

The renowned Columbia, or more commonly named Tiger, Lily appears just after the "cool caves" and continues, mainly on the left of the initial trail until the switchbacks at 500'. However, throughout July and even into August at higher elevations, these huge drooping orange blooms with six petals spotted with burgundy-to-brown spots and growing up to 6' tall, cannot be missed. Unfortunately, some thoughtless hikers have swiped their tops but even without flowers, you can still appreciate their multiple, but spaced whorls of six linear leaves swirling about the towering stems. By late in August, however, their blooms spent, their branches reaching toward the sky are but a botanist's distant memory of their orange glory of earlier summers.

> Deception is as common in plants as it is in the human world. Consider the common daisy. What appears to be the simplest of blossoms turns out to be an intricate assembly of floral parts masquerading as a single flower. The apparent "petals" are in reality each a single flower, with stamens and a pistil all its own; the central disk continues the charade, each tiny point actually its own flower as well, with its own sexual parts. Some Aster Family species (which include all the daisy-like plants in our area) contain only disk flowers (for example Pearly Everlasting and Rayless Arnica) and others have ray flowers only (think Dandelions and Chicory). The combination of both disk and ray flowers is the most common arrangement in this large Family, as in our beloved arnicas and balsamroots. One of the Latin names for the Aster family is the *Compositae*, the term connoting the blended nature of this family's common architecture. Such deception apparently evolved to enlarge the flower head in order to make it more visible to pollinating insects, but it is not unique to the Aster Family. Dogwoods, both in tree and plant form, advertise their presence through bright white bracts, a similar artifice seen in the Christmas Poinsettia's leafy red bracts. Orchids are famous for mimicry as well, rearranging their shape by evolutionary means to entice insects and hummingbirds into thinking what they are not. For example, the *Tolumnia* genus contains one species that is shaped like a certain type of female South American beetle. Male beetles are attracted to the flower in their futile attempts to mate with it! Thus subtle and enthralling anatomy serves as a guise to ensure survival in a highly competitive world.

You emerge from the woods, just after a junction with the connector to the Pacific Crest trail and Herman Creek, onto a series of brief, now largely overgrown openings through which the trail switchbacks as it gradually gains altitude. In these openings, from about 500' to 600', you will find both White-flowered and Scouler's Hawkweeds, the first 2' tall with tiny white daisy-like flowers, the second 3' tall with yellow dandelion-type flowers. Scads of the Scouler's line the trailsides. Both commonly appear on the upslope side. While the Cascade Penstemons of late spring have now mostly vanished by mid-July, Harsh Paintbrush daubs dashes of crimson on the slope while yellow Wall Lettuce and white Woods Strawberry, Queen Anne's Lace and Pearly Everlasting offer their contributions, the last two especially prominent at a right switchback at 550'.

A grassy widening in the trail at about 800' just before it becomes a true road, and where another road enters from the left, holds another Aster Family flower in the summer, one which should be familiar to any urbanite with a lawn – the English Daisy. This inhabitant of city lawns worldwide can range from 1-8" tall and its prototypical white daisy-heads with yellow centers, usually one or two to a plant, resemble a dollhouse version of the Oxeye Daisy, so commonly seen in fields and roadsides in late spring and early summer. Like the Oxeye, English Daisy was introduced from Europe in the 19th Century as a cultivated ornamental plant but has aggressively pursued American real estate ever since. It can bloom as early as April in the city and persist through the summer elsewhere. You can see it popping up like daisy pimples on the faces of lawns in Portland and Seattle, a not un-pretty flower, but one considered a blemish in its usual habitats. Also, crowding this opening are squadrons of Scouler's Hawkweed towering above the omnipresent Self-heal.

Up the road, the tiny flowers of White-flowered Hawkweed compete with the button-like buds of Pathfinder (Trail Plant), mostly on your left. The two can appear superficially similar but the Hawkweed has single stems topped by a diminutive white daisy while the flowers of Pathfinder have multiple stems with a number of terminal nodes masquerading as blossoms.

About 10 minutes after you begin the gentle ascent from this flat open area, openings on your right, secondary to the 2017 fire, reveal a startling display of Fireweed, its deep pink blossoms standing tall atop 4-6' tall stems. The show continues even past **Point 4**, the junction with the Gorton Creek Trail, at least for another 50 yards until you almost reach the left turn starting the trek up Nick Eaton Ridge proper. Undoubtedly the conflagration opened up this area but the heat of the fire paved the way for Fireweed's cottony seeds to germinate and spring forth in such a bounty. These seeds can lie dormant under the earth for over 40 years or more, awaiting such a blaze to activate them and propel them to such a beautiful abundance. It will be ablaze in all the openings from here to the top of the Ridge, its four petals a striking color, as deep pink as the eye can perceive.

By late summer, this tall plant has shed most of its wispy seeds, blown by the rare winds of late August to distant spots of earth far from their mother plant, to settle in and sprout (germinate) without competing directly with their parents.

The trail becomes a road as you hike through the forest, with the Inside-out Flower predominating and an occasional Trail Plant (also called Pathfinder) marking the way. The stems of this species can reach over 3 1/2' tall. Just before and after **Point 4**, at the old Herman Camp, an occasional arc of sunlight arches through this alley of green and brown, highlighting the sometime blooms of Little Wild (Baldhip) Rose. Starting up the Nick Eton Trail at a left turn another 100 yards up the road at **Point 5**, 1,000', you will notice almost immediately Merten's Coral Root, its foot-tall reddish stems and poorly developed reddish-white leaves betraying its saprophytic nature. One fine specimen appears on your left first, followed by another almost immediately on your right. About 20' further, another species with a riveting pattern appears on your right, White-vein Pyrola. This plant has 10" tall stems dotted with greenish-white bell-like flowers. At about 300 yards up this trail, you will find Rattlesnake Plantain, with its mottled green and white leaves and inconspicuous greenish-white flowers appearing almost like knobs rather than blossoms. Notice the difference in the clear delineation of the leaf veins outlined in white on the Pyrola as opposed to the mottled pattern on Rattlesnake Plantain.

Tiger Lily specimens grow to your left beyond the junction and again can be enjoyed just minutes after you turn up the Nick Eaton Trail proper. These, both right and left, may persist at least through late July. They will reappear sporadically along the switchbacks between 1,110-1,900', then again just after a log crossing at 3,000' after the last opening. The common weed, Wall Lettuce grows tall here as well, with stems reaching 4' tall, topped by numerous five-petaled yellow flowers.

Since the 2017 Eagle Creek blaze, Varied-leaf Phacelia has assumed prominence alongside many Gorge trails. It comes into bloom in late June and lasts throughout the summer months. A 3-4' tall shrub-like plant, it is topped by a curving coil of dullish white flowers. A member of the Waterleaf Family, its prominence since the fire may have been caused by the thinning of competing shrubs, the increasing amount of light with the relative absence of greenery in the trees above, and the heat of the fire itself. As with Fireweed, Varied-leaf Phacelia's seeds may be activated by the heat of forest fires. Indeed, the seeds of several species can lie dormant for over 40 or more years, then magically sprout into life once the intense warmth of a forest fire activates their growing cycle.

Progressing up the steepening trail, at **Point 5**, the woods hold examples of Trail Plant, and more samples of Rattlesnake Plantain. To add some color to this green scheme, Broad- leaf Arnica, with its daisy-like yellow heads and scores of the sky-rocket Scouler's Bluebells, with their pale-blue swept- back petals flaring boldly against the brown earth, both add charm to these shady environs, best appreciated on a hot summer's day.

The small openings at 1,700' and 2,000' are filled at this time of year with hot pink Taper-tip onions, white mouse-eared Field Chickweed, a few Herald-of-summer plants and white-topped Yarrow, another Aster Family plant with composite heads of ray and disk flowers, albeit tiny ones. A few lingering Cascade Penstemons hang on through late July, with deep purple tubular blooms and the last vestiges of Rosy Plectritis linger throughout that month as well. An additional surprise is Common Clarkia, with four small flaring pink petals atop 2' stems. Mostly found further east, this Evening Primrose Family member displays unusual petals that start out as small straight projections from the main stem, then flare out into an arrowhead shape at their tips. In between these brief openings, scads of Scouler's Blubells cluster to form, both right and left, miniature blue starbursts.

Beginning at **Point 6**, the first of the six major openings on the Nick Eaton Trail appears. Each will again be listed, with its summer-blooming floral contents, below:

a. This first meadow, at 2,300', shows its colors early in the season, and retains most of them throughout the summer. You are introduced to it by Woolly Sunflower, which is sprinkled throughout all these openings during the summer months. Sky-blue Bluefield Gilia, with ball-heads of gorgeous light blue predominate in July and through mid-August, along with Birch-leaf Spirea. Crimson Harsh Paintbrush, blue Lupine, purple Cascade Penstemon, the thinner tubes of Fine-leaf Penstemon, yellow Wooly Sunflower and St. John's Wort. These all compete for attention like so many noisy siblings, but fortunately in a manner more calm and attractive. Since the 2017 blaze, Varied-leaf Phacelia, about 3' high and topped with a coil of dull white blossoms, will be found in all the subsequent openings described and will continue to dominate the shoulders of the trail in open spots all the way to at least 3,700'.

b. At 2,350', the second opening is a happy repetition of the first, but with the additions of Herald-of-summer's red-flecked pink blossoms scattered along the way, along with hordes of Three-vein Fleabane - shrubby, 2-3' tall and bearing gorgeous purple daisy ray flowers with yellow centers.

c. The third meadow, lying at 2,400', adds to the above the two common Desert Parsleys, Martindale's and Nine-leaf, their yellow cluster of flowers sparkling in the summer sun. Also spy the unmistakable Herald-of-summer, its pink blossoms marked with a deeper red splotch. The Cascade Penstemon will persist throughout the summer months here along with Wooly Sunflower as well.

d. It is a bit of a climb to reach the fourth opening, at 2,550', but well worth the effort, as all the above summer flowers are in bloom here, including the widespread Wooly Sunflower. Masses of Blue-field Gilia and Howell's Daisy (*Erigeron howellii*) compete with the sky for your attention. Birch-leaf Spirea continues to bloom throughout the summer while Sticky Cinquefoil has surfaced, its 2' tall viscous stem crowned by pale yellow blossoms.

e. At 2,625', the fifth meadow contains the beautiful red and blue combination of Harsh Paintbrush and Lupine, so common in our mountains due to their mutually beneficial association. To complement their presence, battalions of Blue-field Gilia carpet the meadow, along with Birch-leaf Spirea. Howell's Daisy shows off its purple daisy-like flowers, while yellow Nine-leaf Desert Parsley bends in the gentle zephyrs of summer at these heights. While not strictly in this opening, a single Little Pipsissewa (Prince's Pine) pokes its modest head up in the woods on your left just beyond this meadow at 2,650', its waxy white swept-back five petals surmounting a 10" stem. It is in this meadow that an unusual plant makes an unforgettable appearance: Salmon (or Giant) Polemonium. About 2-3' tall, each stem is topped by a cluster of light orange tubes. Many appear on your right, some very close to the trail. This is one of the few locations this species blooms in such profusion and this far west.

f. The last opening occurs at 2,750' and offers nothing new but everything splendid, including Tiger Lilies, Gilia, Wooly Sunflower, Cascade Penstemon, Fleabane, Spirea, Paintbrush, Lupine, and Martindale's Desert Parsley and a cache of bright purple Cascade Penstemons. Note that, by summer, Cascade Penstemons appear to have replaced the smaller Fine-tooth Penstemons, which dominated the highlands in May and June. Do not mourn the last of these openings; others will grace the woodlands above and on your loop back down.

The forest above hides another White-vein Pyrola, at 2,700' on the right. This, and the Little Pipsissewa mentioned above, along with the more common regular Pipsissewa, now found above 2,700' here and on to the point where the trail hits the top of the first ridge at 2,850', all belong to the Heath Family. However, many botanists now place them in their very own Wintergreen Family. The pipsissewas and pyrolas are all closely related but their manner of growth can differ. Regular Pipsissewa sends creeping rhizomes (underground stems) in all directions; thus fields of this plant are often found inhabiting shady mid-elevation conifer forests. However, its smaller cousins, like Little Pipsissewa and the Pyrolas – White-vein, Sidebells, and Large – all appear quite asocial, and emerge in isolation here and there, with seeming indifference to whether any relatives live nearby. All these wintergreen-like heaths, however, do share a preference for deeply-shaded habitats; their success in dense forest suggests a partly saprophytic lifestyle.

As you progress the mildly angled trail on top of the Ridge, tons of Tiger Lilies are still in bloom from late July through early August. They are accompanied by Candy Flowers and Birch-leaf Spireas, which do as well in partial forest as in the open.

At 2,850', the trail meets the top of the ridge and, besides Pipsissewa's, many Candy Flowers live here in summer. Look for these plants for ¼ mile before **Point 7**, the junction with the Ridge Cutoff Trail. Here too are the first of the many Foam Flowers you will find in woodlands throughout the Gorge in summer and even early fall. This 12" tall Saxifrage Family plant sends up clusters of tiny white flowers throughout our forests' under-stories from low to high elevations; it deserves to be recognized given its abundance and fragile beauty. Its appearance each summer, as dependable as the ocean's waves and tides, marks the middle of summer's all-too-brief tenure and brightens the canopy of woodlands with its sprays of dainty white bells, vaguely reminiscent of the foam thrown up by the sea. Inhabiting forests from Alaska to the Sierra, Foam Flower is also one of our most widespread and enduring plants. Specimens are often found blooming through October.

In summer, the large opening at 2,960' contains some hot-pink Taper-tip Onions, Wooly Sunflowers, Birch-leaf Spirea, Three-vein Fleabane, and the omnipresent Yarrow. The large white-topped umbrella-like plant just past the saddle at 3,000' is Kneeeling Angelica (*Angelica genuflexa*), its appearance betraying its membership in the Carrot Family. While this moisture-loving plant can grow up to 8' tall, here it is about 5', perhaps a reflection of the relative lack of water at this location. Its hollow stems are topped by a flat cluster of white flowers supported by stokes of its umbrella-like pedicels. The stems of all angelica species can be eaten raw or cooked with sugar syrup to make candied angelica. I would advise against it on conservation grounds alone, but in addition, angelicas look a lot like Water-hemlocks, among the most poisonous plants known to humankind.

Kneeling Angelica

> Saxifrage Family plants carry scientific names that appear similar for a good reason: They are anagrams. Botanists evidently ran out of appropriate Latin-sounding names for these species and, perhaps, as botanists are wont to do, decided to have some fun with words. Their jovial endeavors have given us *Tiarella's, Mitella's, Tellima's,* and *Tolmiea's*. To add to this literal confusion, Tiarella's come in three flavors – Single-leaf, Three-leaf, and Cut-leaf varieties, called *T. unifoliata, T. trifoliata,* and *T. laciniata*. DNA evidence now groups all three in the same species. Our forms are usually the three-leaf variety, although some single-parted leaf forms can be found at higher elevations in the Gorge. Such detailing, however, is cumbersome for us amateurs. By the way, *Tiarella* is the diminutive form of tiara, originally a Persian turban. Our plants do not resemble turbans; even sea foam is a stretch.

The woods at 3,000', from **Points 7**, the junction with the Ridge Cutoff Trail, to **Point 8**, the junction with the Deadwood Cutoff where the trail is fairly level, contain more Foam Flowers and a few pink Little Wild (Baldhip) Roses. Tiger Lilies can be spied, usually on your right, especially after crossing over a log (step cut out in its middle), their orange petals speckled with maroon spots like a predatory feline hiding in ambush in the woods. Scarlet Columbines will persist throughout much of the summer from this part of the trail through the slim steep trail to the crest of the Ridge.

As you start up the steep trail to attain the top of the ridge above, notice the large yet intricate structure of these Columbines, now mostly on your right. Here and up the slope, Foam Flower and Little Wild Rose accompany you as you labor to gain the ridge, along with several Scarlet Columbines on your right. Once on top, another Kneeling Angelica greets you near some rocks. Along the top of the ridge all the way to **Point 9**, the lunch spot by several large boulders, you will also find some Foam Flowers and Little Wild Rose. You could continue along the up and down ridge another ¾ of a mile to another cluster of rocks but you would only gain about another 100' of elevation and no more pleasant views. Resting here at **Point 9** or further at the second rock outcrop, is certainly allowed, considering your elevation gain of about 3,700', considering the ups and downs of this trail, especially so because you now have to lose all that altitude. Nonetheless, it's been a rewarding trip and at least up untill now, hopefully your knees are still functioning; you'll need them on the descent, no matter which trail you take down.

September and October

The Forest Service now keeps the gate open through the autumn months and thus you will not have to walk up the paved road from **Point 1**, the parking area by the turn-off, to **Point 2**, the picnic area at 180'. However, notice the tall sky-blue Chicory as you begin the short and slow drive along with the pretty pink Robert Geraniums adorning the roadside along the way. Unfortunately, blooms are absent as you begin the actual trail at **Point 2** ascending the woods up to the power line at **Point 3**, 400', although here at least some weeds grace the unsightly gash: The well-known and often detested lawn weeds Red and White Clover put in an appearance along with Queen Anne's Lace and Pearly Everlasting. Don't fail to appreciate the multi-hued fallen leaves of the Big-leaf Maples strewn upon the autumn forest floor.

Also here, not that anyone could miss it, is the Common Dandelion. At almost any time of the year, and of far-flung distribution, this Eurasian plant is perhaps the most despised species in the Plant Kingdom, even though scores of other plants do more harm (think of cropland damage from Russian Thistle or allergies to the Ragweeds). Upon close and solitary inspection, however, Dandelion does not lack charms. Its heads of bright yellow rays turn gradually into fluffy balls of individual long-beaked seed heads known as achenes; what youngster, and many an adult, hasn't puffed at a Dandelion gone to seed in order to watch the dainty parachutes float off in the breeze? Perhaps we loathe the Dandelion because it despoils our lawns, but I think we also disdain it as much because it is so common. Just as a fancy dress or piece of jewelry looks all the more valuable displayed as a single featured item in the windows of Hermes or Louie Vuitton than the same item crammed into a discount store window with hordes of similar items, so the scorned Dandelion might be better appreciated were it less common.

At the "cool caves" just as you re-enter the woods after the power line, Self-heal is still in bloom on the left through early October, its purple flowers atop foot-tall stout stems. Even though the flowers are grouped at the top of the stem, examine each small blossom to discover its intricate structure. The sepals are united into a two-lipped tube while the petals are similarly fused into two lips as well, the upper lip shaped like a hood over the lower one, three-lobed and fringed. Such floral architecture is a common indicator of the Mint Family, to which Self-heal belongs.

The switchbacks further on up the trail hold several overgrown openings between 500' and 600' elevation. Here, White-flowered Hawkweed persists throughout September with its tiny white daisy flowers atop 2' tall branching stems. Pearly Everlasting lives up to its name here as well; its white bud-like blooms fringed with a touch of yellow may well last into early November in many habitats. Along the road for about a quarter mile before Herman Camp at **Point 4**, 990', Wall Lettuce continues to show off its yellow flowers, each with five petals, a rare consistency in the Aster Family, where a variable number of petal-flowers forms the rule. Although Wall Lettuce originated in Northern Europe and is considered an invasive weed in Northwest forests, its color here, within this vault of green, is welcome, at least before very much reflection.

The only other flowering plant here in autumn is Pathfinder (Trail Plant). Despite being 2-3' tall, it is often neglected as it has tiny dull white discs for blossoms that practically require magnification for acknowledgement. As you continue up the dusty road, then turn left up the Nick Eaton Trail at **Point 5**, 100 yards from Herman Camp, the saprophyte, Merten's Coral Root, appears first on your left, then on your right. Lacking chlorophyll, its reddish stems and flowers stand out against the green and brown hues of these claustrophobic woodlands. Just a minute or two later, look for White-vein Pyrola still in bloom on your right, at least through mid-September, with its bell-like greenish-white blossoms nodding above its 10" tall stems.

Just past **Point 5**, where the Nick Eaton Trail has yet to steepen, Rattlesnake Plantain remains in bloom throughout September while the other summertime woodland plants here have begun their winter slumber. However, note on your left, just past the stone etched with the names of the trails on your left, about 3-4 minutes up this trail, several yellow Scouler's Hawkweed plants still in bloom. More remarkable are the bright pink blossoms of Fireweed you will encounter a few minutes away, and these will continue in the woods sporadically up to the meadows at 2,200', then beyond in brief openings all the way to 3,100'.

These plants are dwarfs compared to the Fireweed we know from other regions because they have just awakened, in the summer of 2019, after the 2017 conflagration. True to its name, many Fireweed seeds lay dormant under the soil until the heat of a fire promotes their germination. These Fireweed plants are just 2-3' tall now (summer of 2024) but will continue to grow to their eventual, and more typical heights of 5-6'. These should not be confused with Broad-leaf Fireweed (or Willowherb), a 2' tall species of gorgeous deep pink color painting damp hillsides from 4,000' to 7,000' in the North Cascades.

A second species appears to have benefitted from the 2017 fire: Varied-leaf Phacelia. It is bush-like, 3' tall, with curling fuzzy dull white flower stalks. It is far more common here, beginning at **Point 5**, and continuing intermittently in forest and meadows to 3,000', then again on the steep rough trail up to the ridge at 3,500'. I know of no reference to its benefitting from fire but its presence in this trail was hardly noticeable until after 2017.

Continuing along, an occasional Pathfinder and some Wall Lettuce take the hiker to the two small openings at 1,700' and 2,000', where striking pink Taper-tip Onion shares these autumn meadows with Yarrow and Pearly Everlasting, at least early in September. In the forest above, Wall Lettuce, Foam Flower, and Pathfinder predominate, often until mid-October in mild years.

The six openings beginning at **Point 6** now have shed most of their blooms but in the first two, 2' tall yellow St. John's Wort holds steadfast right and left, and at least here, we can forgive its weedy nature so prominent in our urban fields and roadsides. Also in these first two meadows are mats of Autumn Knotweed, a tiny plant growing just 6" tall, with a curl of nondescript flowers sticking up from its basal narrow leaves. It's sort of like a miniature imitation of Lady's Thumb; see the Table Mountain chapter for a more complete description of this late bloomer.

In the last two of these meadows, at 2,625' and 2,750', a different yellow blossom appears - the bushy cool-weather-loving Hall's Goldenweed, so abundant in autumn that we need not forsake it for its weedy name. In fact, this native plant, a member of the Aster Family, is a mainstay of fall heights in the Gorge, and happily so as it graces open meadows at mid elevations through November in many cases, when all other floral life seems moribund. Rub its stem gently to appreciate its glandular secretions, rendering it a resinous feel.

A single specimen of Little Pipsissewa, with its two nodding waxy white bell-like flowers drooping atop a 6" stem, can be discovered hiding in the forest at about 2,700' on your left. In the large stony opening just beyond **Point 7**, the Ridge Cutoff junction at 2,850', Yarrow persists until late September, but in the woods beyond, only Pipsissewa, Foam Flower, and Candy Flower can be found from here to **Point 8**, the Deadwood Cutoff at 3,040'. These pink-striped Candy Flowers began blooming here in the month of April; are these the same plants? Perhaps not. These members of the Purslane family are annuals but may also re-bloom as short-lived perennials (see the North Lake Chapter for a discussion of the differences). It is suspected that the Candy Flowers we see in autumn may not be the same that we see in this spot in the spring. Rather, they may have either re-bloomed from the same plant, or arisen as a different late-flowering variety in the fall. Regardless, they are welcome at these heights, offering a bit of floral color to these often-darkening woods.

> A number of plant species appear to bloom again in the fall after an initial spring flowering. Our bright purple and yellow crocus garden varieties and the colorful dogwood trees of lowland Gorge forests are examples, as are several species of rhododendrons and azaleas. But some species seem to completely disregard our common wisdom about when they are supposed to bloom. I have witnessed Western and Northwestern Saxifrages re-blooming with September's rains and Kittentails showing their pretty purple heads in late December. The same occurs in our city gardens, where

Camellias and Bergenias can begin to bloom in late January and where some Rhododendrons re-bloom in October. These, however, should probably not be thought of as misfits or as delinquent progeny from dysfunctional families. Rather, they are likely genetic variants whose wandering DNA may, with some future and unpredictable catastrophic climatic event, confer some advantage to their kind. Thus Earth's wondrous variety of plant and animal life can take advantage of slight genetic alterations in order to be ready to survive the inevitable changes our planet will incur. One wonders if the same flexibility will be present in our own species

Up the ever-steepening trail beyond **Point 8**, Pathfinder and Rattlesnake Plantain share the woods with Foam Flower and Pipsissewa, with its 10" tall stems topped by four or five pinkish down-turned bells. On top of the ridge at **Point 9**, 3,550', Kneeling Angelica greets you by a cluster of rocks on the trail where I recommend you rest and eat lunch. Beyond this point, up and down along the ridge, Little Wild Rose makes a spotty appearance, but much of the inflorescence is by now altered by the season into a grouping of large red berries called "rose hips", rich in Vitamin C. Also along this stretch in autumn, examples of Foam Flower persist, occasionally until mid-October. Thus ends the fall flower show on Nick Eaton Ridge, perhaps not the most prolific and colorful season to accomplish this loop, but one nonetheless rewarding not only in the amount of cardiovascular benefit it confers, but in the host of unusual plants it offers, further opportunity for amateur botanists to hone their impressive plant identification skills!

Down the Ridge Cutoff and the Gorton Creek Trail

In order to complete a loop, you must retrace your steps down Nick Eaton Ridge to **Point 7**, the Ridge Cutoff, unless you wish to undertake the complete loop by continuing another mile and a half or so on Nick Eaton Ridge beyond **Point 9** to the point where the Nick Eaton Trail meets the Gorton Creek Trail further southeast (I wouldn't). The loop described here, with the re-tracing of your steps back down is not necessarily an onerous task as, once down the steep part of this trial, and below **Point 8**, the grade eases considerably and the remainder of this round trip is a fairly gradual trek downhill with opportunities to view a somewhat different floral array than on the way up. You could also take the Deadwood Cutoff at **Point 8**, but this is a bit rougher a trail and takes just a bit longer. Either way, however, you will combine a great workout with the opportunity to revisit old floral favorites and meet some intriguing new species as well.

Please remember that the descriptions of right and left are given as you are *descending*.

February and March

At these relatively high elevations, at least for the Gorge, not much is happening on the Ridge Cutoff this early in the year. Indeed, snow may still cover even lower parts of this loop. Still, not all is lost botanically as some Trilliums are beginning to bloom as early as March between **Points 7** and **10**, 2,850' to 2,680', along with a few Evergreen Violets, their recumbent yellow petals an enticing herald of spring's commencement. As you turn left at **Point 10**, where the Ridge Cutoff meets the Gorton Creek Trail proper, you may also discover some of these early violets. Should you turn right here instead, in a few yards you will come to the rude path on your left that descends to the famous viewpoint of Indian Point. By all means, take this 15-20-minute detour, even though it forms the definition of the word "steep".

Here, at the Indian Point side trail, you will drop down quickly at first to a plateau overlooking to your left the sheer east wall of a subsidiary crest off Nick Eaton Ridge. Then you proceed along the narrow but level humpback to a field of talus and, straight ahead, the spire of Indian Point itself, **Point 11** at 2,450'. If competent on somewhat unstable basalt, climb the crumbly pinnacle (a Class 3 scramble) for the most spectacular viewpoint on the Oregon side of the Gorge. If not, satisfy your soul by admiring the almost equally stupendous views out to the river and amuse yourself by yodeling silly sounds at the wall to your left and listening to how inane they are as they echo back to you.

Back on the main trail (yes, it's a struggle to climb back up, but you could have left your pack at the junction), head down the Pacific Crest Trail-like broad switchbacks and search for the few Trilliums beginning to emerge and the occasional Evergreen Violet, these mostly below 1,200' in late February and above it by mid-March. Just beyond **Point 12**, a large left switchback at 2,240', several Evergreen violets are blooming by late February. At **Point 13**, just after a right indentation, a winter-only stream may be flowing. Here, Coltsfoot can be found as early as late February, particularly on your right, its cluster of up to 12 disk-like dirty white flowers atop 2' high stems. This water-loving plant hardly looks like it belongs in the Aster Family; its main distinction, aside from its extremely early time of flowering, is its habit of sending up its inflorescence before its leaves even get a chance to emerge. By the time the leaves catch up, the flowers have all but withered, leaving everyone other than early-season explorers to wait and wonder when the plants will bloom!

Between this point and **Point 4**, Herman Camp, where the Gorton Creek Trail rejoins the Nick Eaton loop, you may find sparkling white Trilliums and a few more Evergreen Violets, but that is all. No matter – this loop still has offered plenty of exercise and a taste of the floral banquet to come later in the spring.

April

Now Nick Eaton loopers will truly experience some of the best the Gorge has to offer, with views as fine and flowers as diverse and colorful as most any other trip hereabouts. Down the Ridge Cutoff, between **Points 7** and **10**, you will find many Trilliums, now in gleaming satiny white. Just past the turnoff from the Nick Eaton trail, scan the right bank for a choice selection of Evergreen Violets, their yellow ground-hugging flowers atop dark green heart-shaped leaves. Adding color to the background tapestry of light spring green, a number of intense blue Oregon Anemones line both sides of the trail between 2,800' and 2,700'. The trail here is relatively gentle, causing one to wonder if it will ever descend to the Gorton Creek Trail. However, after about ⅓rd of a mile, the way begins to drop a bit more steeply, and here, at approximately 2,740' in two brief partial openings, enough sunlight reaches the forest floor to encourage the growth of two small fields of Glacier Lilies, their nodding yellow flowers replete with six reflexed petals above a brace of oblong leaves clasping the 1' tall stems. You will have to climb higher to find its

Avalanche Lily

sibling, the white Avalance Lily (*Erythronium montanum*), rare in the Gorge and not observable from the trails described in this book. Try the Vista Ridge or Paradise Park Trails on Mt. Hood in June and July or the parklands around Mt. Rainier; fields of these white beauties sparkle in the open woods from 4,000' to 6,000' in these locations.

As you greet the Gorton Creek Trail, **Point 10**, another side trip beckons, the steep descent (and necessary re-ascent) to Indian Point. This 15-20-minute detour provides some of the fiercest scenery in the Gorge. As you first turn right, then quickly left, look on your right for several examples of Oregon Anemone, and, on both sides of the trail, Trillium. Watch your step as you drop down to a landing and proceed across a humpbacked sandy ridge to reach the talus and a view straight ahead of Indian Point, our **Point 11** at 2,450'. On the ridge just before the talus, Martindale's Desert Parsley grows in these decidedly non-desert-like conditions, with umbrella-shaped clusters of yellow compound flowers thriving off the stony substrate. Here too, you will find Manzanita just coming into bloom with its white berry-like blossoms and, beneath this 5-6' tall shrub, its younger and shorter sibling, Kinnikinnick, a low-lying shrub with similar berry-flowers, these more pinkish than its taller brother. You may climb the Point itself, if competent on easy but loose rock, but you will not see many blossoms up there until early May, although the views are spectacular.

As you turn left at **Point 10**, 2,680', to descend the gentle Gorton Creek Trail, note the many white Trilliums decorating these largely open woods. These are a brilliant white above about 1,500', but below that, they begin to turn first pink, then burgundy, as April progresses. The second large sweeping switchback, this a left turn, is our **Point 12**, 2,240'. From just beyond to about 2,150', Oregon Anemones tint the forest blue, then reappear just before and after a small seasonal stream at **Point 13**, 2,000', then again at 1,550' and 1,125'. Clumps of Evergreen Violet and Candy Flower are prominent at 2,200' after crossing this stream as well. Below 1,200', Hooker's Fairy Bells add their cheerful voices, though hiding in white underneath their broad parallel-veined leaves, a sign of allegiance to the Lily Family; these continue to **Point 4**, Herman Camp, especially on the right of the trail at first, then on its left. Big-leaf Sandwort spreads its rather small lance-shaped leaves at the bases of tall conifers and tops its 6" tall stems with tiny white blossoms at 1,750', then again between 1,300' and 1,100'. Meadowrue is frequent below 1,300', particularly at 1,150' on your left.

This completes the Nick Eaton-Gorton Creek circuit, although you will still have to descend (gently) another 1½ miles down the Herman Creek Trail to reach your vehicle. This round trip in April repays your efforts many-fold, not only with its bounty of both woodland and meadow blooms and a spectacular display of Grass Widows, but with its inspiring views, especially off the Indian Point side trip. But even should you chose not to undertake those detours, this loop will reward with treasured flowers and sufficient exercise to render it a Gorge favorite at any time of year.

May

Could May exceed April in the splendor of this loop's floral displays and the diversity of its blossoms? Perhaps, for many of April's flowers continue to flourish throughout May, while entirely new species now begin their mid-spring emergence. Rather than worry about which is better, make this loop in both months to savor the similarities and distinctions. Beginning from **Point 7**, the Ridge Cutoff, a right turn as you return *down* the Nick Eaton Trail, Trilliums, Evergreen Violets (the low-lying yellow flowers on your right just as you embark on this descent), and deep blue Oregon Anemones appear in the woods, particularly during the first half of the trail down to the Gorton Creek Trail. Here too, between the start of this cutoff at 2,850' and down to about 2,775', Oregon Boxwood now is in bloom, although its flowers may be more difficult to detect than its leaves. This 2' tall shrub carries its finely-serrated leaves in pairs, the base from which appear nearly microscopic maroon four-petalled flowers in small clusters.

As you gently descend this cutoff, note the Vanilla Leaf and Inside-out Flowers now in bloom. Between 2,800' and 2,650', Star-flowered Solomon's Seal appears, first on you right in clumps, then on your left between 2,750' and 2,700'; it is also on your left in shining carpets. Its broad but tapering leaves form a virtual ground cover here, surmounted by its starry white flowers with six twinkling petals. Hooker's Fairy Bells, Scouler's Valerian (especially at 2,760' on your right), and Miner's Lettuce, with its strangely beautiful single-leaf collar pierced by its stem at 2,700' on the right, complete this portrait of white on green. Fortunately for the May traveler, the Glacier Lilies at 2,740' on the trail, described under **April**, above, are still in bloom through the early days of this month, although not quite as brilliant in display.

Botanists often change both Latin names and species and genus status, not just to confuse us amateurs, but partly because new information arises in all the sciences; indeed, that is the nature and value of the scientific method. New DNA testing and closer scrutiny of plant genera has led to new attributions for several of the species described in this text. For example, Star-flowered Solomon's Seal, referred to in this book as *Smilacina stellata*, is now assigned to a different genus by some botanists, and called, in Latin, *Maianthemum stellatum* (now evidently a male species instead of a female one). *Disporum hookeri*, our very own Hooker's Fairy Bells, is, according to some, now in a different genus, and named *Prosartes hookeri*, while our beloved Candy Flower, (my *Montia sibirica*) is at present assigned to the *Claytonia* genus, as is Miner's Lettuce, *Montia perfoliata* in this text. Big-leaf Sandwort, *Arrenaria macrophylla* here, is assigned to the genus *Moeringia*. I have chosen to blissfully ignore these modern alterations for three reasons: 1.The older designations are still broadly employed and in wide use by the majority of amateurs. 2. Not all botanists agree with these recent nomenclatural changes. 3. These recently-issued designations may well change in the near future. because of newly acquired knowledge.

As you reach **Point 10**, the junction with the Gorton Creek Trail at 2,680', note several Little Wild (Baldhip) Roses to your left, perhaps just beginning to bloom by mid-month. Here, by turning right, and locating a rough side trial on your left, you can make the 15-20-minute detour to Indian Point, **Point 11**. Although a steep descent and re-ascent, this side trip, as described above under **February** and **March**, is especially worthwhile in May. Trilliums and Oregon Anemones are still in bloom until mid-month and can be found just as you begin the precipitous descent. Hairy Manzanita, a tall shrub with white berry-like blossoms appears as you approach the talus after this drop, and its sibling, the low shrub, Kinnikinnick covers the stony ground just beyond. Yellow Martindale's Desert Parsley is common on the rocks once you reach level ground just before the talus as well. By all means climb to the actual summit of Indian Point if you are comfortable on Class 3 rock, although beware of some loose stones masquerading as solid holds in mid-climb.

Here, at the top, you gain not only an overarching view north, west, and east over the river and out to Mt. Adams, but the discovery, on the north-facing sheer cliff, of the almost otherworldly shimmering magenta-pink of Rock (Cliff) Penstemon (*Penstemon rupicola*). The glorious mat of tubular flowers surmounts grayish-green wooly leaves, the whole viewed from afar appearing as a blotch of almost artificial color to rival any cultivated garden scene. You can now disregard the fabricated and lurid purples

Rock Penstemon

and magentas of grocery store asters and orchids for here, in the wild, nature has provided as rich a color display as any in the city.

Spotted Saxifrage

One other new species appears here, on the north slopes off the summit of Indian Point, although not as spectacular as the Rock Penstemon. It is the Matted (Spotted) Saxifrage (*Saxifraga bronchialis*). "Saxifrage" means rock-breaker in Latin since members of this species were thought to actually split the rocks upon which they grew. Instead, many stone-loving plants in the Saxifrage Family, and the Stonecrops as well, grow off the soil and debris of cracks in the rocks and stone walls. They are no less remarkable, however, for their ability to thrive in such apparently lifeless habitats. The Matted Saxifrage here has 6" wiry stems with alternate small leaves and is topped by several white flowers with five petals – nothing unusual here until you bend down for a closer look. Each petal is marvelously spotted with maroon dots fading to yellow as they progress inward toward the center of the flower. As usual in the floral world, searching with a small eye amply compensates for minimal effort.

Descending from **Point 10** Gorton Creek, the Trail swoops in two broad and gentle switchbacks accompanied by many yellow Evergreen Violets, a plant which is particularly partial to mossy banks. Trillium is scattered from here down to about 1,500', now mostly turning a wine-red shade. Around the second large switchback, this a left turn at **Point 12**, 2,240', Small-flowered Alumroot, with its clustered sprays of white bell-like flowers, inhabits open rocky areas; these will continue to about 1,500'. Candy Flower flourishes after a right-hand dip in the trail at 2,200' near the stream (now drying) at our **Point 13**, 2,000', but also can be found wherever the trail crosses a moist area. Yellow Columbia Gorge Groundsel (Senecio) is frequent on both sides of the trail between 2,550' and 2,400', then again at 1,900' to 1,800'.

Here too, you can find its close cousin, Broadleaf Arnica, also a yellow daisy flower in the Aster Family, but far taller and more robust than the 1' tall Groundsel. At least early in May, the Oregon Anemone is still displaying its deep-blue flowers between 2,600' and 2,100', while Meadowrue and Star-flowered Solomon's Seal grace the heights here and there all the way from 2,100' to 1,100'. Plume-flowered Solomon's Seal makes an appearance as well, particularly between 1,500' and 1,200'. The bright yellow bunched flowers of Cascade Oregon Grape are scattered throughout these woods between 1,900' and 1,100', although do not expect to see every such bush adorned with blooms. These deep-forest dwellers rarely blossom all at once, and some bushes do not flower at all in some years.

Certain plants bear borrowed names because of a supposed resemblance in some way to an already-discovered species. Such is the case with the oddball plant Wild Ginger (*Asarum caudatum*), quite common on this trail, particularly on your left between 2,000' and 1,900'. Don't expect to see its weird flowers without a bit of work as you must bend down to discover, beneath its heart-shaped glossy ground-hugging 4" leaves, the solitary 3" bell-like flowers with three flaring petals of an indescribable brownish-purple hue, each petal looking for all the world like a Salvador Dali tapering moustache. Unrelated to true Ginger, Wild Ginger does nonetheless smell like the tropical spice, especially when crushed (but don't be tempted – take my word for it). The roots have been employed as a ginger substitute. But why would this plant's flowers conceal itself under its leaves, hiding on the forest floor like some floral refugee banished from its own habitat? Probably because these earthbound blooms are pollinated by creeping ants and have no need to hold their heads up high; their status, however, is not diminished by their lowly dwelling.

From the switchback at **Point 12** back to level ground at **Point 4**, Sweet Cicely, Broad-leaf Arnica, Fairy Slipper, and Baneberry are all plentiful. Especially notable on this rather same-same part of the trail are a clutch of Bunchberries on your right at 1,300'; a number of Scouler's Valerians just a short distance further on your left; Red-flowering Currant at 2,100', 1,500' and again at 1,350' on your left; Broad-leaf Arnica between 1,250' and 1,100'; the Columbia Gorge Groundsel between 1,300' and 1,100'; many Star Flowers between 1,100' and 950'; and the rose-purple flowered Nevada Pea between 1,200' and 900'. Single examples of certain species are notable at these lower elevations as well. A Little Wild (Baldhip) Rose is in flower now at 1,300' on your left, complemented by Leafy Pea, while a single Honeysuckle trails its bright orange blossoms on the ground on your right at about 1,000'. This particular specimen, normally a vine, is apparently in search of something to climb. The Little Wild Rose, when in fruit, produces a bright red seed rich in Vitamin C. In fact, many plants contain compounds similar to those biologically active in humans as well.

Back at **Point 4**, your starting point for this round trip, take satisfaction in having achieved a rigorous loop and in having discovered more blossoms than most amateurs can remember in an entire month. As you descend the road, then the trail back to your cars and SUV's, savor the best that May has to offer in the Gorge.

Wild Ginger

June

Plenty of wildflowers continue to bloom on this round trip in the month of June, and by completing this circuit down the Ridge Cutoff, then the Gorton Creek Trail, you'll not only be experiencing a subtle difference in ecological neighborhoods than if you descended the way you came up, you'll be saving your knees a bit as well, as this loop is less steep on the way down.

From the beginning of the Cutoff, at **Point 7**, you will encounter not only the familiar May blooms of the Inside-out Flower, Vanilla Leaf, and Star-flowered Solomon's Seal, all still doggedly persisting on the upper, mostly level reaches of the trail until at least mid-month, but also the newly emerging blooms of Bear-grass. Common from 2,800' to 2,750' in open woods, it is especially prominent on your right, its striking white plumes here up to 5' tall. Don't fail to inhale; the delicate aroma, faintly reminiscent of vanilla and spice, belies this plant's giant stature. Color is found with the presence of the yellow Columbia Gorge Groundsel (Senecio), here a 1' tall plant with daisy flowers, between 2,700' and 2,600', and with the similar yellow Broad-leaf Arnica, particularly abundant on both sides of the trail between 2,750' and 2,695'. The latter species can be distinguished by its taller and more robust structure, growing up to 2½' tall, and having oval leaves in pairs opposite each other; the groundsels (senecios), one of the largest genera of plants in the world, have leaves which progress up its stem in alternate fashion rather than in pairs, and, in this particular species, the leaves are lobed and toothed.

As you begin your descent to the Gorton Creek Trail, look to your right at 2,790', about 500 yards before the junction. The elegant maroon-spotted orange blossoms of Tiger Lily appear, here 3' to 4' tall, with its six tepals (combined petals and sepals) swept backwards and re-curved like some tropical bird in flight. Should you choose to turn right at the junction (**Point 10**, 2,680') and then left soon thereafter to descend the wickedly precipitous Indian Point Trail, note the blue-purple tubular flowers of Fine-tooth Penstemon on your left once you reach the level sandy area just before the talus-choked ridge. Here too, you will continue to find Martindale's Desert Parsley, with yellow sprays of clustered blooms lying low against the stony ground, and a single specimen of Wooly Hawksbeard on your left just before the Point itself, our **Point 11**. This fuzzy 1½' tall plant has yellow dandelion-like flowers and, look as hard as you may, you will not discover this species anywhere but here, on rocky bluffs and lowlands of the Columbia Gorge. Ascend the pinnacle of Indian Point itself, if you are comfortable on Class 3 slightly unstable rock, for a most awe-inspiring view of the Gorge.

Between **Points 10** and **12**, the large left switchback at 2,240', the Inside-out Flower is prominent, while Candy Flower and Meadowrue add interest at the drying but still-damp areas between 2,200' and 2,100'. A right in-cut on the trail brings one, in about a mile from the junction, to **Point 13**, a by-now dry streambed at 2,000'. The Columbia Gorge Groundsel is present here and there between 1,950' and 1,900'. Broad-leaf Arnica occurs sporadically between 2,600' and 1,500', although you will find its elliptic opposite pairs of leaves and 2' tall flowerless stems often. Note that many plants inhabiting the deep forest, such as Arnica and Oregon Grape, do not bloom all at once. Some of these species flower successively, perhaps out of consideration to their neighbors, while others may bloom one season then not the next. Seeing the leaves of Arnica emerging in early spring, the aspiring naturalist may be led to believe that a carpet of yellow will soon blanket these woodlands, but such an image turns out to be an illusion, as only a handful of these plants will actually blossom. As opposed to open meadows, with their abundance of light, forests are more frugal in their allotment of blooms. We should not blame nature for this disappointment; it lies instead in our own fallible expectations.

Coursing down the gentle switchbacks below **Point 13**, the Little Wild (Baldhip) Rose displays its cheery blossoms frequently between 1,500' and 1,200', especially on your left. Arnica and Groundsel add their yellow daisy-like flowers all the way down to 1,000'. Sweet Cicely, its coarsely toothed leaflets twice divided into three's, and Candy Flower, with its five dainty white petals lined like peppermint with pink stripes, decorate these low-lying woodlands as you approach your ending point for the loop at **Point 4**. Head right as you pass the old Herman Camp to return, in a mile and a half, to your vehicles.

Completing this loop in June serves not only as an opportunity to review your knowledge of old floral favorites but to keep up your conditioning program as well; over 3,500' of elevation gain is good but tiring, while over 3,500' of descent can be torment for the joints. Such are the rewards and travails of the amateur botanist.

July and August

The Nick Eaton Ridge Cutoff Gorton Creek loop is a splendid idea in the summer months as you escape the city heat, being in the cool forest much of the way, but with plenty of open areas to enjoy the sunshine and the flowers. Beginning right at **Point 7** at the Ridge Cutoff, 2,850', look to your left, then right to see several fine specimens of Tiger Lily. Fragrant Bear-grass accompanies you as you hike the mostly-level trail here, preferring these open woods at mid-elevations. Another companion, however, is the notorious weed, Wall Lettuce, Although a member of the Aster Family, this 2-3' tall plant lacks the neat disk and ray flowers of a daisy, and instead, unlike any other species in its huge family (perhaps some 22,000 species), consistently displays five yellow "petals" (actually five separate ray flowers) on its many wiry stems. Notice its weirdly-shaped leaves, lance-like in outline but with jagged lobes and flanges clasping their base. Its sinister introduction from Northern Europe has allowed this shade-lover to spread from urban fields and roadsides to Cascade forests as high as 4,000'. We probably should not blame the plant for taking advantage of our welcoming woodlands, but Wall Lettuce is unseating a number of forest-dwelling native plants, and insidiously working its European way up our trails to ever-higher elevations. Thus far, no means of control is at hand.

Along the Ridge Cutoff, between **Points 7** and **10**, Broad-leaf Arnica, Pipsissewa at 2,800' to 2,700', Foam Flower, Twin Flower at 2,800' to 2,700', and an occasional lingering Inside-out Flower, can all be found. Here too, are several more examples of the glorious orange Tiger Lily at about 500 yards from the start of the Cutoff on your right, putting many cultivated garden flowers to shame with its size and brilliant color. Yet it is a new summer flower which captures our interest here, arching over the banks on the right side of the Cutoff Trail in its first quarter mile – the 2-3' long-stemmed Turtle Head, or Woodland Penstemon (*Nothochelone nemorosa*). Our only penstemon not averse to partial shade, Turtle Head is a pink-to-purple tubular flower so like other penstemons that many botanists consider it to be included within that genus. Its saw-toothed opposite leaves look like a penstemon's, and its flowers, inside and out, are certainly penstemon-like. Only slight technical differences from other penstemons cause most taxonomists to place it in its own genus, possibly a rank trick of professionals to confuse us amateurs. Fortunately, calling it "Woodland Penstemon" will not endanger your status as a flower lover. *Chelone* means "turtle" and *notho* means "false"; I can't see much of a resemblance but more imaginative plant hunters were evidently impressed.

It's not a bad idea to turn right at **Point 10**, the junction with the Gorton Creek Trail, 2,650', to take advantage of the crude path soon appearing on your left which abruptly descends to fabled Indian Point, our **Point 11**. Here, dramatic views of temple-like pinnacles and the mighty swell of the Columbia below make the effort well worthwhile. As you descend, note additional pink Turtle Heads peeking at you on the right. In July, blue-purple Fine-tooth Penstemon appears on your left as you finally reach level ground on a sandy ridge. Also note yellow Martindale's Desert Parsley hugging the rocky soil and a single

Turtle Head

example of the narrow endemic, Wooly Hawksbeard, a Gorge-only 1½' tall plant with fuzzy stems and yellow dandelion-like flowers on your left as you approach the fearsome-looking spire that is Indian Point. Scramble up the Class 3 loose pinnacle if competent on rock to savor additional far-flung views.

As you return to the main trail and turn right to begin the lazy, looping descent between **Points 10** and **12**, yellow Broad-leaf Arnica greets you early in the summer but by August, its daisy-like flowers are mostly gone to fuzzy seed. In a similar disappearing act, the Inside-out Flower has mostly vanished by mid-July. Nonetheless, a number of all-summer blossoms dot the woods here, including white sprays of Foam Flower from 2,650' to 2,400'; Rattlesnake Plantain on your left at 2,520', with its mottled leaves and snake's rattle of a flower; Queen's Cup, with glistening white flowers with six petals, especially on your right between 2,500' and 2,200'; Wall Lettuce, pretty much all the way down; and Little Wild (Baldhip) Rose here and there but particularly prominent on the left between 2,600' and our **Point 12**, the second large switchback at 2,240'. A trickle of moisture may still spill from the stream at 2,000', **Point 13**, and in the transient dampness, some Candy Flowers are in bloom here throughout the summer.

Between **Points 13** and **4**, where the Gorton Creek Trail intersects the Nick Eaton Loop, some Arnicas remain in flower throughout early August, while Candy Flowers continue here and there as well. Also, many Pathfinder plants (*Adenocaulon bicolor*) are now in bloom. This species, also called Trail Plant, hides its membership in the Aster Family well, as there appears to be no connection to our familiar and colorful daisies, asters, and arnicas. However, this species *is* an Aster Family member, although its miniscule dirty whitish floral heads look like the product of some urban weed, disproportionately small atop its 2' tall thin stems. What distinguishes this plant, however, is its heart-shaped coarsely toothed leaves. Old-time explorers would walk through a patch of Pathfinder, then look back from whence they came: The top sides of the leaves are bright green while the undersides are all fuzzy and silver in hue. Once walked upon, the leaves, now outlined by their upside-down position because of their square stems, pointed the way back. In this way, old-time game hunters were said to always know their path home by the conspicuous upturned leaves left in their path. However, it's best not to wander off-trail to test this theory. No need to trample the vegetation just to test history.

You turn right here, our **Point 4** at Herman camp, really just a widening in the Herman Creek Road, and continue back about 1½ miles to your vehicles. Take pleasure in the accomplishment of this strenuous loop, with as much elevation gain than almost any other in the Gorge except, of course, for the Mt. Defiance/Starvation Ridge marathon further east. Meadows, flowers, views, and cool forest have all combined for as wonderful a summer outing as the Gorge has to offer.

September and October

Fall certainly means fewer wildflowers, but the cooler weather and changing colors bring their own rewards. After you've retraced your sometimes-steep steps back down the Nick Eaton Trail to the junction with the Ridge Cutoff, **Point 7** at 2,850', turn right to descend this cut-off to the Gorton Creek Trail. Here, still coloring the woods just after the junction, several Tiger Lilies appear on your left in early September, their orange hues matching the maturing colors of autumn. About half-way down, as the trail starts to descend more rapidly, note several Merten's Coral Roots at about 2,775', on your right. These plants lack chlorophyll, having come to depend on fungi in the deep forest soil, and thus are a reddish color with not a trace of green. Wall Lettuce continues its invasion of these shady woods, with

Pathfinder

small yellow five-petalled flowers atop wiry stems. Pathfinder (Trail Plant) inhabits this environment as well, although its tiny white button-like flower heads are inconspicuous. In the first brief opening, at 2,745', where arches of sunlight can finally penetrate the forest gloom, White-flowered Hawkweed raises 2' tall thin stems topped by small white daisy-like flowers with orange centers.

Point 10 is the junction with the Gorton Creek Trail at 2,680'. An obligatory side trip for many hikers to **Point 11**, Indian Point, requires a right turn here for a few yards to discover the crude path on your left offering a steep descent to this viewpoint. The rugged trail is brightened by several examples of Turtle Head, with pink-purple flowers looking like two-lipped tubes with a longer lower than upper lip. Wall Lettuce inhabits the precipitous slope as well, along with a few long-lasting Little Wild (Baldhip) Roses. While no blooms remain at Indian Point itself in the fall, the views continue to make this side trip worthwhile.

From **Point 10** all the way through to **Point 4**, the junction with the Herman Creek/ Nick Eaton Loop, much Wall Lettuce remains in flower through at least early October, its yellow flowers with five petals each trying to hide its identity as an Aster Family species. Appearances are not as revealing as we'd like them to be in plant identification. Most Rose Family plants barely resemble our common garden roses, while most Buttercup Family species bear little resemblance to our notion of what a buttercup should look like. Remember, when dealing with plant affiliations, they are assigned families by dint of their reproductive parts and seed characteristics, a taxonomy verified by recent DNA analyses.

There are occasional openings down the gentle switchbacks of the Gorton Creek Trail. The brief ones at 1,800' and 1,600' contain examples of White-flowered Hawkweed and Pearly Everlasting, while a lower opening at 1,100' holds this latter only. Through these low woodlands, clumps of Pathfinder, especially at 1,350' on your right, and some Candy Flower persist through the autumn months, the latter perhaps as a re-bloom. Otherwise, no flowers are brave enough to face the oncoming colder weather, or are simply smart enough to know better than to expose themselves to the fierce gray of late fall. Nonetheless, true Gorge hikers know that the views; the exercise (about 3,500', considering the ups and downs of this trail); the yellows, reds and browns of fallen leaves marking the paths; and the remaining flowers; all combine to render a late-season hike around this loop a pleasure no matter the month, temperature, or weather. Congratulate yourself on a task well-achieved and perhaps one to be repeated again after winter has evicted most casual hikers from these heights

The Flower Finder: Nick Eaton Ridge/Gorton Creek Loop

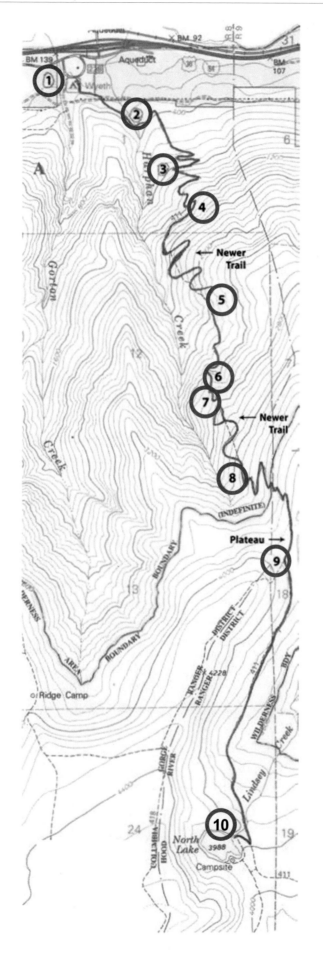

North Lake (Wyeth Trail)

Many early Columbia Gorge trails were built in the late 19th to mid-20th centuries by folks who wanted to reach a destination as directly as possible, be it a viewpoint, a stand of particularly desirable timber, or a favorite fishing lake. When a dam was constructed in the 1930's to contain the waters of Lindsey Creek, creating North Lake, the Forest Service was unaware that the straight-up track to it would be so popular that, in a few years, the vertiginous trail would become a badly-eroded rut. Drainage efforts throughout the years proved so unsuccessful that, in the late 1980's, to the dismay of hikers hoping for a severe conditioning workout, the Forest Service created a new trail, the Wyeth Trail (#411), complete with more civilized and leisurely switchbacks, thus adding an additional 7/10s of a mile to the 5½ mile trek to the Lake. Grizzled hikers, climbers, and conditioning fanatics complained but to no avail; they were referred to Munra Point for a more vertical ascent. Normal-minded hikers simply enjoyed the more relaxed pace to this unexpected, if artificial, body of water high up on the eastern flanks of Green Point Mountain. The more than 4,000' of gain still grants a more-than-decent workout, while the flowers displayed along the way remain, happily, the same.

CAUTION: As with all Gorge trails that venture above 3,000', snow often lingers throughout April and into early-to-mid-May. This becomes important on the North Lake Trail in the area between **Points 9**, the junction with the Green Point Mountain Trail and **10**, the lake itself. The trail between these points is relatively level, crosses several marshy areas and is easily lost in the snow. Instead of mushing around looking for footsteps or open trail, it is best here to retrace your steps and descend. After all, you've already gained all the elevation available, the lake will be covered in snow and ice if you can actually find it, and there are no wildflowers out beyond this point at those times of year anyway.

February and March

To hike the Wyeth Trail to North Lake in late winter and early spring, you will have to park at the gate blocking the road to the campground and walk the short level distance (about 3 to 5 minutes) up the paved road to the trailhead. Even in these colder months, you may find Robert Geranium along the sides of this road, and, beyond, just past the trail sign, more of this pretty pink flower should be out, lining the old logging road that marks the beginning of the trail. Adaptable to a fault, Robert Geranium, the ubiquitous Eurasian weed, can bloom continuously through some winters. Charming it may be, but as testimony to its invasive nature, consider that a 1994 text considers it "rare in western Oregon." In fact, Robert Geranium has now invaded huge tracts of lowland forest in the Gorge as well as lining city roadsides and invading urban gardens. Studies are underway to discover effective eradication measures as this beautifully dangerous plant is displacing native flora with dispatch.

Here, on the eastern side of the Crest, blooming times are a bit behind those on the milder western side. Winter brings colder temperatures east of Cascade Locks, while conversely, it's warmer there in the summer, all a result of the moderating effect caused by proximity to the ocean. Nonetheless, by late February and early March, Oaks Toothwort can usually be found along both sides of the ancient earthen road where the trail originates at **Point**

1. This, one of our earliest bloomers, has lightly-colored lavender flowers of four petals each. Usually three or four flowers bloom atop each 6" tall stem. This precocious flower, along with another lowland February bloomer, blue-purple Snow Queen, benefit from early-season blooming, when the deciduous trees at these low elevations are leafless, thus allowing more light into the wooded realms where these flowers thrive. In fact, many Snow Queens can be found in these woods, although not on the Wyeth Trail itself. To see a colorful garden of these low beauties, walk west up the access road beyond the gate towards the large animal farm at the end of this road (please do not disturb the nice folks running the farm) and glance into the woods to your left; here, the 4" tall blue-purple flowers form a continuous understory from late February through late March. A number of Indian Plum shrubs are also in bloom, beginning right at the trail signpost on your right and continuing on both sides up the dirt road leaving the parking area.

As you make the sharp left turn at the trail sign 100 yards from the parking lot, a few more Toothworts appear. The cleared brushy area beyond contains no flowers in these early months, but just before and after the power line, bunches of these lovely and delicate Toothworts, in the Mustard Family, can be found, especially on well-drained slopes, from the crossing of Harphan Creek at **Point 2** to **Point 3**, the start of the steeper switchbacks at about 750' elevation. The last of the Toothworts at this early time of year occur just beyond the right turn at a large rock face at 1,300', **Point 4**, just at a stony clearing. Also by mid-March, Evergreen Violets are blooming at low elevations and can be found in abundance just beyond the power line, then again beyond the stream crossing, from 400' to 700'. These yellow violets bloom a bit earlier than their larger yellow siblings, the Stream (Wood) Violets, and are most often found growing off mossy banks; Stream Violets have more heart-shaped and thinner deciduous leaves with a taper tip and they prefer wetter habitats.

Where are the harbingers of spring, the Trilliums? They are present and accounted for, though a bit later than west of the Crest. A few can be seen by the power line emerging in mid-to-late March; more will be apparent later in that month through April and May at higher elevations. Another early flower, also a Mustard Family member, Hairy Bittercress, is also present by late February and early March at the power line. You will be excused for failing to notice this non-native weed as its 4-10" tall stems are topped by the tiniest of white four-petalled blossoms; perhaps "blossom" is a misnomer, as the microscopic 3 mm flowers are often closed tightly. Propagation is by crawling insects, obviating the need for showy blooms.

Another premature bloomer is the Shining Oregon Grape. By late February in milder years to mid-March under colder, darker conditions, its bright yellow clusters of flowers are borne on woody stems up to 10' tall in urban cultivated forms grown in Portland gardens and highway medians. Here at the power line, however, the native plants are about 1½ -3' tall. There are several differences between this Oregon Grape and the more common Cascade Oregon Grape. The shiny-leaved species prefers sunny locations at lower elevations, blooms earlier than the forest-dwelling and more ubiquitous dull-leaved Cascade species, and has seven to nine leaflets compared to 9 to 19 for the Cascade plant. Both are members of the exotic-sounding Barberry Family, which, although harboring many tropical plants, also contains several common Northwest species: Vanilla Leaf and the Inside-out Flower.

Early spring travelers will largely need to content themselves with these blooms; indeed, in many years, snow covers the trail above 2,500' until middle to late in March. However, one last bit of colorful joy may be discovered by late March just above the power line and up to about 600' – the Calypso Orchid or Fairy Slipper. This is the pink jewel of early spring woodlands, with its slipper-shaped lower lip spotted in red-purple and single egg-shaped dark green leaf lying on the forest floor. It can often be found growing atop a mossy surface, and indeed, thrives from its many thread-like and delicate roots which gather nutrients from the damp earth below. Persisting throughout the spring, the well-named Fairy Slipper brings a sense of cheer to every winter-weary mountain traveler, announcing the coming of warmth and sunshine. While we long for the sun of spring and summer, and may curse the winter rains, this plant helps us remember that where the sun always shines, there's a desert below.

April

You may still have to park at the gate in early April; never mind – it's a short walk to the trailhead. Anyway, the flower show begins right at your cars with the burning bush of Red-flowering Current by Harphan Creek, just west of the parking area. Garden cultivars of this 6' tall shrub, in pink and white, are planted on roadsides and public gardens in cities for early-season color, but here, the native plant for once actually seems to outshine its city cousin in height and brilliance. Walking up the paved road beyond the gate, the ever-present Robert Geranium appears, but at the grassy areas by the trailside parking area at **Point 1**, note two different geraniums: Robert's occupies the lawn on the south by the trail sign, but, turn around and you will see a similar but smaller pink geranium, Dovefoot Geranium (*Geranium molle*) carpeting the lawn to the east and north of the parking strip. This is smaller but just as pretty a plant, with kidney-shaped three-lobed downy leaves (*molle* means "soft"). Unfortunately, it's also an invader from Europe and Asia, although far less aggressive than Robert Geranium. It flourishes by early spring in damp grassy depressions. Travelers on I-84 can spot its rosy glow from mileposts 18 to 20 in the median strip, where it almost appears like an artificial pink rug from afar. Its seeds, like all geraniums, are spear-shaped capsules whose exact length and shape distinguish the species. These capsules are the fused styles (the female parts) which form a central beak; in some species, like these two geraniums, this beak is relatively short, about ¼ -1" long but in others, including the common Filaree

found mostly east of the Crest (Catherine Creek east of Bingen in Washington contains entire gardens of this plant), the style can be up to 2" long. The beak of Filaree is spirally twisted and is the source of a fascinating tale: The twisted style drops to the ground and, as it alternately dries, then is dampened by spring rains, it straightens out and re-twists, thus screwing itself into the ground to be germinated.

> The anatomy of flowers and seeds are a bit too technical for an amateur field guide, but some basic information hopefully won't prove too tedious and might even be helpful: The functions of a flower are to contain the reproductive parts of a plant and to attract pollinators (although some plants are wind- or self-pollinated). The male parts inside the flower are called the stamen, composed of a filament (a long stalk) topped by an anther bearing the pollen, which will fertilize the female parts, usually of another flower of the same species. The female parts are called the pistil, consisting of a basal ovary where the seeds are produced; a style, which is generally a long tube conducting the seeds upwards; and the stigma, the top part of the pistil, usually a bulb-like enlargement which receives the pollen from the male parts of another flower of the same species. Occasional plants, such as Meadowrue, Colt's Foot, and Goat's Beard, are dioecious, meaning they come in either male or female versions; most species, however, bear both female and male parts on the same flowers and thus have evolved mechanisms to prevent self-fertilization, which might produce nonbearing or inferior offspring. These mechanisms are sometimes architectural, as for example in plants where the anthers greatly exceed the stigma, or temporal, as in species in which the pollen of one plant matures before its seeds, but in tune with those of a neighboring plant of the same species. As with humans, anatomy and timing are everything in reproduction.

Up the road that serves as the start of the trail, Robert Geranium and yellow Cascade Oregon Grape accompany you up to the abrupt left turn, where pale lavender Oaks Toothwort takes over. Just below the power line, at about 300', three Serviceberry shrubs, these about 15' tall and on your left, display their raggedy white blossoms to brighten this weedy slash, which also contains some Dandelions, white

Woods Strawberries, Hairy Bitter Cress, and Shining Oregon Grape, in flower at least early in the month. Up the gentle woods beyond, many Fringe Cup, white Trillium, and lavender Toothwort are out, the first clumping especially above the creek crossing at **Point 2,** the last more spotty but prominent on hillsides up to about 1,500'. The shy Fairy Slipper (or Calypso) Orchid cannot hide here; its pink flowers with spotted pouches are at 750', 900', and 1,200', on the left thriving off the mossy undergrowth. The yellow Evergreen Violet also makes love to the moss, and can be found just after **Point 2,** and up to 700', then again in profusion after the steeper switchbacks beyond **Point 3** and on your left at 1,400'. Just before the open rocky areas at **Point 4,** find a field of Trillium in the open woods on both sides of the trail.

As you approach the first sighting of the small stream at **Point 5,** 2,200', a Red-flowering Current bush is on your left along with several Trilliums, a nice contrast of red and white. Several more Red-flowering Currents mark a pleasant change from the omnipresent green at 2,300' as well. Occasional Trilliums occupy these open woods, particularly at 2,350' on your right. Just before the talus and stream area at **Point 6,** 2,700', a pretty congregation of Candy Flowers resides, while at the stream, several more examples of Red-flowering Current cheer up this somewhat drab shrubby area. Just beyond, however, the view opens up to the north over the river and the sun is finally allowed entry. Further along, at **Point 7,** the series of openings begin; these make the previously constant woodlands seem a distant green blur. For here, at 2,750' in early April, the glorious purple Columbia Kittentails cover the north-facing bluffs and lead the way to a combination of yellow Glacier Lilies, Martindale's Desert Parsleys and more Kittentails which grow in colorful profusion up to the top-most opening at 3,200'.

Beyond these stone-filled meadows, little else is in bloom this early. No matter; the exercise and the lake compensate while your trip amply prepares you for the greater floral glories to come in May. Amateur botanists cannot be greedy in this young spring season, but will probably always appreciate any hints of color after the angry grey skies of a Northwest winter.

May

By mid-May, there are two additional benefits to the hike up to North Lake: The gate is down so you can park at the actual trail sign, thus avoiding the brief walk up the paved road to the campground; and, more importantly, the flowers on this trip are at their best. Even at the parking area, **Point 1,** color appears in the form of the two foreign geraniums mentioned above in April. The more common and larger Robert Geranium covers the area by the sign on the south side of the lot, while its Eurasian sibling, Dovefoot Geranium, can be seen growing in swaths of pink on the grassy lawn to the east of the cars.

Indeed, Robert Geranium, the more adaptable of the two, also frequents the crude road which inaugurates the trail. This coy but ingenious invader prefers partially shaded locales and has taken readily to our temperate Pacific climate. It can be found here, and after the left turn at the second trail junction 100 yards ahead, along with Sweet Cicely. Although this latter species lacks distinctive flowers (they are short-lived, microscopic, and a dull white), its leaves, extending from 1' tall stems, are as unmistakable as the plant is common. These are twice divided into three's, nicely and regularly toothed, and have a spring-green softness that embraces the forest floor on many shaded plateaus at all elevations in our Northwest forests.

Up the sharp left switchback just after the trail junction with the Gorge Trail (#400), many Oaks Toothworts are in bloom early in the month, while the Trilliums which graced this part of the trail before and after the stream crossing at **Point 2** are now turning a deep and aged purple. Just before an unexpected dip in the trail as you enter the power line slash, note three beautifully-formed Mock Orange shrubs on your left. These, fit enough for a cultivated garden, emit an odor reminiscent of citrus fruits, hence the name.

Leaves are classified in several ways by botanists, the chief being whether they are deciduous or evergreen. Leaves can be alternately arranged on a stalk, opposite each other; or whorled at the stalk's base; or at mid-stem. Leaves can be simple, with a single leaf emanating from a stem, or compound, with a secondary stem (called a petiole) supporting a number of leaflets. Leaf architecture can also be classified as pinnate, in which leaflets are arranged in two rows from a central stalk (think of Oregon Grape or peas and vetches) or palmate, in which the leaflets extend out from a common point like the fingers from a hand (think of lupine leaves). Moreover, leaf shapes have been described as ovate, oblong, elliptic, cordate (heart-shaped), sagittate (arrow-shaped), linear, lanceolate, etc. Leaf bases have been described as cordate, auriculate (with "ears"), or sagittate. Margins can be said to be entire (untoothed) or lobed, serrated, or dissected (deeply lobed). For us amateurs, it may well be sufficient to simply recognize basic shapes associated with our favorite flowers; indeed, several nearly identical species can be most easily distinguished by their leaves, as for example Stream and Evergreen Violets, or Woods and Broad-petal Strawberries. More important than anatomy is purpose: Leaves perform the vital function of photosynthesis by taking up carbon dioxide from the atmosphere and emitting oxygen as a by-product, a fortunate occurrence for all us land-dwelling animals. In each leaf's cell (and indeed, in all animal cells as well), resides structures called mitochondria which supply the energy needed for the cell to perform its major functions. Indeed, these specialized, tiny organelles originated billions of years ago when a free-living bacteria invaded a simple single cell, eventually allowing cells to agglomerate to form all multi-cellular forms of life. Strange but true, we are all the result of an ancient infection! They also capture water from the air, although most hydration is performed by a plant's roots. Leaves also can sprout new baby plants (think of Youth-on-age), can lend support to a plant through modifications called spines, or protect a plant through thorns (also modified leaves) and even feed a plant by capturing prey (think of the sundews, and pitcher plants). Without leaves, we would all be absent.

The Inside-out Flower is much in evidence here and continues to about 700', at least into late May. Cascade Oregon Grape is now displaying its bright yellow flowers in bunches and its holly-like leaves above its woody base, particularly between **Points 2** and **3**, the beginning of the steeper switchbacks at around 750'. At times here, as elsewhere in the Gorge, this Oregon Grape forms a continuous ground cover, particularly under second- and third-growth Douglas Fir forests logged 75 to 150 years ago. It may favor the areas where some disturbance has occurred; early Native Americans did not believe it was as common before massive logging operations began in the Industrial Age. Indeed, the preponderance of Douglas Firs and Western Hemlocks in this forest raises an interesting question: With so many trees all over the world, wouldn't the Kingdom of *Plantae* comprise the largest of all earth's kingdoms, more than, say, *Animalia* or *Fungi*?

Scientists have often wondered and debated which Kingdoms of organic life dominate our planet in terms of mass. Some say it's the fungi, because these organisms, now awarded their own Phylum distinct from plants, spread their filaments, known as hyphae, underground to inconceivable distances. Indeed, the largest organism believed to exist is a mushroom in Eastern Oregon whose various extensions both above and below ground, occupy many square miles of territory! Others, however, are equally convinced that is the lowly bacteria which comprise the greatest biomass, given their ubiquity in sea and on land, in the atmosphere, in the driest deserts, and in the deepest recesses of earth yet plumbed by man. More recently, researchers have calculated biomass by proxy: the amount of carbon stored in the various kingdoms known to humans. Certainly, although mammals are large (think elephant and whale), while bacteria, archaea, protists and viruses are ubiquitous, no organisms match the biomass of plants (the kingdom of Plantae). Indeed, the combined biomass of plants, at 450 gigatons, account for a majority of this total (think the Amazonian forests and the temperate forests of North America and Russia), even though water covers almost three quarters of the planet. The study also uncovered some uncomfortable facts: Humans and their livestock have come to outweigh all other forms of vertebrates while the mass of other animals and fish have been declining during the past 150 years. If grasses and plants have conquered the earth, it is we who have transformed it probably for our planet's eternity.

Just before you reach the power line, the papery white flowers of the 5'-tall Thimbleberry shrub are now in bloom. Note also the three specimens of Serviceberry on your left. Their straggly white flowers appear to have a multitude of petals, but appearances can be deceptive for there are really only five petals per bloom. There only appear to be more because the many blooms at the ends of the branches bunch together and frequently overlap. These 4-25' tall shrubs (here, about 10' tall) checker open woods throughout the Gorge in spring at almost all altitudes.

The power line itself contains the expected weed, Dandelion, but native plants are in evidence here as well, benefiting from the sunshine this anthropogenic gash has created: white Woods Strawberry, Thimbleberry, whitish-pink Miner's Lettuce, the tiny blue Small-flowered Forget-me-not, and the pale purple of Small-flowered Tonella (*Tonella tenella*), with its diminutive broad-lipped flowers.

This last, an often-unnoticed species within the Snapdragon Family, is frequently found at low elevations on open ground throughout the Gorge. Its tiny flowers deserve a closer look. Atop 4-8" tall single stems, they have five strangely-shaped petals, the lowest formed into a seemingly oversized broad lip, the middle two swept back and upward, framing the two small nubbin-like upper petals. Perhaps this outlandish appearance is meant to attract some pollinating insect, but from a close inspection, many human eyes regard these blooms as odd, yet in a uncommonly charming way.

Small-flowered Tonella

The gently-sloping woods before and after **Point 2**, the stream crossing, are filled with Vanilla Leaf, Inside-out Flower, Sweet Cicely, the large white blooms of Columbia Wind Flower, pink-to-white Star Flower, and, late in the month, the cheerful pink of Little Wild (Baldhip) Rose. A graceful specimen can be found at about 700' of elevation on your left. Between **Points 2** and **3**, the white bushy-topped blooms of Scouler's Valerian are now popping up, particularly on the right between 750' and 900'. These 1-2' tall plants have pointed, toothed, and lobed leaves, and resemble their somewhat larger cousin, Sitka Valerian, a plant associated with high mountain meadows in the Cascades.

Here at around 500', the trail rises steadily but in graceful arcs, accompanied by Vanilla Leaf, Inside-out Flower, Sweet Cicely, the arching Plume-flowered Solomon's Seal, its close relative the mat-forming Star-flowered Solomon's Seal, along with the occasional Hooker's Fairy Bells. However, as you begin to ascend more steeply, at **Point 3**, about 750', the yellow Evergreen Violet appears, blooming just opposite the ruin of an old rusting metal lumber-working tank in the woods to your left just after a brief verdant plateau at about 1,000'. (Rumor has it that this ancient bin was part of a whisky still, but rumor often mistakenly confounds sin with innocence.) Here, there is a right turn and, just before it, a clump of reddish-purple Spotted Coral Roots can be found hiding in the forest on your right. Further up the switchbacks, the welcoming lapis blue of the Oregon Anemone comes into view at about 1,200' elevation, then reappears sporadically from 1,300' to 1,700'.

Just following this right turn, several examples of Bitter Cherry (*Prunus emarginata*), a small tree about 15' tall, poke through the woodlands between 1,000' to 1,200'. This relative of the common cherry (hybrids and cultivars of *P. avium*), sports white flowers with five petals and multiple golden stamens, a trait shared by many in the crowded Rose Family. Unfortunately, these cherries are almost inedible because of their bitter taste, although birds seem to relish them. The *Prunus* genus, however, is responsible for many of the fruits we cherish, such as cherries, apricots, pears, and plums. Apples are closely related, being of the *Malus* genus. Even blackberries and raspberries (*Rubus*), and those delicious strawberries (*Fragaria*), are

Bitter Cherry

all contained in genera within the Rose Family, whose fruits and rosehips have been relished by animals and humans since Eve took her first bite. Although that may be because these rose products harbor Vitamin C and are therefore good for our health, I suspect it's rather due to the fact that they also contain plenty of that sweet fruit sugar, fructose.

Just before the rocky openings and right switchback at 1,300', **Point 4**, three red and yellow Columbines decorate the trail on your left. In both of these openings, two Naked Broomrapes can be found growing in proximity to the rocks; in the first opening, these are on your right while in the second, they are on the left. Six inches tall, and pretty as a perfect purple flower should be, this plant holds a dark secret: Its existence depends upon "raping" (hence its name) nearby species, particularly the stonecrops and saxifrages which grow in these openings. This innocent-looking flower obtains its energy through dastardly predation and outright parasitism. However, we should probably not be too judgmental about the livelihood of plants. At any rate, we can conveniently ignore what goes on underground, (although we are learning more about the complex interweaving and mutual dependencies amongst fungi, bacteria, and the roots of all plant species) and simply appreciate the beauty top-side. For now, let the professional botanists worry about the rest.

Residing in these two openings, which afford pleasant views out to the river and north beyond, are two small plants which deserve notice. Little-leaf Candy Flower can be found growing off the talus blocks in both openings, while the diminutive purple Small-flowered Tonella appears in the first opening, at least to those who look closely for these 4-6" specimens. Prairie Star, with its raggedy cleft pink-to-white petals, also graces these rocky breaches in the forest cover. In-between the openings, Scouler's Valerian, with 1-2' tall stems and fuzzy white compound flower tops, and ground-hugging Woods Nemophila, with tiny blue flowers, are both frequent companions.

At these openings, there is a right switchback and on the north-facing rock at this point, Broadleaf Stonecrop plasters this wall with its jade-green fleshy leaves and starry yellow flowers. This plant appears to grow directly off the stone and, indeed, its roots do invade rocks and create tiny crevices which enable it to flourish. You may notice another stonecrop, or at least its leaves, growing here, the Oregon Stonecrop. While its yellow blooms won't appear until early summer, its succulent yet radiant deep-green leaves distinguish it readily from the blue-green leaves of the Broad-leaf species. It also thrives on stone and, like all plants in this *Sedum* genus, its leaves hold enough water to see it through a hot, dry summer. Compare the leaves of these stonecrops to the Jade Plant of indoor offices and living rooms; they are members of the same family. *Sedums*, and their close cousins, the *Sempervivums*, commonly called Hens and Chicks, are also familiar to most Northwest gardeners and prosper outdoors as well as in indoor potted plants. You can see the resemblance to their distant relatives, the Cacti, in their water-storing capabilities.

Also at this opening, and close to the rock, look to your left to see many examples of Little-leaf Candy Flower. Just past this mossy cliff, the rare Suksdorf's Mist Maidens grow off the dripping face to your left at about 1,350-1,400'. Small white flowers with five petals dangle from the moss-covered face like stars against a dark gray sky, their precisely scalloped leaves adding to their allure. These damp and generous gardens beg for close scrutiny, even if your botanically-challenged hike-mates are scurrying on. Please also note that Jolley equates this species with Sitka Mistmaidens, while others, such as Hitchcock et al., disagree. The Sitka species look similar but differ in root shape and distribution, hiding in high alpine rock crevices, generally above 7,500' in the North Cascades and Olympics.

Beginning at about 1,600', and continuing up to about 2,800', Cascade Oregon Grape is now in flower, with its bunches of yellow blooms presaging the dusty blue berries yet to come. Note that not every bush bears its blooms at the same time, so as not to exhaust pollinating insects. In addition, Red-flowering Current puts in a nice show at about 1,600' on your left. At 1,650', and continuing for about another 200' of elevation, the Fairy Slipper, or Calypso, Orchid remains in bloom. Stop and admire it here, as it will soon retire for the remainder of the season.

Also at 1,650', a clump of low-blooming Bunchberry, with four white "petals", actually leafy bracts, can be found on the left. Just before, then between the two points where the trail approaches the streams (**Point 5** is the first stream approach), much Candy Flower is now sprouting, with its paired leaves mid-way up its 8-10" tall stems and its five white petals candy-striped with pink. Look for this species from here all the way to the wet area at **Point 6**. On your left at 1,950', note the strange Rayless Arnica, several of which cluster here but are only rarely seen elsewhere in the Gorge. They can be considered the opposite of Dandelions in that all their individual flowers are centered in the disk, with no flaring "petal flowers" to be seen.

Early in May, Trillium is still in bloom above the streams, but by late in the month, it has turned burgundy with age. At a relatively flat area at 1,850', then just before and after each stream approach (the second approach is at 2,350'), the shiny, vein-streaked leaves of Wild Ginger recline on the often-damp forest floor. These heart-shaped and varnished dark green leaves are the only evidence of this plant's existence unless you gently lift them up to reveal the strangest flower in all the Northwest. The blooms of Wild Ginger are a dark purple-brown which blends in well with the duff upon which the leaves rest. Each plant has three petals which end in flaring tapering lobes which look for all the world like a Salvador Dali moustache, or at least like something this surrealist would have imagined as a flower. This species is pollinated by crawling ants and thus has no need to advertise its presence with gaudy color or on a lofty perch.

In the forests between **Points 5** and **6**, Small-flowered Alumroot and Fringe Cup are in bloom up to about 2,700'. Here, at **Point 6**, where a seasonal stream still flows, yellow Stream Violet and Candy Flower line the muddy trail; strategically placed rocks and sliced tree trunks help avoid some of the mud and water here. As you hop across the moss-covered rocks, note the leaves of Brook Saxifrage (*Saxifraga arguta*), with cogs spaced as regularly as if a careful mechanic had crafted them. This saxifrage sends up sprays of white flowers intermittently throughout the spring. Just beyond this crossing, on your right, a Red-flowering Current shrub is now in full splendor, with drooping pinkish-red clusters of flowers to brighten this semi-open area. A more unusual shrub can be found here as well, also on the right: the Clasping-leaf Twisted Stalk. This is a 3' tall Lily Family bush with pointed parallel-veined leaves which clasp the stem. These are accompanied by dainty pinkish-white flowers pendent beneath. Look closely at the pedicel, the flower's stem. These stalks have a kink which allows the bloom to hang straight down. Its close cousin, Rosy Twisted Stalk, used to grow here as well and at the trail junction with the Gorge Trail far below, but has been absent in recent years. With non-clasping leaves, a height of only 1-2', and pink flowers suspended from non-twisted stems, it can be easily distinguished from its larger relative. Although gone now, it may well return to these areas in the future and can be found alongside many trails around Mt. Hood in damp areas.

There are four meadowy openings beginning at **Point 7**, 2,750', and in May they are filled with color:

a. Just before the first opening, at 2,750', at a left turn as you begin to emerge into the open, blue-purple Kittentails may still be in bloom, at least early in the month. Also early, yellow Glacier Lilies grace this meadow along with two specimens of scarlet Harsh Paintbrush at its start on your right. On both sides of the trail, but particularly on your left, Cliff (Menzies') Larkspur holds its deep purple-blue flowers above 2' tall stout stems. Colonies of Small-flowered Blue-eyed

Spreading Phlox

Mary inhabit the opening along with the strangely mottled brownish-purple flowers of Chocolate Lily and the low-growing yellow Martindale's Desert Parsley.

A new yet common species is present here as well, the winsome Spreading Phlox (*Phlox diffusa*), a mat-forming perennial plant whose needle-like leaves are often completely hidden under its showy pastel pink-to-blue-to-white five petalled flowers. A closer look behind each flower finds that, as is typical in the Phlox Family, the sepals form a long tube from which the petals appear to sprout. Lean down, even if it's a chore with a pack on, to sniff the ineffable aroma of these plants. Some describe the odor as spicy but to me, it smells like licorice candy, its perfume only adding to its allure. Mats of this Phlox cushion rocky slopes and outcrops throughout the west, from British Columbia south to the Sierra. Fortunately, it's no stranger to the Gorge.

b. The second meadow, at 2,900', might be considered more of a rock-strewn opening than a grassy plain, but an opening it is, replete with Glacier Lily, Small-flowered Blue-eyed Mary, Phlox, Harsh Paintbrush, Larkspur, white Large-flowered Cryptantha with an orange "eye", purple Fine-tooth Penstemon, and Chocolate Lily. Check out the view north across the river to Dog Mountain and beyond.

c. At 3,100', the third meadow is also fairly rocky, but still holds the Blue-eyed Mary's and Larkspurs noted above. Moreover, it also contains several flowering examples of Shining Oregon Grape and blue Broad-leaf Lupine, this last in masses from the middle to the end of the opening. Fine-tooth Penstemon lends its purple tube-shaped flowers to this array. A less spectacular species, Miner's Lettuce, is also in bloom here on the left, notable for its single round leaf split by its piercing stem. Yellow Nine-leaf Desert Parsley is also present, a 2' tall plant with umbrellas of yellow compound flowers, contrasting with Martindale's Desert Parsley, lower and smaller, but with broad finely-dissected blue-green leaves. To complete this balanced garden, Grass Widows wave their satin-flowered reddish-purple flowers in the May breezes so common at these open heights. Orange-yellow stamens, protruding from their centers, accentuate these blossoms' beauty.

d. Yes, Lupines, Shining Oregon Grape, Grass Widows, Harsh Paintbrush, the two desert Parsleys, and Fine-tooth Penstemon are all here in the fourth meadow at 3,200', but joining them is a most unusual and thoroughly delightful plant found near western Gorge trails only here and on Dog Mountain – Yellow Bells (*Fritillaria pudica*). This close relative of Chocolate Lily is relatively rare, especially this far west, preferring as it does dry grasslands. Another example can be found at about 2,450' on the right of the trail on Dog Mountain (though earlier in the spring) in the large meadow before the top of the Puppy (Windy Point). Atop 6-10" stems, one to two bells of limpid yellow hang shyly downward toward the grassy ground, almost surmounted in height by this plant's grassy leaves. The yellow shades into orange as the blossom ages, adding even more grace. Cultivated fritillaries are sometimes seen in Northwest gardens but these are special cultivars; this yellow delight should never be disturbed as it cannot be successfully transplanted, no matter how charming it would look bordering your lawn.

Above the meadows, **Point 8**, 3,280', marks the beginning of ten switchbacks (the Lowe's must have skipped two as they insist there are eight) in the woods up to the plateau and finally the junction with the Green Point Mountain Trail, #418, **Point 9**, at 3,920'. Here, Red-flowering Current will occasionally be seen coloring this otherwise forested section, but no further flowers are in bloom except several early-flowering Little Wild Rose and Bear-grass specimens. These last are especially sweet-scented. Fortunately, the plateau itself, after the left turn at the trail junction with the Green Point Mountain Trail, contains Vanilla Leaf and the Inside-out Flower, at least after mid-month, and an occasional Little Wild Rose along with some Red-flowering Currents about halfway to North Lake, as well as yellow Stream Violets surrounding the two stream crossings just before the Lake. At the Lake itself, **Point 10** at 3,988', late in the month a large specimen of Serviceberry can be seen on the rock outcrop while Bear-grass perfumes this stone-strewn neighborhood with its fragrant lily-like odor.

Adventurous and hardy hikers can make loops here to Rainy Lake or backtrack and climb Greenpoint Mountain, then descend to the Gorton Creek Trail (with the aid of a car shuttle). Most of us are content to retrace our steps and savor again the glories of those meadows, the deep shade of majestic Douglas Firs, and the emerald green of moss-plastered rocks and forest floors. The Wyeth Trail in May is at its best.

June

The trailhead, **Point** 1, is marked by a signpost and surrounded by Himalayan Blackberry and Robert Geranium. We have one native blackberry, the Trailing Blackberry (*Rubus ursinus*), which, though rare along trails in the Gorge, can be found on your left just after you begin the trek up the dirt patch right out of the parking lot. It, unfortunately, carries sharp thorns the equal of the Himalayan sort. Up the dirt road to the junction with the Gorge Trail, pink Robert Geranium joins pinkish-white Star Flower and, early in the month, white Plume-flowered Solomon's Seal and Woods Strawberry. Following the abrupt left turn here, at the beginning of the trail proper but before the power line, white Columbia Wind Flower begins and continues in woods below and above the power line gash. This flower could be mistaken for a very large Woods Strawberry as both are white, have three toothed leaves, and carry yellow stamens at their centers. But note that the Strawberry prefers open areas, is low-lying and its stamens are much deeper in yellow hue than those of the Wind Flower.

Tiger Lily

You actually descend to the power line, and, in an open area just as you enter this gash, note White (or Birchleaf) Spiraea, a vanilla-white relative of our bushy pink garden spiraea's. At the power line itself, between 250' and 300' elevation, reddish-pink Nevada Pea trails above other vegetation; somewhat weak-stemmed, it is partially supported by other plants but we should not demean this habit as it helps many vines in the Pea Family, such as cultivated beans and weedy vetches, to survive.

> Most flowers wither with age, much as people do. However, some flowers, unlike people, change color as they age. In our mountains, Trillium and Fringe Cup are two prime examples. Since the color change is from white to purple or pink, it is difficult to intuitively understand any evolutionary advantage to this mutation, as it would not appear that the plant is conserving energy by turning a different hue. Botanists may have studied the chemical processes involved, but we amateurs can simply appreciate the different shades that allow such flowers to age gracefully. More apparently beneficial are the color changes which leaves of certain species undergo as they first develop. Many high-elevation plants, such as Spreading Stonecrop (found at higher altitudes than the Gorge) and several Aster Family plants of high mountain meadows, develop leaves which at first are a reddish-purple, then, when mature, become green. Certain stonecrop leaves, exposed on rocks to the constant glare of the sun, stay red throughout their life spans. The reddish-purple color actually protects these leaves from the sun's ultraviolet rays, which might damage them in their fragile youth. While this safeguard undoubtedly benefits those plants, other species living right next door lack these protective mechanisms yet appear to do just fine in their sun-drenched habitat, a puzzle as yet unsolved. By the way, the autumn color changes of deciduous leaves, particularly prominent in our maples, is not a metamorphosis as much as a loss: When these leaves lose their chlorophyll in the fall, they simply reveal the oranges, reds and browns that were hiding underneath. Thus green doesn't change to orange; it simply disappears.

Other plants benefiting from this artificial opening include the native Thimbleberry, with its papery large white flowers; Woods Strawberry; yellow Scouler's Hawkweed, with its dandelion flowers; and the weed Dandelion itself; Oxeye Daisy; Hairy Cat's-ear, with its close impersonation of a Dandelion flower; fuzzy-flowered Rabbit-foot Clover; and pink Common Vetch. This last, a universal weed of field and roadside, is allied with the peas (*Lathyrus* genus) and also has tendrils which cling to other vegetation. However, vetches are generally smaller and have bristles completely surrounding the style, whereas peas have their bristles on one side only, sort of like a toothbrush.

Here, just under the power line on the left, then just past it as well, the ravishing Tiger Lily *(Lilium columbianum)* appears. With stems 4' tall and radiant-orange nodding 2"- wide flowers, this lily is unmistakable. Its petals are purple-spotted in the center and recurved as they taper to a pointed tip, making this plant hard to miss. Preferring thickets, open forests, and meadows, Tiger Lily has a number of close relatives on the West Coast, but none as showy and thriving this far north – it is as recognizable as it is beautiful but please leave it be – it is usually impossible to transplant it to city gardens. By the way, the spots on the petals were thought by Native American children to give you freckles if you got too close; don't let that deter you – it's probably a superstition, although no scientific tests have yet been conducted.

Red Elderberry

You now enter deep woods, where Star Flower and the Inside-out Flower predominate. Proceeding up the gentle path to **Point 2**, the crossing of Harphan Creek, masses of Star Flower twinkle against the green sky below while Nevada Pea lends color to the forest, here often more purple than its usual pinkish-red. The Solomon's Seals, Plume- and Star-flowered, are both present, along with Hooker's Fairy Bells early in the month and later, Queen's Cup, lining the left side of the trail a bit before the creek. As you cross this stream, now quite low, and begin at first gently up the mossy trail, watch for Scarlet Columbine immediately after the stream, then the many Columbia Wind Flowers populating this forest early in June, followed here by Fringe Cup, now turning pink with age. The ubiquitous Robert Geranium, pink above its ferny leaves, grows here and there up to 600'. The weedy Cleavers clings to vegetation just after the creek crossing on the right side of the trail, its linear whorl of leaves barely surmounted by tiny white flowers with four petals. You may be all too familiar with this common weed of urban fields and gardens.

Several other plants appear in this cozy valley of green after crossing the stream, where the trail becomes steeper as it ascends to the old metal logging equipment seen to your left at around 1,000'. Scouler's Valerian appears beginning at around 500', a 1-2' tall plant with clusters of fuzzy white flowers which occasionally are pink or red at their centers. Look for the bright pink of Little Wild (Baldhip) Rose at a right turn at 750', and for Columbine, which graces the forest just before and after a left switchback at around 800'. Between 700' and 800', several specimens of a small but distinctive flower can be found growing low in the forest, perhaps only 4" high. Its name, Single Delight, describes its unique beauty: A single white waxy bloom tops a whitish stem with basal oval to elliptic leaves. The nodding flower itself is the picture of modest ornament, with five frilly petals centered by a green bulb (the ovary). From this orb springs a single straight style capped with a five-tipped stigma looking for all the world like a chess rook. If you can find this diminutive gem, stoop to inhale its fragrance, as delicate as the plant itself.

In the rocky areas in these woods, Small-flowered Alumroot is common, along with Trail Plant (Pathfinder) late in the month. By the time the going gets steep after **Point 3** at 750', you will at least be encouraged upward by the presence of much Star Flower, Plume-flowered Solomon's Seal, and even a few Phantom Orchids (Ghost Flower), a spooky waxy-white plant appearing as an apparition of some disembodied spirit. It can be found in the dark woods at around 1,100' on your left. Here too, find some color with the Columbia Gorge Groundsel (Senecio), a yellow daisy-like flower atop 1-2' tall stems with deeply in-cut leaves. Several examples are on your left at about 1,250'. Early in the month, the Inside-out Flower is still in bloom, while by mid-June, Little Wild Rose appears from 1,000' to 1,400'.

At 1,200', then again at 1,250', two examples of Red Elderberry (*Sambucus racemosa*) appear, both to your right down the hill about 50' below the trail. This 10' tall shrub in the Honeysuckle Family has alternate leaves divided into five to seven lance-like and toothed opposite leaflets and, at this time of year, creamy blue-white pyramidal clusters of flowers. The red berries will follow later in the summer but you would be best advised to leave them be; their raw taste is insipid. I would not recommend gathering quantities to concoct wine or jam; the celebrated elderberry wine of poetry and song comes from a European species whereas our native elderberries – this species and the Blue Elderberry (*S. cerulea*) are used for jam but it is rather inferior to the old-world product. Native American children would peel the pith just underneath the stems of elderberries to make pea shooters and straws. It's probably best not to repeat this process as we have adequate commercial substitutes.

As you progress up the steepening trail, note the unusually beautiful crimson and yellow flowers of Columbine on your right just before the opening and right turn at **Point 4**, 1,300'. At this point, the jade green leaves and starry yellow flowers of Broad-leaf Stonecrop are plastered on the rock face at the right turn. The small stony openings just before and past this turn feature white Field Chickweed, with its raggedy-cleft petals; a miniature forest of Harsh Paintbrush at the first opening on your right; several more examples of Broad-leaf Stonecrop right and left; a single naked Broomrape on the right, its small purple flower with orange stamens masquerading innocence while it parasitizes the roots of its neighboring stonecrops; Little-leaf Candy Flower, white with pink stripes, growing off the moss-covered talus; purple Nevada Pea, leaning on and borrowing strength from other vegetation, as some neighbors are wont to do; Broad-leaf Lupine, blue and tall; and the hardy pink Spreading Phlox, seeding the Gorge air with its sweet licorice-candy bouquet.

Between **Points 4** and **5**, the first stream approach at 2,200', the North Lake Trail now enters deeper forest and ascends more gently, passing typical late spring/early summer woodland plants. Hooker's Fairy Bells can be found early in June up to about 2,300' while the two Solomon's Seals, Plume- and Star-flowered, are also in bloom. The Columbia Gorge Groundsel is present here and there from 1,300' to 1,750', with lyre-shaped leaves and yellow daisy flowers. Baneberry appears between 1,400' and 1,900', while tall yellow daisy-like Broad-leaf Arnica brightens the woods at these elevations and up to 3,000' as well. Two fine specimens of Striped Coral Root are on your right at 1,650'; note that these strange all-pink saprophytes lack any green chlorophyll and make their living off fungi in the duff of the deep forest floor.

At 1,560', then again at 1,900', specimens of White-vein Pyrola can be discovered on your left hiding in the moss. Note that the white lines outline the veins on the dark-green roundish leaf and are not mottled like the more elliptical leaves of Rattlesnake Plantain. Twin Flower is now out and especially prominent between 1,650' and 1,850' on the left where the trail levels out. Its ground-hugging bifurcating stem holds two clones of a pinkish-white bell-shaped flower that defies any descriptor other than the overused "lovely". From 1,750' to 2,150', you can find several examples of Pathfinder (Trail Plant), a species more renowned for its bicolored leaves (top vs. bottom) than for its 2' tall stem sprinkled with almost unnoticeable flowers. The usual suspects, Inside-out Flower, Star-flowered Solomon's Seal and Vanilla Leaf, are frequent inhabitants up to 2,500'.

The small stream is first approached at a left switchback at 2,200', **Point 5.** Here, and at the second approach to this same stream, also at a left switchback at 2,350', yellow Stream (Wood) Violet and pink-striped Candy Flower thrive in the moist habitat. Just beyond **Point 5,** Common Monkey Flower prospers in the damp area to your right, while above, occasional examples of the Inside-out Flower and Broad-leaf Arnica present themselves, although the forest here is thick and often devoid of flowering plants except for an occasional Black Huckleberry bush, with its small pendant and globular pinkish flowers barely noticeable until, later in the summer, they burst forth as sweet juicy black fruit. Twin Flower is in bloom from 2,350' to 2,450', frequently found on your right. Also out now, despite the shade, is Bunchberry; a clump occupies the base of a huge Douglas Fir at 2,400' on your right. By the time you reach the squishy damp area at the second approach to the stream, **Point 6,** 2,700', Youth-on-age is clearly in evidence, with 2'-tall stems adorned with brownish tubular flowers ending in frilly tips. This moisture-loving relative of the similar Fringe Cup can be distinguished from that plant by its darker flowers which do not turn pink with age, and by leaf-buds at the base of many of its leaves. These will drop off, as the season progresses, to germinate in the damp earth and form new plants of their own. Piggyback Plant is another name for this species, emphasizing its mother-daughter relationships.

Much larger than most any other species in our forests, Cow Parsnip is now in bloom in this wet area, along with another, but much less gentle giant, Devil's Club. Although both these monsters of our Northwest forests bear clusters of white flowers, and both here can top 7', Cow Parsnip's blooms are in flat-topped umbels while Devil's Club holds its flowers in an upright spike. More to the point (excuse the pun), Devil's Club bears fiendishly-sharp spines all along its stem and even along the petioles (stems) of its leaves. While the derivation of the species name, *horridum*, is a no-brainer, the genus name is a bit confusing: Appropriately, *oplo* means armor in Latin but *panax* means curative. In fact, this brutally armed Frankenstein of a plant has been used by native peoples for centuries for everything from arthritis to diabetes. Please do not grab hold of this plant should you slip; to add injury to injury, its spines are not only sharp and barbed, but are also tipped with a toxin irritating to the skin. There are many clubs to join, but please do not join this Devil's Club.

> It appears that what occurs underground in the plant world, while less attractive than what we can see above, is however, no less intriguing. Aphids are tiny insects that love to eat bean plant greens. Wasps kill aphids. Wasps should therefore appear to be friends of bean plants and this has been known for some time. This protection occurs even in bean plants living in proximity to the infested plant but not attacked directly. Indeed, the mechanism was believed to be airborne: As aphids attack the plants, the flowers emit an odor which attracts wasps and repels aphids. However, even when these volatiles are blocked, the bean plants continue to attract wasps when attacked by the pesky aphids. By what mechanism could this occur? Bean plants often share root systems but when these are cut, the infested plants still share this property of protecting itself and its neighbors. However, these plants also share fungal root systems called *hyphae* and when these are cut, no protection occurs to the attacked plant or its neighbors. It thus appears that bean plants, and probably many other flora, can communicate with their neighbors, and thus likely kin, that they are under attack and must protect themselves. An underground internet? Perhaps.

Much as savannah animals, many other species congregate at this watering hole, including Common Monkey Flower; Red-flowering Current, just past the stream; Stream Violet; Fringe Cup; Salmonberry, with its delicate crinkly-magenta flowers; and, also just past the wettest part of **Point 6**, Varied-leaf Phacelia. Salmonberry is a moisture-loving member of the Rose Family, closely related to the blackberries and raspberries. Here, it forms a shrub 4' tall, with mild thorns. By mid-summer the yellowish-red fruits will be out, mushy but, to some misguided tastes, preferable to Thimbleberry. Try some anyway – you may be in the group that appreciates their lack of character.

Above the stream, Star-flowered Solomon's Seal predominates from 2,700' to 3,000', in places forming a carpet across the forest floor; its root systems can crowd out and eliminate other plants. Such competition is not limited to the animal world – even such dainty plants as this one can be frightening to other species. Less so is the tiny-flowered White Hawkweed. Despite its name, it is a native plant, one with 2' tall stems and diminutive white daisies at the tips of its several terminal branches. It prefers sunshine and can be found in scattered openings in the woods between **Points 6** and **7**. Also just out toward the end of June is Foam Flower, its white sprays of flowers brightening the woods above 2,500'.

By the time you reach the first major opening, **Point 7**, you may wish to take a break or at least slow down to appreciate the fine flower show upcoming. Blooms here, in the four following meadows, are as eye-filling in June as in May:

a. The first opening, at 2,750', displays Cliff (Menzies') Larkspur, here deep blue and prominently spurred; yellow Broad-leaf Stonecrop, running amok on the abundant talus walls on your left; the two yellow Desert Parsleys – low-lying Martindale's with ferny blue-green leaves and more upright Nine-leaf, with grass-like bright green leaves; tall blue Broad-leaf Lupine; scarlet Harsh Paintbrush; Small-flowered Blue-eyed Mary; white Woods Strawberry on the dry hillsides; pink and white Spreading Phlox, carpeting the ground to your right; yellow Wooly Sunflower, its bright daisy heads lifted above its fuzzy-grey cleft leaves; several remaining Chocolate Lilies, especially on your right; and a lone white Northwestern Saxifrage in mid-meadow on your left.

b. Just a bit above the first opening, the second meadow, at 2,900', proudly boasts many of the same blooms, including the Lupines, Phlox, Larkspurs, Sunflowers, and Desert Parsleys. But here, Fine-tooth Penstemon, with its purple tubular flowers also appears, along with tall now-white Western Groundsel and Yarrow. In addition, not just one, but two species of Paintbrush are now present: the Harsh species, more common early in the Gorge, and Common (Tall) Paintbrush, a later-blooming and taller species but no less colorful, with its bracts dipped in what appears to be the brightest orange-red imaginable. This species can be distinguished from its cousin, the Harsh Paintbrush, by its lance-shaped leaves, which do not end in lobed tips.

c. The third meadow, at 3,100', although smaller than the others, contains much the same floral display. Both Paintbrushes thrive here, along with Lupine, Chocolate Lily, Groundsel, carpets of Phlox, Larkspur, Fine-tooth Penstemon, and Small-flowered Blue-eyed Mary. Although the Woolly Sunflower is absent, its companions adequately compensate.

d. As you enter the last meadow, at about 3,200', Phlox and Lupines dominate, along with the outlook over the Columbia far below. While white daisy-like Groundsels and the two Paintbrushes compete for attention, a little-noticed pale yellow flower on your left half-way through should also garner some regard as it is not often identified in the Gorge, nor frequent at these heights. Sticky Cinquefoil, a 1½'-tall plant with glandular stems and green bracts surrounding its small pale yellow flowers, is a member of the Rose Family, but do not look for thorns or perfume. Again, family members are classified based upon their reproductive organs (stamens and pistils) and their seeds, and more recently by their DNA, not by any resemblance to each other's leaves or flowers. There are probably family members you may think lack your kin's common traits, but, fortunately or not, DNA testing is now upon us.

In the dark woods between and above these meadows, Queen's Cup, with its pure white flowers of six tapering petals held on a single stem above its basal rosette of elliptical leaves, covers parts of the forest floor between 2,900' and 3,000'. Trillium can be found, scattered and now purpling with age, along with Baneberry, the two Solomon's Seals, an occasional Spotted Coral Root (look for several on your left at 3,250'), the gratifying deep blue of the Oregon Anemone (a clump can be found at 3,280' on your left) and just beyond, its white cousin, the Columbia Wind Flower. Evergreen Violet and a few long-lasting Fairy Slipper (Calypso) Orchids remain out between 2,900' and 3,500'. At the switchbacks, beginning at **Point 8**, 3,280', Little Wild Rose is now in bloom, especially at a right-hand turn at 3,450'.

Sticky Currant (*Ribes viscosissimum*), a 3-4' tall shrub, can also be found at these switchbacks. The entire plant is covered with viscous hairs and emits a strange chemical odor in the summer heat. Its white tube-shaped flowers are also sticky and hang from the tips of branches beneath its maple-shaped leaves. This shrub and its sibling, Red-flowering Currant, serve as hosts for a fungus called Blister Rust, which is decimating pine populations in the eastern parts of the Northwest. Do not blame the currants; I'm sure they had little to do with the massacre, although they themselves do not seem to be weakened by the fungus.

Nearing the plateau, again note Red-flowering Currant, a more common relative of Sticky Currant. It has unmistakable drooping clusters of blooms and is particularly prominent around 3,400' and 3,500'. Here too, Bear-grass is now in flower, spreading its delightful cologne throughout these dim woodlands. At 3,725', watch for a prime example of White-vein Pyrola on your right. Here and there, yellow-flowered Cascade Oregon Grape dots the woods and Foam Flower is just starting to make its summer-long debut, its dainty white spray of flowers above a forest covering of maple-like leaves.

Sticky Current

On the plateau itself, just before the trail junction with the Green Point Mountain Trail, #418, our **Point 9** at 3,920', Stream Violet pokes its yellow flowers up through the forest duff and two specimens of the oddball Pinedrops (*Pterospora andromedea*) poke boldly up on your right about 50 yards after reaching level ground. This large saprophyte cannot be mistaken for anything else, with its 3'-tall stout stems and drooping reddish bells masquerading as flowers. Little Wild Rose puts in a credible appearance just after the left turn at the sign to Green Point Mountain, along with Queen's Cup and Hooker's Fairy Bells, especially at a dip in the trail about 100 yards after the sign.

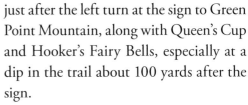

Pinedrops

Along this part of the trail, with its many ups and downs, Little Wild Rose, Trillium (now purple), Bear-grass (particularly in the opening half-way to the lake looking out to Mt. Defiance to the east), Vanilla Leaf, and Foam Flower are all apparent. Just at the mucky areas around the two stream crossings, look for Stream Violet and the rare Rosy Twisted Stalk (*Streptopus roseus*). This Twisted Stalk, a 1½'- tall shrub-like plant resembles Clasping-leaf Twisted Stalk but with several differentiating features: Easiest to observe, it has pink, not white, flowers, and these have only mildly flaring tips. Moreover, its leaves are not as clasping as its larger sibling in the Lily Family. Also, it is usually un-branched and the pendant flowers do not bear the kinked pedicels (stalks) of the Clasping-leaf species. However, both Twisted Stalks share a preference for moist locations and both discretely hide their blooms underneath their leaves. As usual, a close look reveals the delicate and fragile beauty of these blossoms.

At North Lake, **Point 10**, you can relax, eat lunch, and admire not only the view across the placid waters but also the Serviceberry bushes growing on a rock outcrop near its west bank. You can also breathe in the perfumed air of the Bear-grass growing all around the lake (actually a reservoir – note the earthen dam that you traverse to get to the rocky lunch site). Also take note of the emerging red flowers and towering spires of Fireweed by the west shore.

The rest will do you and your joints good while you contemplate all the marvelous flowers which will accompany your journey back down. Rest assured that many of these will continue to grace this trail throughout the summer months as well.

July and August

Although the display of blooms tapers off somewhat during the summer months on the trail up to North Lake, this trip in summer still offers many floral species to discover and adds a bonus of cooling shade and a bit more solitude than you'd find on more well-known trails closer to Portland, especially on a weekend. As you walk up the old overgrown road from the parking lot at **Point 1**, many rose-purple Robert Geraniums greet you, seemingly thriving in the shade. Just at the trail sign at the junction with the Gorge Trail, there is a sharp left turn and here, as you walk up above the road, Woods Strawberry is out on your left, with its white flowers adorning the sandy banks where enough light can reach its toothed triplet of leaves.

The opening at the power line soon appears as you descend a bit, then cross this breach in the forest. The trampled ground here affords an opportunity for all manner of urban summer weeds to intrude, although perhaps they should not be maligned for being opportunistic immigrants. Ox-eye Daisy (in July), Dandelion, and Queen Anne's Lace (in August) are all present, along with the natives Yarrow, Pearly Everlasting and Woods Strawberry.

When children are asked to draw a flower, they most often unknowingly sketch an Ox-eye Daisy *(Chrysanthemum leucanthemum)*; they may color its petals other than their natural white although they usually get the yellow disk flowers correct. The proto-typical simple "flowers", 1-2' tall Ox-eye Daisies camouflage their complexity with a superficially elementary appearance of pure white "petals" circling a central yellow disk. In

Ox-eye Daisy

fact, as in most members of the Aster Family, this simplicity is deceiving as the "rays" are actually each a separate flower in itself and the central disk is also a composite of many disk flowers. Both ray and disk flowers each contain a set of stamens and pistils, the full reproductive system of a separate bloom.

> We often mention stems of plants but barely give them much careful thought except the brief recognition that they support a plant's leaves and flowers. However, stems accomplish much more than provide support and points of attachments for leaves, the inflorescence, and fruits. They also transport, through osmosis and a venous system much like that in animals, vital nutrients and water up the stem to the leaves, where photosynthesis can convert those ingredients into energy in the form of carbohydrates. These are then distributed to all other parts of the plant as well. Moreover, through cell division and multiplication, stems produce the growth necessary to allow the plant to reach towards the light which makes photosynthesis possible. Stems also serve as storage vessels, along with roots, bulbs, and tubers, holding nutrients in reserve until needed in leaner times. Finally, their green color also means that stems are as capable of photosynthesis as are leaves, and often contribute a goodly share of energy for the plant's overall survival. Please do not disregard stems; there's clearly more to them than meets the casual eye.

Thus this apparently single blossom on a single stem is actually composed of hundreds of individual flowers, much as a coral colony in the ocean is a composite of multiple individuals. This has led to the Aster Family being called not only the *Asteraceae*, but also the *Compositae*. This adaptation has undoubtedly contributed to the family's prominence as one of the largest in the world. Hundreds of tiny flowers congregating to mimic a single large bloom can obviously attract more pollinating insects. Anyone driving the freeways of the Northwestern states in late May through early July is familiar with whitened fields of this not unattractive daisy, which is said to compete for some crops and hence is classified as a bothersome weed. Ox-eye Daisy is certainly related to our garden Shasta Daisies but it is not, as some have claimed, simply a smaller cultivar.

Although our Ox-eye Daisy is a distant relative of fancier chrysanthemums cultivated for centuries and worn as a compulsory adornment at prom night, there's another chrysanthemum that's even more interesting and potentially more useful: *C. cinerariifolium*, a finicky plant which grows best in the volcanic soil of Rwanda. While all chrysanthemums produce small amounts of a natural insecticide, pyrethrin, *C. cinerariifolium* happens to contain an abundance of it and has enriched previously dirt-poor farmers in that war-torn nation. These farmers now grow this plant in huge fields to sell to giant fertilizer companies. The UN's Food and Agriculture Organization has sponsored training for Rwandan farmers, who can now profit from a humble relative of the Ox-eye and can proudly claim they are doing their part in providing a sustainable and natural insecticide which hopefully can be increasingly employed to replace harsher chemicals now in use.

A native plant, just as manifest in its original habitats as Ox-eye Daisy along roadsides, now appears both just before and at the openings created by the power line, Fireweed (*Epilobium angustifolium*). It cannot be avoided, not that you'd want to, because this plant is really not a weed in any sense. A native of the Northwest from the Alaskan coast to Northern California, beloved Fireweed colors huge tracts of lowland and mountain meadows with 4'- to 8'-tall spikes of rose-purple 1-2' four-petalled flowers *en masse*. A member of the Evening Primrose Family, row upon row of these lovely plants are often seen in the foreground for photos and posters of our most famous snow-capped mountains. Beautiful as it is, especially massed as if planted by a grand gardener, in fact this lushly-hued species harbors a secret to its grouped success: Fireweed was named not for its blooms of blazing color but for its habit of growing in areas devastated by forest fires. More recently, it has appeared by the thousands in burned-over logging tracts. Scientists have discovered that Fireweed's seeds may lie dormant in the soil for up to a century until activated by the heat of a forest fire or the burning of lumber slash.

Fireweed

Heat-activated seeds are well-known in several plant species, but this cannot explain the entire distribution of Fireweed as it is also common in upland meadows absent a fire for many centuries. This tall and resplendent species may harbor several mechanisms of germination, a fortunate trait for flower lovers everywhere, and for bees who manufacture a delicious honey from its vibrant blooms. It may also be pollinated by moths, a common means of reproduction in the Evening Primrose Family, named because, unlike Fireweed, many of its species open as the sun sets, and only then emit a fragrance which lures nocturnal flying insects to its blooms. By conserving energy during the sundrenched daylight, these plants can be said to truly be living on the edges of dawn.

In the woods before and after **Point 2**, the crossing of Harphan Creek, The Inside-out Flower, Queen's Cup (especially at 550' and 900'), and Pathfinder (Trail Plant) can all be found in frequent locations. As you progress gently up the broad swath rimmed with bright green forest, several examples of Rattlesnake Plantain grow off the moss in the first 100 yards beyond the creek, both left and right. Do not confuse their mottled elliptical leaves with the white vein-outlined, rounder leaves of White-vein Pyrola. Both may prominently advertise their presence off moss in dark woodlands by attracting the attention of crawling insects to their leaves rather than to their somewhat unimpressive flowers.

Within this green monopoly, spots of color enliven the woods with the rose-to-purple Nevada Pea, sometimes trailing on the ground and often sprawling over other vegetation. Little Wild (Baldhip) Rose and a few remaining pink Robert Geraniums persist here as well. Later, by early August, Wall Lettuce adds some yellow through its small but distinctive spray of blooms with five petals each, a rarity in the Aster Family. By 750', just after a left turn and a brief plateau, the going gets steeper at **Point 3**. Soon you reach a green plateau and can spot the old rusted logging relic back in the woods on your left at 1,050' as you approach a right switchback. Here, Little Wild Rose and Pathfinder reside. Up the trail, these same two plants continue to thrive, along with an occasional Rattlesnake Plantain.

On the rock face at the right switchback at 1,300', **Point 4**, and at the openings just before and thereafter, the starry yellow blooms of Broad-leaf Stonecrop, with its jade-green leaves, are fading, but the Oregon Stonecrop, draped upon the same talus faces, is now in full flower, with equally attractive yellow blooms and much greener, yet still stout and succulent, leaves. At 1,560' and 1,900' White-vein Pyrola displays not only its showy leaves but its flowers of whitish-pink drooping bells. A clump of Pipsissewa has begun to bloom by mid-July at about 1,800' and, in the more level area between 1,750' and 1,900' several examples of Wild Ginger continue to be in bloom, although you will have to identify it by its glossy heart-shaped leaves plastered against the frequently damp ground. Lift these gently to gain a peek at the weirdly-colored brownish-purple flowers with three petals ending in tapered tips resembling nothing more than a waxed moustache.

The trail approaches, but does not cross, the creek, first at **Point 5**, 2,200', then again at 2,350'. Wild Ginger appears near both approaches, while the weedy Wall Lettuce continues pretty much constantly up to about 2,450'. At this point, you will begin to find the dainty white panicles of Foam Flower above maple-shaped leaves; these only increase as you climb higher and will persist throughout the summer and even into fall. At times covering large tracts of forest floor, these common denizens of the deep forest do seem to spread ocean-like foam across the dark reaches of our mountain woodlands. They are particularly prominent on this trail at 2,500' to 3,750'. A few Little Wild Roses add cheerful pink notes to the woods between 2,200' and 2,500', along with Rattlesnake Plantain. A single clump of Queen's Cup thrives at 2,300' on your right.

By this time of year, when you reach the rocky open area at **Point 6**, 2,700', there may be only a little moisture left. Still, water-loving plants continue to bloom throughout much of the early summer. Herculean Cow Parsnip may still be out, its monstrous leaves appearing to bear the weight of the world, and you may still join the Devil's Club if you are not careful. Just before this all-too-brief opening, watch for a specimen of Ocean Spray on your right; this 6' bush sports panicles of brownish-white flowers at the tips of its arching branches. All four openings between **Points 7** and **8**, from 2,750' to 3,200', contain fine examples of blue Broadleaf Lupine, red Harsh and Common Paintbrush, and the vivid pink of Taper-tip Onion, all now blooming throughout the summer months. The odor of this last is unmistakable and, even though the natives dug these bulbs for food, and despite the fact that the intrepid explorer David Douglas devoured them, please do not uproot these plants to use as a condiment. You do not, in all likelihood, enjoy the excuse of starvation nor are you the first of your race to explore these parts.

Above these meadows, you enter the seclusion of giant Douglas Firs and occasional true firs and hemlocks. Short openings afford sufficient light here and there to support Bear-grass while deep woodland plants such as Little Wild Rose and Foam Flower remain in bloom throughout the long, dry summer. Above **Point 8**, however, where a series of steep switchbacks begins at 3,280', several other forest species can be found despite the deep shade. Yellow daisy-flowered Broad-leaf Arnica appears in early July while an occasional Turtle Head (Woodland Penstemon) arches out from the forest with its tubular rose-purple flowers. Pipsissewa bears its cheery-pink nodding bells above its toothed dark-green leaves all the way from 2,800' to 3,600'.

On the plateau, Queen's Cup reappears, while Pipsissewa and Rattlesnake Plantain both continue, and the weirdly-wonderful Pinedrops, with reddish 3' tall stems and pink bulblets as "flowers", appear in two examples on your right about 50 yards after reaching the plateau and about 100 yards before reaching the Green Point Mountain Trail sign, our **Point 9**, 3,920'. Here a clump of trailing Twin Flowers appears, still in bloom through much of July. Along the crest after the left turn at the sign, Little Wild Rose, Bear-grass, Queen's Cup, Pipsissewa, and Sidebells (One-sided Pyrola) are all common. About ¼ of a mile after the sign, look for specimens of orange and maroon-spotted Tiger Lily on your right, a glorious sight in these often-shaded woods.

At the marshy areas and stream crossings ¼ mile before the lake, look for a 3-4' tall plant with alternate toothed arrow-shaped leaves and a tassel of yellow-headed daisy-like flowers. Arrow-leaf Groundsel (*Senecio triangularis*), is a member of the world's largest genus (about 1,500 species), the *senecio's*, in the second largest plant family, the Aster Family, with over 23,000 members. (The Orchids are a slightly larger group.) Although relatively rare in the Gorge except at high elevations not often approached by these trails, it is common in the Cascades in wet high alpine meadows. The senecio's have a bevy of confusing common names, including groundsels and butterweeds. Senecio derives from the Latin meaning old man ("senile"), a reference to the developing seed heads, which look a bit bald, sort of like your author.

At the Lake, **Point 10**, 3,988', Bear-grass perfumes the air through late summer, while an unusual spirea, White (or Birch-leaf) Spirea, grows on your right on top of the earthen dam that you cross to reach a lunch spot and resting area by the rocks on the northwest shore overlooking the Lake. Since most spireas are rose-colored, this one merits notice. You may observe folks fishing in the lake, but, by and large, an outing to North Lake on the Wyeth Trail is marked by solitude and shade, not a bad combination for a steaming summer's day. Take a plunge to cool off if you are brave enough, then dry off for the descent, a not-too-steep return and a day more than well-spent in the wilderness.

September and October

Autumn is an appealing season to explore the Gorge. Colors are changing the green and brown scene into an array of yellows, reds, and orange hues. The air is crisp and, with leaves dropping from the deciduous trees, the forests are a bit more transparent. Moreover, flowers are still present; what they may lack in diversity, they compensate for in their hardiness and unique character.

For example, at the parking lot, **Point 1**, the two geraniums, Robert and Dove-foot, hold steadfast through September and, for the Robert Geranium, even throughout the winter. This pretty, if invasive, species dots the woods up the old dirt road, and even the trail beyond, up to 1,200' in places, with its charming tiny pink blooms above frilly leaves. Yellow Wall Lettuce, another highly invasive European weed, also is present, both in the campground by the parking area and, later, up the trail from 400' to almost 3,000', at times forming an almost continuous boundary on both sides of the path. Its foreign origins can be detected by its absence anywhere but trailside, having been dispersed by passers-by, and not naturally occurring in the deeper woods away from the main trails.

As you briefly descend to the power line, Queen Anne's Lace persists through mid-to-late November (note its bird's-nest cup as it goes to seed), along with Hairy Cat's Ear, Pearly Everlasting, and a few white Woods Strawberries, the latter out through September. In the woods before and after the stream crossing at **Point 2**, Pathfinder (Trail Plant) is present, but you may notice its triangular leaves, fuzzy white beneath, before you recognize its single thin stalk ending in several secondary stalks (pedicels). These are topped by tiny nondescript dull-white buttons for flowers but more likely, by this time of year, by seeds with linear protrusions spreading from a central point.

Little else is in bloom by fall from here to past the beginning of the steeper switchbacks (**Point 3**) except for the omnipresent Wall Lettuce. However, at the huge rock face where the trail turns right, **Point 4** at about 1,300', two stonecrops, Oregon and Broad-leaf, may still display their yellow, sharp-petalled blooms growing right off the stony wall. The Oregon species sports thick bright-green leaves while the Broad-leaf variety boasts jade-green foliage. Just past this spot, immense specimens of Wall Lettuce have taken up residence, thriving off the rocky slope to your left. These are the largest examples of this species I have seen in the Gorge; they evidently prefer this often-dank shaded site for their new home in America.

Aside from Wall Lettuce, not much of interest is growing for the flower fancier between this point and the meadows at **Point 7**, except for some Pathfinders and Rattlesnake Plantains along the level areas between 1,650' and 1,900'. In the meadows above, between 2,750' and 3,200', however, Taper-tip Onion lends a bit of welcome hot-pink color to the scene early in September, along with its appetizing odor, while Yarrow will persist here through October.

However, a new plant, one that should be learned well by all Gorge travelers, comes into bloom in early September, and will last throughout the fall – Hall's Goldenweed (*Haplopappus hallii*). This Aster Family member with straggly yellow daisy-like flowers, is actually a shrub – note its woody base – the sole Aster family shrub in the Gorge. Rising 2-3' tall, though often arching its stems in the winds of its commonly-exposed sites, Hall's Goldenweed is not a weed at all, but a native species. It is also a narrow endemic, although it occurs outside the Gorge in the middle Oregon Cascades as well. This plant, named after H. M. Hall, an early western botanist, is part of a large genus of mostly desert shrubs which may have begun to inhabit the Gorge mistakenly during the summer months without realizing the deluges to come in winter. Its Greek genus name means "simple seed down", referring to the position of the pappus, the hairs on top of the seed-head in dandelion-like plants.

Hall's Goldenweed

In the third opening, at about 3,100', an example of the large shrub, Ocean Spray, can be found. Here it is about 4' tall, while at lower elevations, such as in the Willamette Valley, it can grow up to 10'. It features sprays of dull white-to-brown flowers on multiple branches. Above these openings, Foam Flower begins at about 3,200' and, through September, continues to dapple the dark forest with enchanting sprays of tiny bell-like white flowers all the way to the plateau and junction with the Green Point Mountain Trail at **Point 9**, 3,920'.

It continues beyond the up and down trail to North Lake, **Point 10**. In early to mid-September, Little Wild (Baldhip) Rose is still in bloom, especially evident between 3,200' and 3,700'; a good example is at the right switchback at 3,300'. Bear-grass may persist at these elevations as well, and is present above 3,200' and on the plateau. Still out in September on the plateau and along the trail to the Lake are some examples of Twin Flower sprawling over the forest floor on both sides of the path. At North Lake, Bear-grass perfumes the autumn air and reminds us that not all flowers bloom in spring and summer. The descent marks an opportune time to try to recognize species from their leaves alone, absent any inflorescence – the mark of the well-informed amateur botanist!

Epilogue

There is no doubt that there are errors in this book. I would appreciate learning of them through readers' comments. Despite notions to the contrary, plants do travel, or at least their progeny do. Annuals can be especially tricky to pin down to a single location. However, in the Gorge, even these, as well as perennials, do grow in about the same spots each year. Nonetheless, let me know if I have erred in any of these descriptions, locations or elevations. Comments can be sent through the Publisher or to me at wldflr@comcast.net.

Taxonomy is as tricky and variable as plant locations may be. Updated nomenclatures will certainly confuse us amateurs but that is not their sole intent. Rather, as new evidence about origins and relationships among species and genera emerge, change is guaranteed. This is especially true as recent DNA findings shuffle the naming cards faster than a Las Vegas croupier. I have therefore provided an appendix with more familiar names as found in the majority of texts alongside newer designations in several more recent works. I have preferred the older names in this book as they are far more familiar to Oregon's and Washington's amateur plant enthusiasts. For example the Solomon's Seals have been known for the past century as belonging to the *Smilacina* genus, but some botanists have now placed them in the *Maianthemum* genus. Similarly, the attractive Grass Widows adorning our low to mid-elevations prairies in earliest spring have long been placed in the *Sisyrinchium* genus but have more recently been assigned to the genus *Olysnium*.

I have also shortened some common names as their formal names are tedious. "Plume-flowered False Solomon's Seal" ("Western Solomon's Plume" in Jolley – you can see how this gets confusing) has been abbreviated to "Plume-flowered Solomon's Seal" for example and "Spreading Phlox" to simply "Phlox" most of the time as there are no other large mats of phlox in the western Gorge. Knowledgeable ordinary folk most often use these simpler terms in the field. Botanists will protest but they also abbreviate as well. I hope amateurs and simple flower lovers will approve.

I have tried to avoid technical jargon wherever possible and thus have not provided a glossary but these are available in most references cited in the bibliography. Thanks to readers everywhere – I hope you have learned something about our panoply of wildflowers and enjoyed the read as well.

For those seeking evidence of climate warming in the blooming dates of Gorge wildflowers, in the main, they must look elsewhere for longer-term studies and perhaps in more exposed and extreme locations. Although we can all see the shrinking of glaciers in our Northwestern mountains, over the more than 60 years or so of noting blooming times, from the late 1960s to 2024, I cannot say overall that the times of flowering have markedly changed. Nonetheless, I have noted Broad-leaf Arnica's blooming a bit earlier, from mid-May as opposed to June, along stretches of the Nick Eaton Trail between 800-1,200', as well as a number of Long-spurred Violets in flower as early as the first week in April in meadows on both the Ruckel Creek and Nick Eaton Trails.

It should be borne in mind, however, that I also have not visited the exact same locations on the exact same dates year after year. Thus, the best I can hope for is that my observations can serve as a baseline for future botanists and climate scientists but cannot be taken as even close to a controlled study. Nonetheless, the increasing acceptance among politicians and the public that climate change is not only real and accelerating, but that it is created by human beings, should serve as both a sign of progress and also an opportunity for future attempts to ameliorate it.

There is no denial of the overwhelming scientific evidence of human-caused climate warming. In fact, scholars now believe we are entering a new epoch, the "Anthropocene", a time when sapient beings such as us have the power and capacity to alter the Earth's total environment. Humans, as opposed to all other organisms on our planet, plant or animal, now have the foresight and power to choose to influence our environments and ecosystems rather than to simply accept, and try to adapt to, catastrophic changes that surely will occur in the future. We are stewards and wardens of our home, our Earth. When a tree falls in the forest, everyone can hear, and feel, it.

Of importance, countries such as Brazil and India are now reducing deforestation, recognizing that algae, plants and forests are the earth's lungs. These chlorophilic species sequester far more carbon dioxide than all attempts at the capping and taxation of carbon we can muster. We are thus stewards and wardens of our home, our Earth, and must live up to our Latin species designation as *Homo sapiens*. Our country, which gave birth to the notion of wilderness, should not be the first to destroy it. Denying that a battle must be fought is the first step in losing it.

Appendix

RECENT SCIENTIFIC NAME CHANGES OF SELECTED WILDFLOWERS

Most changes involve generic, rather than specific names. Common names have not changed, although these are quite a bit more varied and may differ depending on whom you talk to and where you have your discussion. The following list cannot be considered complete due to the ever-evolving nature of scientific taxonomy:

Common Name	Old Scientific (Latin) Name	New Scientific Name
Big-leaf Sandwort	*Arenaria macrophylla*	*Moehingia macrophylla*
Bunchberry (Canadian Dogwood)	*Cornus canadensis*	*Cornus unalaschkensis*
Candy Flower	*Montia sibirica*	*Claytonia sibirica*
Chocolate Lily	*Fritillaria lanceolata*	*Fritillaria affinis*
Columbia Kittentails	*Synthryris stellata*	*Synthyris missurica, suspecies stellata*
Common Sweet Cicely	*Osmorhiza chilensis*	*Osmorhiza berteroi*
Cow Parsnip	*Heracleum lanatum*	*Heracleum maximum*
Fireweed	*Epilobium angustifolium*	*Chamerion angustifolium*
Grass Widows	*Sisyrinchium douglasii*	*Olsynium douglasii*
Hall's Goldenweed	*Haplopappus hallii*	*Columbiadoria hallii*
Miner's lettuce	*Montia perfoliata*	*Claytonia perfoliata*
Mountain Oregon Grape	*Berberis nervosa*	*Mahonia nervosa*
Oaks Toothwort	*Cardamine pulcherrima*	*Cardamine nuttallii*
Ookow	*Brodiaea congesta*	*Dichelostemma congestum*
Oregon Boxwood	*Pachystima myrsinites*	*Paxistima myrsinites*
Pineapple Weed	*Matricaria matricariodes*	*Matricaria discoidea*
Plume-flowered Solomon's Seal	*Smilacina racemosa*	*Maianthemum racemosum*
Rock Penny Cress	*Thlaspi fendleri*	*Thlaspi montanum*
Rose Campion	*Lychnis coronaria*	*Silene latifolia, subspecies coronaria*
Rough Wall Flower	*Erysimum asperum*	*Erysimum capitatum*
Shining Oregon Grape	*Berberis aquifolium*	*Mahonia aquifolium*
Star Flower	*Trientalis latifolia*	*Trientalis borealis*
Star-flowered Solomon's Seal	*Smilacina stellata*	*Maianthemum stellatum*
Wall Lettuce	*Lactuca muralis*	*Mycelis muralis*
White Campion	*Lychnis alba*	*Silene latifolia, subspecies alba*

Selected and Annotated Bibliography

Alden, P. et al. 1998. **National Audubon Society Field Guide to the Pacific Northwest.** New York: Alfred A. Knopf. A total guide, including birds and animals, very brief descriptions of our flowers but gorgeous, though small, photos.

Clark, L. 1976. **Wildflowers of the Pacific Northwest from Alaska to Northern California.** Edited by J.G.S. Trelawny, Seattle: University of Washington Press. Technical descriptions for the truly committed; this is a complete compendium, with illustrations.

Gilkey, H.M. and La Rea, D.J. 1975. **Handbook of Northwestern Plants.** Corvallis: Oregon State University Press. A more reader-friendly text, though still quite a bit of a read; quite comprehensive.

Hitchcock, C.L., Cronquist, A. Owenby, M., and Thompson, J.W. 1955-69. **Vascular Plants of the Northwest.** Seattle: University of Washington Press. Illustrations are sometimes more informative than photos and this five-volume set has the best. Too heavy to carry and too dense to thoroughly read, it is only for the serious. Still, if you can get to a library that owns a copy, do so, just for the drawings and for an understanding of how botanists describe plants.

Jolley, R. 1998. **Wildflowers of the Columbia Gorge.** Portland, OR: Oregon Historical Society Press. The definitive and only localized guide to the Gorge. It shows photos, some a bit unclear, and list frequency, blooming times, and general, though not specific, habitat. By including the middle and far eastern sections of the Gorge, it is comprehensive, but a bit excessive for those seeking to learn about flowers west of Hood River. However, it remains invaluable as a reference and field guide.

Kozloff, E.N. 2005. **Plants of Western Oregon, Washington and British Columbia.** Portland, OR: Timber Press. A technical guide to ferns and flowering plants by the dean of Northwest botanists, this volume also contains a color photo section, though it is separated from the descriptions, necessitating annoying page-turning. It covers some areas far from the Gorge, but it is still a valuable resource as it provides amateur-friendly keys and descriptions.

Lowe, D. and Lowe, R. 1988. 35 **Hiking Trails Columbia Gorge; 2nd Edition.** Beaverton, OR: Touchstone Press. The hiking bible for the Gorge.

Matthews, D. 1988. **Cascade-Olympic Natural History.** Portland, OR: Raven Editions and The Audubon Society of Portland. If you are serious about the Northwest and its ecosystems, read this book. Yes, it includes a page-turning color photo section and yes, it provides descriptions of far more than wildflowers, including animals, birds, trees, and reptiles. Still, it remains the best-written natural history text about our outdoors. Read it for its content, but also for its prose, a heady mix of folksy charm and academic wisdom.

Pojar, J. and MacKinnon, A. (Eds.). 1994. **Plants of the Pacific Northwest Coast.** Redmond, WA: Lone Pine Publishing. Indispensible and not limited to shores and sea-sides. Clear photos are paired with nontechnical descriptions. Though it provides some far-flung species, all the main Gorge flowers are present and accounted for. Brief and easily-followed keys are presented for all the major flower families. Trees and ferns are here as well, along with, for those interested, mosses, lichens, and fungi.

Ross, R.A. and Chambers, H.L. 1988. **Wildflowers of the Western Cascades.** Portland, OR: Timber Press. Large photos help with identification, although they are in small sections separated from the nontechnical descriptions. Worth it for the pictures alone.

Schneider, R. Revised by Yuskavich 2014. **Hiking the Columbia River Gorge.** Helena, MT: Falcon Guides. Current descriptions of all the major Gorge trails.

Strickler, D. 1993. **Wayside Wildflowers of the Pacific Northwest.** Columbia Falls, MT: The Flower Press. Despite its title, this volume covers territory from the coast inland all the way to the Rockies. Nonetheless, most of our Gorge species are here and the book is worthwhile for its large and gorgeous photos.

Sullivan, W.L. 2006. **100 Hikes in Northwest Oregon and Southwest Washington, 3rd Edition.** Eugene, OR: Navillus Press. One section covers many of the Gorge trails, with clear trail directions and illustrations.

Turner, M. and Gustafson, P. 2009. **Wildflowers of the Pacific Northwest.** Portland, OR: Timber Press. The photos are small, but clear and the descriptions nontechnical. The text arranges flowers by color and provides habitats and blooming times for each species. Of its several advantages, not the least is its modern nomenclature. It is the single volume that I would recommend first to amateur botanists.

Printed in the United States
by Baker & Taylor Publisher Services